Handbook of Sensory Physiology

Volume III/2

Muscle Receptors

By

D. Barker · C. C. Hunt · A. K. McIntyre

Edited by

C. C. Hunt

With 128 Figures

Springer-Verlag Berlin · Heidelberg · New York 1974

ISBN 3-540-06891-0 Springer-Verlag Berlin · Heidelberg · New York
ISBN 0-387-06891-0 Springer-Verlag New York · Heidelberg · Berlin

Library of Congress Cataloging in Publication Data. Hunt, C.C. Muscle Receptors. (Handbook of Sensory Physiology, V. III/2) Bibliography: p. Contents: Barker, D. The Morphology of Muscle Receptors. Hunt, C.C. The Physiology of Muscle Receptors. McIntyre, A.K. Central Actions of Impulses in Muscle Afferent Fibres. 1. Muscle Receptors. 2. Neuromuscular Transmission. I. Barker, David. II. Hunt, Charlton C., 1918. III. McIntyre, Archibald Keverall, 1913. III. Title. IV. Series. [DNLM: 1. Mechanoreceptors WL700 H236 v. 3 pt. 2.] QP351. H34 Vol. 3, no. 2. [QP369] 591. 1'82'08s. [596'. 01'852] 74-13983

Typesetting, printing: Joh. Roth sel. Ww., München. Binding: Brühlsche Universitätsdruckerei, Gießen

Preface

This section will consider the structure and function of muscle receptors, as well as the central nervous system mechanisms with which they are concerned. In volume I of this Handbook, receptor mechanisms are discussed in detail. Also, the crustacean stretch receptor and the frog muscle spindle have been considered. The present section will be concerned with vertebrate muscle receptors with an emphasis on mammals.

Muscle receptors provide interesting examples of specialized mechanoreceptors. The muscle spindle is a striking case of a receptor which is regulated in its function by the central nervous system in efferent neurons. Muscle receptors have long been known to play a crucial role in the reflex regulation of movement. In recent years it has become apparent that these receptors are also important in sensory phenomena such as the perception of position and movement.

St. Louis, July 1974 C.C. Hunt

Contents

List of Contributors

BARKER, David

Department of Zoology, University of Durham, Science Laboratories, South Road, Durham DH1 3LE, Great Britain

HUNT, Carlton C.

Department of Physiology and Biophysics, Washington University, School of Medicine, 660 South Euclid Avenue, St. Louis, Mo. 63110, USA

MCINTYRE, A.K.

Department of Physiology, Monash University, Clayton, Victoria, Australia 3168

Chapter 1

The Morphology of Muscle Receptors

By

DAVID BARKER, Durham (England)

With 214 Figures

Contents

I. Introduction

A skeletal muscle is such an obvious example of an effector organ that it generally comes as a surprise to learn that most of the nerve fibres innervating it are sensory. Studies of the composition of cat hindlimb muscle nerves (Boyd and Davey, 1968; Stacey, 1969) have shown that sensory axons account for about two thirds of the total somatic component; most of them are non-myelinated. All the non-myelinated sensory axons, as well as some of the myelinated ones, terminate as free endings on blood vessels and in fat and connective tissue (Stacey, 1969). Some of these function as nociceptors responding to deep pressure or to squeezing the muscle, and compare with similar receptors found in skin. The rest of the myelinated sensory axons supply two types of mechanoreceptor responding to stretch, namely, the tendon organ and the muscle spindle; a few paciniform corpuscles, such as occur more commonly in joint capsules and periarticular tissue; and occasionally one or two of the much larger Pacinian corpuscles. The small muscle fibres that compose the spindles receive their own supply of motor (γ) axons, which on average constitute 43% of the total somatic motor component in cat hindlimb muscle nerves (Boyd and Davey, 1968). The receptor equipment of a cat hindlimb muscle thus receives a very large share (81%) of the total supply of somatic nerve fibres, while the muscle fibres that execute its contractions and constitute the bulk of the organ receive the small remainder as skeletomotor (α) axons. A schema of the innervation of mammalian skeletal muscle is shown in Fig. 1.

The sections that follow describe the structure and innervation of receptors in vertebrate striated muscle, and also, where known, their distribution and development. The largest and most complex of these is the muscle spindle.

II. The Muscle Spindle

A muscle spindle is composed of a bundle of small ('intrafusal') muscle fibres that receive both a motor and a sensory innervation. The receptor lies in parallel with the ordinary ('extrafusal') muscle fibres, and for part of its length is enclosed within a fluid-filled capsule. It is a mechanoreceptor that responds specifically to active and passive changes in muscle length, this sensitivity being modifiable by the activation of its motor supply and the contraction of its muscle fibres. Spindles occur in tetrapod vertebrates and are distributed within the majority of the somatic muscles. They occur singly, or may be variously combined in groups, or

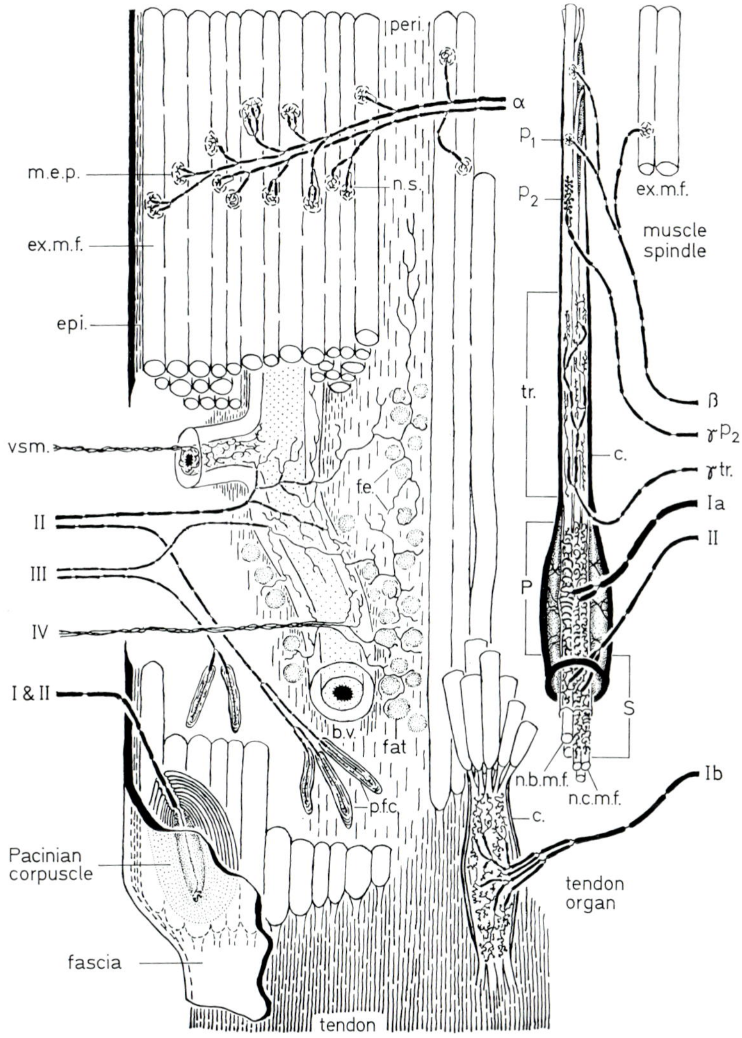

Fig. 1. Schema of the innervation of mammalian skeletal muscle based on a study of cat hindlimb muscles. Those nerve fibres shown on the right of the diagram are exclusively concerned with muscle innervation; those on the left also take part in the innervation of other tissues. Roman numerals refer to the groups of myelinated (I, II, III) and non-myelinated (IV) sensory fibres; Greek letters refer to motor fibres. The spindle pole is cut short to about half its length, the extracapsular portion being omitted. *b.v.* blood vessel; *c.* capsule; *epi.* epimysium; *ex.m.f.* extrafusal muscle fibres; *n.b.m.f.* nuclear-bag muscle fibre; *n.c.m.f.* nuclear-chain muscle fibre; *n.s.* nodal sprout; *m.e.p.* motor end-plate; *P* primary ending; p_1, p_2 two types of intrafusal end-plates; *peri.* perimysium; *pf.c.* paciniform corpuscle; *S* secondary ending; *tr.* trail ending; *vsm.* vasomotor fibres. (Modified from BARKER, 1967)

intimately associated with tendon organs. The functional significance of the single encapsulated receptor with its sensory and motor innervation is sometimes stressed by using the term 'spindle unit', this being interchangeable with, and sometimes preferable to 'single spindle'. A 'tandem spindle' has two or more encapsulated regions spaced widely apart along the length of a common intrafusal bundle, whereas a 'compound spindle' is formed by two or more spindle units bound together side by side with variable degrees of fusion of intrafusal bundles and capsules.

The muscle spindle was first recognized as a discrete entity by WEISMANN (1861). BATTEN (1897) was of the opinion that MIESCHER (1842) had observed them in the mouse, and other authors have given priority to HASSALL (1849) and MUNK (1858), but there is very little in these early writings to substantiate these claims. Weismann's description and figures of the small bundles of thin muscle fibres that he found in teased frog muscle leave no doubt that they were muscle spindles. He supposed that they were ordinary muscle fibres in process of splitting up to form additional ones. In the following year KÖLLIKER (1862) described similar structures in the frog's pectoral-cutaneous muscle, elaborated Weismann's growth theory, and called the structures 'muscle buds' *(Muskelknospen)*. The many protagonists of the muscle-bud theory who followed added considerably to knowledge of the morphology of the spindle, but were chiefly preoccupied with describing the stages that were supposed to occur in this process of muscle growth. KÜHNE's (1864) suggestion that the spindle might be an "apparatus with a peculiar, still unknown, physiological function for the contraction of muscle" was ignored, though it was his term *Muskelspindel*, alluding to its fusiform shape, that was ultimately to be adopted.

The identification of the spindle as a sense organ rather than a stage in muscle development, or, as others had suggested, some kind of pathological product, was finally established by KERSCHNER (1888a, b). His evidence was circumstantial and largely depended on pointing out the similarities that existed between the spindle and the tendon organ, which had been discovered earlier by GOLGI (1880). The sensory nature of the tendon organ had not been proved, but it was generally believed that Golgi was correct in assuming that it was a receptor recording muscle tension. Moreover, CATTANEO (1888) showed that cutting the ventral roots in dogs left the supply to the tendon organ intact, so that with the sensory nature of this receptor established, more weight was given to Kerschner's argument when he elaborated it further in 1893.

Quite apart from the similarity of the muscle spindle to the tendon organ, both KERSCHNER (1888b, 1892, 1893) and RUFFINI (1893b) concluded that the spindle was a sense organ because of the form and disposition of the complex innervation distributed to the intrafusal muscle fibres within the capsule, which they were the first to describe in any detail in the mammal. The sensory nature of these endings was demonstrated by SHERRINGTON (1894) in his classic paper on the sensory innervation of muscle, for he found that they remained intact in muscles of the cat and monkey whose motor supply had degenerated after ventral root section. He also introduced the term 'equatorial region' to designate the fusiform, fluid-filled part of the capsule that encloses the area of sensory innervation, and distinguished between this and the 'polar regions' of the intrafusal muscle fibres that extend on each side.

Among these attempts to establish the sensory nature of the spindle only ONANOFF (1890) determined the effect of cutting the dorsal roots, but he did not specify the endings that degenerated. It was not until the much later work of

HINSEY (1927), HINES and TOWER (1928), and TOWER (1932) that the results of this type of experiment were fully described. It was also left to them to demonstrate that the polar regions of the spindle were innervated by somatic motor axons. The majority of observers had assumed this to be so because of the similarity of the endings to extrafusal motor end-plates, and also because in lizard muscle branches of the same motor axon had been seen to terminate in both intrafusal and extrafusal plates (PERRONCITO, 1901). RUFFINI (1893b, 1897, 1898), however, had supposed the endings to be sensory, and later KULCHITSKY (1924a) and HUNTER (1925) had claimed that they were derived from sympathetic motor axons.

Recognition of the fusimotor component in muscle nerves came slowly. The bimodal nature of the motor fibre-size histogram was demonstrated by ECCLES and SHERRINGTON (1930) in deafferented cat nerves, but they assumed that the group of small motor axons simply supplied small motor units, evidently being unaware of LANGLEY's (1922) suggestion that such small axons seen by him in cat ventral roots were probably concerned with the motor innervation of spindles. It was ultimately HINSEY (1934), reasoning from the data of SHERRINGTON (1894) and ECCLES and SHERRINGTON (1930), who proposed that the small motor axons in muscle nerves innervated the intrafusal muscle fibres. Proof was provided by LEKSELL (1945), who showed that there was a large increase in the afferent discharge from a cat muscle under some initial tension when these 'γ efferents', as he named them, were selectively stimulated, conduction in the larger α axons being blocked by pressure.

The years since Leksell's paper was published have seen increasingly intensive work on muscle receptors; 137 papers appeared on their morphology, pathology and physiology during 1946–1955, and 575 during 1956–1965, as compared with only 51 during 1936–1945 (ELDRED, YELLIN, GADBOIS and SWEENEY, 1967). Most of this work has been concerned with the mammalian muscle spindle, though knowledge of the amphibian spindle has also advanced considerably, and some attention has recently been paid to spindles in reptiles and birds. Histologists have greatly extended knowledge of spindle structure and innervation, taking full advantage of the introduction of electron microscopy and histochemical techniques, and being encouraged by the progress of parallel investigations carried out by neurophysiologists. The best review of work on spindles in the last century is that by REGAUD and FAVRE (1904), and there are excellent reviews by HINSEY (1934) and TIEGS (1953), which together cover the years that led up to the recent period of intensive activity. More recently the best account of morphological work on spindles is included in the review by MATTHEWS (1964); in a later book (1972) he gives his own view of the field in the early chapters. There is also some useful histological information in a chapter by COOPER (1960). An account of mammalian spindle innervation is given by BARKER (1967), who has also (1968a) reviewed knowledge of vertebrate fusimotor innervation.

The muscle spindle has been the central topic of four symposia held during the past decade; in Hong Kong (BARKER, 1962a), Stockholm (GRANIT, 1966), Dundee (ANDREW, 1966) and Cleveland (BANKER, PRZYBYLSKI, VAN DER MEULEN and VICTOR, 1972). The proceedings of these meetings contain a number of histological contributions which record the conflicting evidence and controversies that have at times attended the emergence of truth. The present (December, 1972) state of knowledge about spindle morphology is described below.

A. The Mammalian Muscle Spindle

Most of the information about mammalian spindle morphology has been derived from studies of the receptor in hindlimb muscles of the cat. This has been

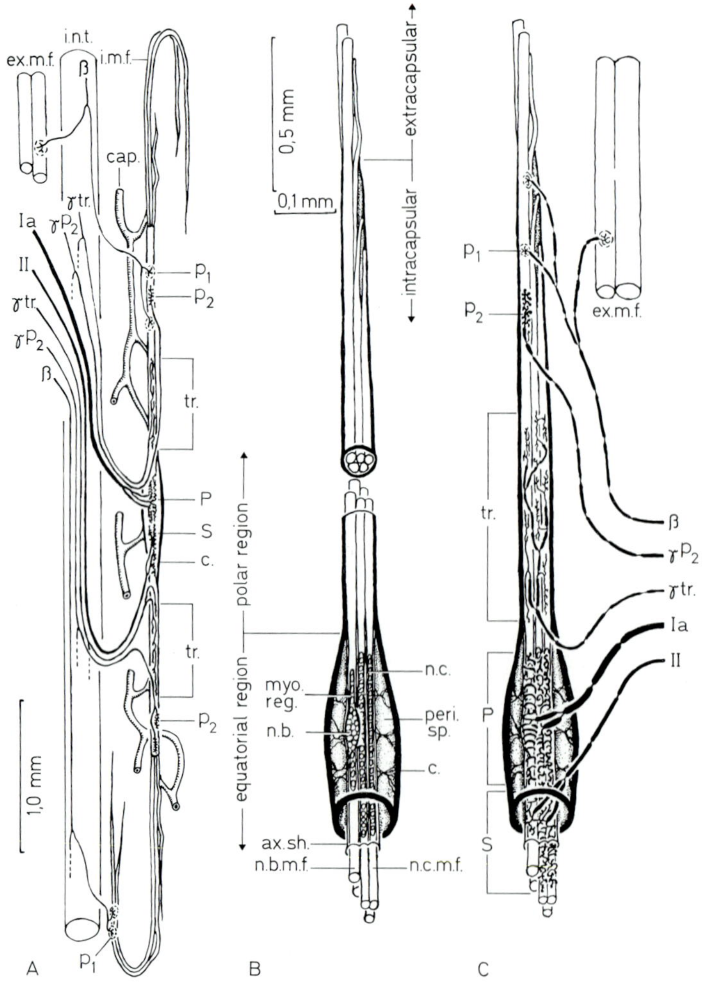

Fig. 2. Schematic diagrams illustrating the structure and innervation of the mammalian muscle spindle as found in cat hindlimb muscles. A shows the general proportions of the receptor; the ends of the poles are curled round so as to fit into the figure. B shows the morphology of the equatorial region and about half of one pole, and the same figure is used in C with the addition of the sensory and motor innervation. *ax.sh.* axial sheath; *c.* capsule; *cap.* capillary; *ex.m.f.* extrafusal muscle fibres; *i.m.f.* intrafusal muscle fibres; *i.n.t.* intramuscular nerve trunk; *myo.reg.* myotube region; *n.b.* nuclear bag; *n.c.* nuclear chain; *n.b.m.f.* nuclear-bag muscle fibre; *n.c.m.f.* nuclear-chain muscle fibre; *P* primary ending; p_1, p_2 two types of intrafusal motor end-plates; *peri.sp.* periaxial space; *S* secondary ending; *tr.* trail ending. Sensory nerve fibres indicated by Roman numerals, motor fibres by Greek letters

the natural consequence of the classic descriptions of cat spindles by RUFFINI (1893b, 1898) and SHERRINGTON (1894), and because muscles of the cat's hindlimb have so frequently been chosen for neurophysiological experiments. Apart from the cat, the information has been derived mainly from human spindles and those of the rabbit and rat.

Schematic diagrams illustrating the general features of the structure and innervation of a cat spindle are shown in Fig. 2. The intrafusal, or axial, bundle consists of 4–12 muscle fibres in which those of small diameter and short length predominate over those that are thicker and longer. The total length of the bundle, as attained by the longest fibres, usually falls between 7 and 10 mm, and part of this length, generally the middle third, is ensheathed within a lamellated capsule. A major feature of the encapsulated part is the equatorial region where the capsule dilates to enclose a fusiform fluid-filled space within which the axial bundle is suspended as it courses through and receives the sensory innervation. A thin sheath invests it in this region (the axial sheath or inner capsule) and supporting strands of trabecular connective tissue arise from it and traverse the periaxial space to fuse with the capsule wall, here at its thickest (see Plate 1, Fig. 6). The extent of the equatorial region corresponds with the length of the periaxial space, which varies according to the number of sensory endings present. The juxta-equatorial regions on either side of it mark the beginnings of the poles, which extend from the limits of the periaxial space to the furthest extremity of the origin or insertion of the muscle fibres composing them. There is no constant feature characteristic of one pole and not the other, but it is convenient to designate them as proximal or distal with respect to their orientation with the origin and insertion of the extrafusal muscle bundles. The encapsulated part of a polar region usually occupies about a third of its length. As the capsule thins down and finally disappears, the number of intrafusal muscle fibres decreases, and those that are left tend to spread out and pursue individual courses to the site of their attachments to intramuscular connective tissue, tendon, or extrafusal endomysium.

Within the equatorial region the intrafusal muscle fibres undergo a slight decrease in diameter and a marked change in nucleation. The longest and thickest fibres possess a short region full of round vesicular nuclei which tapers off on either side into a single row of elongated nuclei within a central core of sarcoplasm. These are the 'nuclear bag' and 'myotube regions' described by BARKER (1948). The shortest and thinnest fibres contain a single central row of nuclei only (COOPER and DANIEL, 1956; BOYD, 1956), and are called 'nuclear-chain fibres' (BOYD, 1960). The length of these is such that few extend beyond the limits of the capsule. The nuclear bags, myotube regions, and nuclear chains together form the site of the rings and spirals of the primary ending, which is supplied by the thickest sensory fibre to enter the equatorial region and is distributed to both types of muscle fibre. In some spindles this is the only sensory ending present, but most also have another type, the secondary ending, which is supplied by a medium-sized afferent. Usually, as in Fig. 2, only one of these is present lying adjacent to the primary ending, its less regular rings and spirals being distributed predominantly to the nuclear-chain fibres, those closest to the primary wrapping round parts of the chains themselves. Secondary endings may lie on each side of the primary ending, and the most complex spindles may receive as many as four or

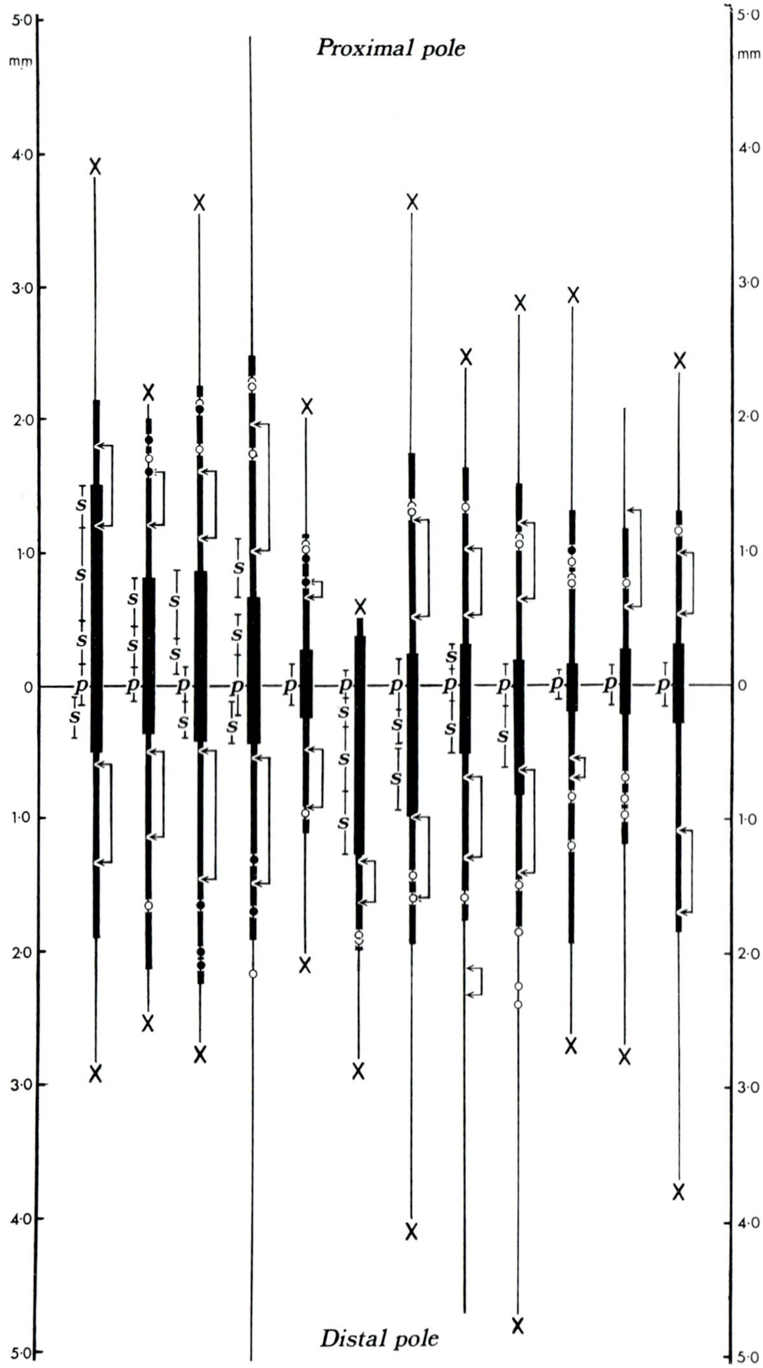

Fig. 3. Schematic representation of the distribution of sensory and motor endings to twelve spindles teased from silver preparations of the tenuissimus muscle of the cat. The bundle of intrafusal muscle fibres composing each spindle is represented as a single line, the capsule as a thickening of this line, further thickening indicating the length of the periaxial space. Each

five. In some instances the terminals assume an irregular 'flower-spray' (RUFFINI, 1898) form.

The motor innervation is distributed to the polar regions and is usually restricted to their intracapsular portions. There are three kinds of motor ending; a diffuse, multiterminal trail ending (BARKER and IP, 1965), and two types of end-plate designated as p_1 and p_2 (BARKER, 1966, 1967). The trail endings and p_2 plates are supplied by the exclusively fusimotor γ axons, whereas the p_1 plates are supplied by β axons, which have a skeleto-fusimotor distribution. Trail endings are almost invariably present and typically occupy an area just under 1 mm long on either side of the equatorial region. The plate endings occur further along the poles and either one or both types may be present. Most of the plate innervation is located on the nuclear-bag muscle fibres, there sometimes being several plates of each type per fibre. The trail innervation is most frequently distributed to both the bag and the chain fibres.

Most of the nerve fibres innervating the spindle travel together with the primary afferent from a nearby intramuscular nerve trunk to enter the equatorial region as a spindle nerve, but there are often other small bundles that enter the poles separately, and the p_1 collaterals of β axons often enter singly, particularly when their plates are located towards the extreme ends. BARKER, STACEY and ADAL (1970) found that the average number of motor axons entering single spindles in various cat hindlimb muscles varied from 8 to 15. Added to this there are the primary (1) and secondary (0–5) afferents, and the sympathetic axons concerned with the innervation of the spindle's rich vascular supply. Fig. 3 is a schematic representation of the distribution of sensory and motor endings to twelve cat tenuissimus spindles.

Such are the general features of a mammalian muscle spindle as exemplified by those in cat hindlimb muscles observed in light microscope preparations. Plate 1 illustrates some of the features seen in transverse section. Various aspects will now be considered in more detail taking account of ultrastructure, histochemistry, and information about spindles in other mammals.

1. *The Capsule, Axial Sheath, and Periaxial Space*

As seen with the light microscope the length of the periaxial space corresponds with the proximal and distal limits of the fusiform dilatation of the capsule, but electron micrographs show that it extends as a narrow cleft for a considerable distance along the poles. The equatorial length of the periaxial space and the length of the capsule vary according to the number of sensory endings present. Thus in teased, silver preparations of cat tenuissimus the equatorial length of the periaxial space was 0.35 mm in a spindle with one primary ending, as compared with 1.9 mm in another spindle with one primary and four secondary endings (see Fig. 3). The capsule lengths of these spindles were 2.2 mm and 4.3 mm, respectively. At their widest, mid-equatorial point the capsules in these preparations measured

spindle is aligned with the length scale so that the zero line passes through the middle of the primary ending (p). Secondary endings indicated as s; p_1 plates as filled circles; p_2 plates as hollow circles; linked arrows indicate areas occupied by trail endings. X indicates pole cut during teasing; such poles may have received a plate innervation beyond this point

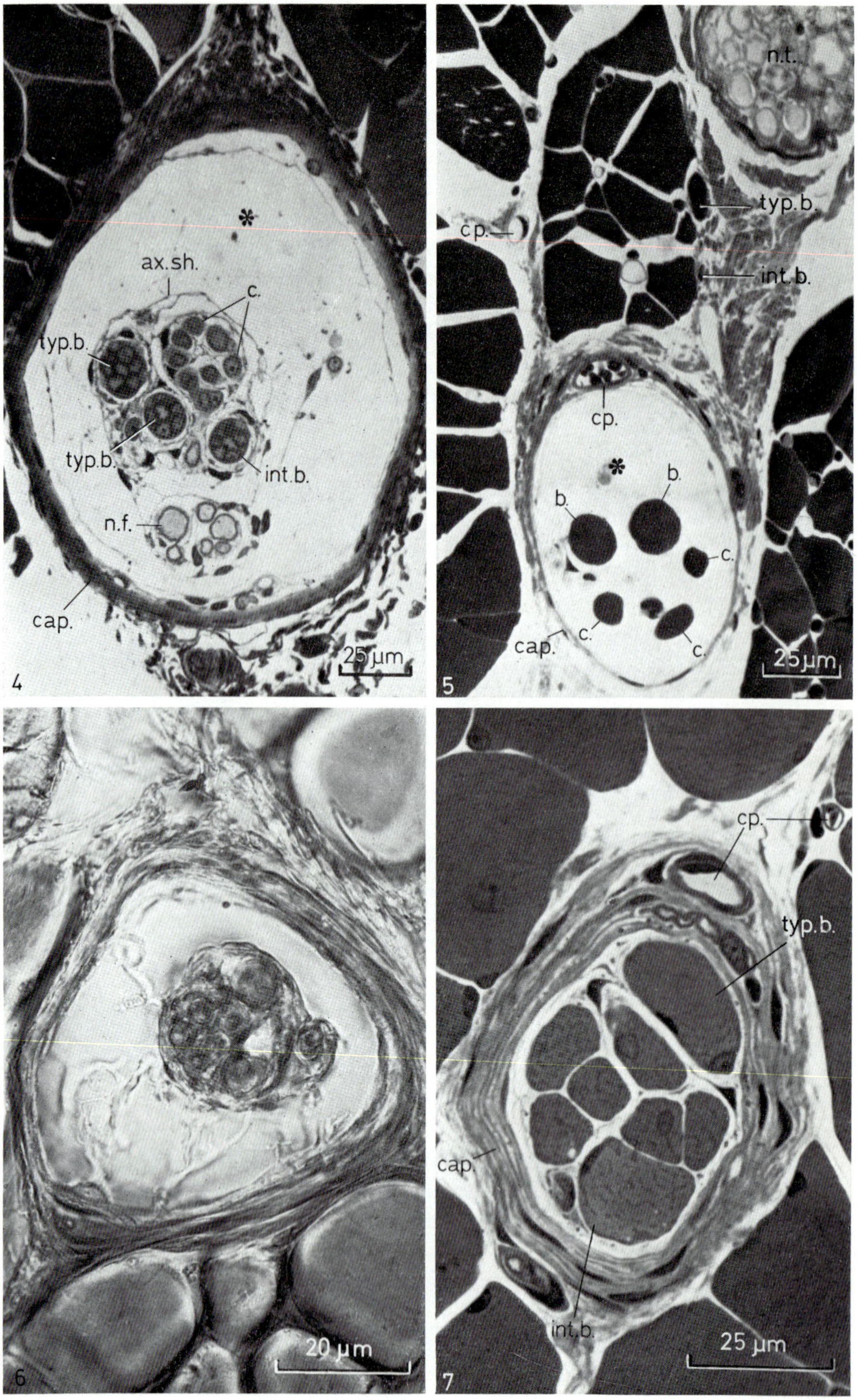
ax.sh.
c.
typ.b.
typ.b.
int.b.
n.f.
cap.
25 µm
4
n.t.
typ.b.
cp.
int.b.
cp.
b.
b.
c.
c.
c.
cap.
25 µm
5
20 µm
6
cp.
typ.b.
cap.
int.b.
25 µm
7

Plate 1

100–150 μm across with walls 10–15 μm thick. These measurements apply to single spindles; as already noted, the capsule may be variously elaborated in compound spindles. It may also occasionally extend to include, in a separate compartment, one or two neighbouring skeletal muscle fibres, or even a small bundle of them (Cooper and Daniel, 1963; Thompson, 1970).

The capsule has a lamellated appearance which is due to layers of very thin flat cells arranged in a concentric tubular fashion alternating with layers filled with collagen fibrils. The ultrastructure was first described by Merrillees (1960, rat) and there are subsequent studies by Gruner (1961, man), Landon (1966, rat), Corvaja, Marinozzi and Pompeiano (1969, cat), Goglia (1970, cat, rabbit, rat), and Banker and Girvin (1971, dog). The cytoplasm of the 'capsular sheet cell', as Merrillees called it, contains a large number of pinocytotic vesicles, small mitochondria, polyribosomes and glycogen granules. Each cell is surrounded by a basement membrane and closely interdigitates with its neighbour so as to form a continuous layer one cell thick. Here and there the edges of the cells overlap, and their cell membranes occasionally form regions of intimate contact variously described as desmosomes (Merrillees, 1960), terminal bar tight junctions (Landon, 1966; Banker and Girvin, 1971), or *zonulae adhaerentes* (Corvaja et al., 1969). Banker and Girvin (1971) describe the layers of the sheet cells as mostly being arranged in pairs, each pair enclosing a channel, the lumen of which is clearly differentiated from the spaces containing collagen fibrils that alternate with the pairs. Vesicles of various sizes are observed in all stages of release into the channels which open into the periaxial space. In such channels the basement membrane is confined to the outer surface of the sheet cells (see Plate 2).

The number of layers of sheet cells in the thickest part of the capsule varies slightly as between different spindles in the same muscle, and there are also variations as between different muscles (e.g. in the cat the capsules in tenuissimus

---

Plate 1, Figs. 4–7

Fig. 4. Toluidene-blue-stained 1 μm transverse section through the equatorial region of a spindle from cat hindfoot lumbrical muscle. The intrafusal bundle, surrounded by an axial sheath (*ax.sh.*) and suspended in the periaxial space (*), consists of two typical bag (*typ.b.*) and one intermediate bag (*int.b.*) fibre and six chain fibres (*c.*) *cap.* capsule; *n.f.* nerve fibre

Fig. 5. Toluidene-blue-stained 1 μm transverse section of cat tenuissimus muscle showing two spindles. The more obvious of these (at the bottom of the figure) has been cut through at juxta-equatorial level and is composed of two bag fibres (*b.*) and three chain fibres (*c.*). On the righthand side of an extrafusal muscle-fibre fasciculus above this spindle, another spindle has been cut through close to the extremity of one pole. Two bag fibres are present at this level, which on being traced through to the equatorial region and examined ultrastructurally (both in transverse and longitudinal section) proved to be a typical bag fibre and an intermediate bag fibre. *cp.* capillary; *n.t.* nerve trunk

Fig. 6. Transverse frozen section (25 μm) through the equatoral region of a cat rectus femoris spindle showing the intrafusal bundle suspended in the periaxial space and supported by trabecular connective tissue (fibrocytes; cf. Plate 2, Fig. 8). Aqueous haematoxylin stain. (From Barker and Cope, 1962)

Fig. 7. Toluidine-blue-stained 1 μm transverse section of a cat hindfoot lumbrical spindle, which has been cut through at the level of the junction between juxta-equatorial and polar regions where the periaxial space is reduced to a narrow cleft. The intrafusal bundle consists of one typical bag and one intermediate bag fibre, and six chain fibres. Small grey patches in the muscle fibres are mitochondria

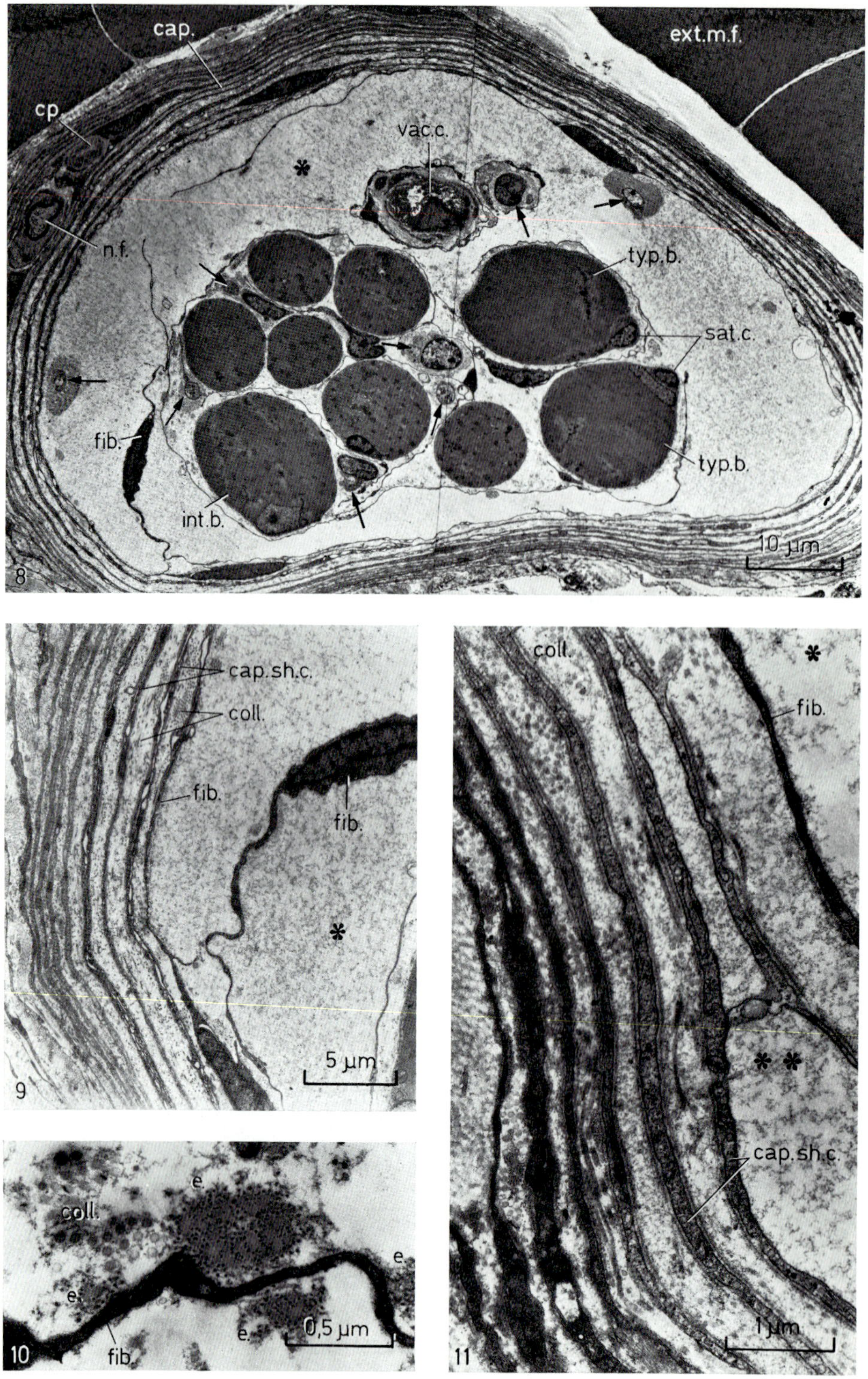
cap.
ext.m.f.
cp.
vac.c.
*
n.f.
ty.p.b.
sat.c.
fib.
int.b.
typ.b.
10 µm
8
cap.sh.c.
coll.
fib.
fib.
*
5 µm
9
coll.
e.
e.
e.
e.
fib.
0,5 µm
10
coll.
*
fib.
* *
cap.sh.c.
1 µm
11

Plate 2

are much thinner than those in rectus femoris) and different species. The number of layers in rat lumbrical spindles is 6–9 (LANDON, 1966), in cat lumbrical spindles 9–12 (CORVAJA et al., 1969), in dog gastrocnemius spindles 8–13 (BANKER and GIRVIN, 1971), and in human spindles sampled in forearm biopsies 12–15 (GRUNER, 1961). The collagen fibrils in between the layers are arranged so that they either encircle the capsule or run longitudinally. The outermost layer of the capsule is composed of thicker collagen fibrils similarly arranged among which there are scattered fibrocytes. The innermost layer of the capsule is composed of a lining of similar fibrocytes, some of which cross the periaxial space to join other cells of the same type which form the axial sheath and endomysial enclosures of the intrafusal muscle fibres (see Plate 2). Similar cells form the endoneural sheaths around incoming myelinated axons. Such cells resemble the capsular sheet cells, but have less cytoplasm, fewer vesicles, and no basement membrane. Their adjoining edges may form regions of close apposition described as terminal bar tight junctions by LANDON (1966) and as *zonulae adhaerentes* by CORVAJA et al. (1969). In the polar regions the endomysial cells generally form a more or less complete envelope around each muscle fibre, but towards the equator two or three muscle fibres may come to share the same envelope. The polar limits of the capsule occur where its inner lining of fibrocytes merges with those in the axial sheath, and the innermost layer of capsular sheet cells disappears. The space between endomysial envelope and muscle fibre contains non-myelinated preterminal axons, collagen fibrils, elastic fibres (see Plate 2, Fig. 10), and, as noted by ADAL (1969), convolutions of folded and branched basement membrane, which may also occur in the periaxial space.

Elastic fibres are more numerous around intrafusal than around extrafusal muscle fibres, and their distribution in spindles has been studied in detail by COOPER and DANIEL (1963, 1967) and GLADDEN (1972a, b) (see Fig. 12). A longitudinally orientated network of thick elastic fibres surrounds the extracapsular portions of the intrafusal muscle fibres. Within the capsule this network continues around the nuclear-bag fibres as the equator is approached, but over the region of the bags themselves the elastic fibres become very sparse. By contrast the number of elastic fibres around the chain fibres becomes greatly reduced as they enter the periaxial space, but

Plate 2, Figs. 8–11

Fig. 8. Low-power electron micrograph of a transverse section through the juxta-equatorial region of a spindle from a cat's hindfoot lumbrical muscle. The intrafusal bundle is composed of two typical bag (*typ.b.*) and one intermediate bag (*int. b.*) fibres and six chain fibres. *cap.* capsule, *cp.* capillary, *ext.m.f.* extrafusal muscle fibre, *fib.* fibrocyte, *n.f.* nerve fibre, *sat.c.* satellite cells, *vac.c.* vacuolated cell in periaxial space (*). Arrows point to non-myelinated axons (mostly preterminal trail axons) surrounded by Schwann cells and collagen fibres. Longitudinal sections through one of the typical bag fibres, the intermediate bag fibre, and one of the chain fibres are illustrated in Plate 8, and transverse sections of one of the typical bag fibres, the intermediate bag fibre, and one of the chain fibres are shown in Plates 7 and 10

Figs. 9–11. Illustrate the structure of the capsule (Figs. 9, 11) of the same spindle, and the differences between collagen (*coll.*) and elastic (*e.*) fibres (Fig. 10). In Fig. 11 a pair of capsular sheet cells (*cap.sh.c.*) enclose a channel (**) and lack a basement membrane on their inner surfaces. *fib.* fibrocyte. (Unless otherwise stated these, and all succeeding electron micrographs in this chapter on Muscle Receptors, were prepared by Dr. M.J. STACEY using cacodylate-buffered glutaraldehyde and osmium fixation followed by uranyl acetate and lead citrate staining)

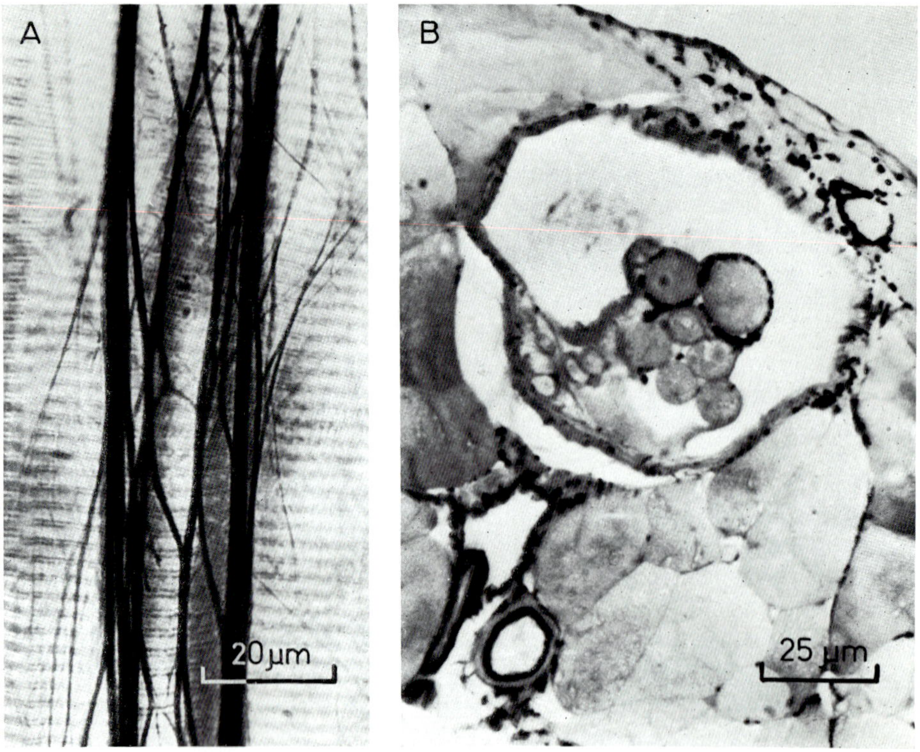

Fig. 12A and B. A Part of the extracapsular polar region of a spindle lying between extrafusal muscle fibres. Elastic fibres have been stained black with resorcin-fuchsin. The specimen was teased from cat tenuissimus muscle, overlying extrafusal muscle fibres being removed. B Transverse section (10 μm thick) of cat tenuissimus spindle stained for elastin with resorcin-fuchsin. The section was cut midway between the beginning of the periaxial space and the equator. Elastic fibres appear as black dots. Two bag fibres are surrounded by elastic fibres, but there are scarcely any around the five chain fibres. (From GLADDEN, 1972b)

increases by two or threefold around the region of the nuclear chains. A few elastic fibres are also found in the outermost layer of the capsule.

Capillaries course for long distances between the lamellae of the capsule, and may occasionally penetrate it and travel close to the axial bundle (see COOPER and DANIEL, 1963; CORVAJA et al., 1969). In rabbit tenuissimus spindles their presence in the periaxial space is a constant feature (R. W. BANKS, personal communication). Nerve fibres also travel within and penetrate the capsule, and where the spindle nerve enters there is continuity between the periaxial and subperineural spaces and fusion between the perineurium and capsule wall. According to SHANTHA, GOLARZ and BOURNE (1968) the perineurium is composed of an outer layer of fibrocytes and an inner layer, or layers, of epithelial cells. The perineural epithelial cells are regarded by them as histologically identical and continuous with the capsular sheet cells (see Fig. 13), a conclusion substantiated by demonstrating that the sheet cells share various oxidative and phosphorylating enzymes in common with perineural epithelium, that are absent from epineurium, perineural connective tissue or endoneurium. Bourne and his co-workers have also shown that the lamellae of Pacinian and Herbst corpuscles, the perisynaptic (teloglial) cells of motor end-plates, and the choroid of the eye are all likewise derived from a continuation of perineural epithelium, which they regard as an extension of the pia-arachnoid mater from the central nervous system. They conclude that the spindle capsule serves as both a mechanical and metabolically active diffusion barrier protecting the sensory endings and separating them from surrounding body fluids.

This raises the question of the nature of the fluid that fills the periaxial space. Until recently it was customary to call this a lymph space since Sherrington (1894) claimed to have injected it via the lymphatics of the leg of a cat or monkey (which of the two is not specified). Supporting evidence was

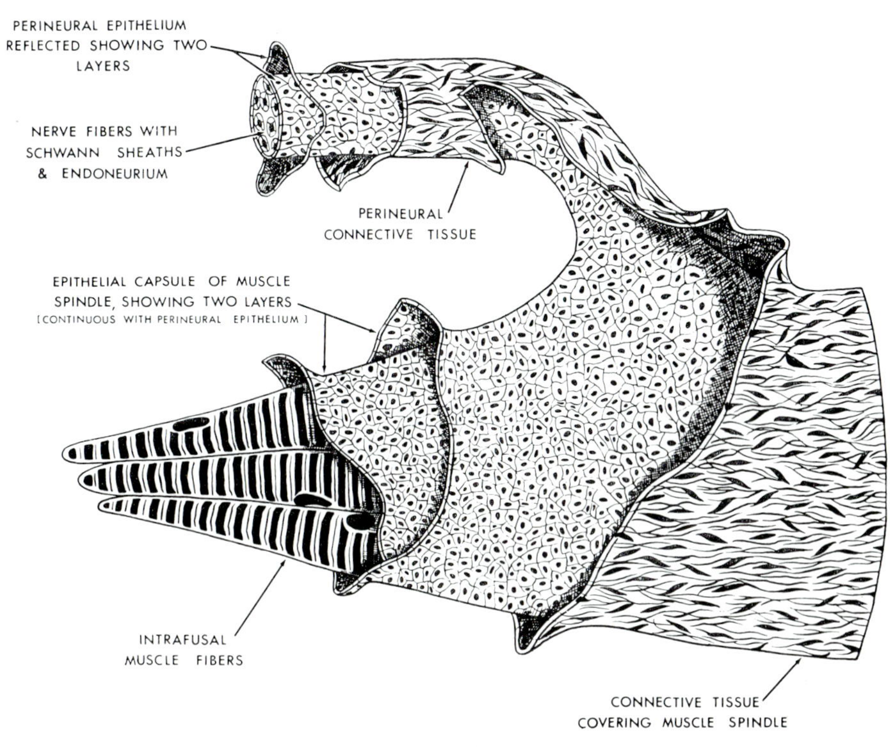

Fig. 13. Diagrammatic representation of a part of a neuromuscular spindle showing in detail the relationship between the coverings of the extrafusal nerve which supplies the muscle spindle as well as the covering of the muscle spindle itself. The perineural epithelium covering the nerve of supply continues to cover the muscle spindle as its capsule. This covering is reflected and shown to be made up of multiple layers of cells lying one on top of the other. Immediately surrounding the perineural epithelium is the perineural connective tissue covering which extends on the capsule of the muscle spindle. This layer of connective tissue is cut and reflected back on the nerve as well as on the muscle spindle to show the continuity of the perineural epithelium of the nerve fasciculus as a capsule of the muscle spindle. (From Shantha, Golarz and Bourne, 1968)

provided by JAHN (1959), who observed round cells in the periaxial space of living frog spindles, which she identified as lymphocytes by comparison with similar cells in blood smears. Also VOSS (1962a) observed many leucocytes, including lymphocytes, in the periaxial space of many spindles in the flexor pollicis longus and brachioradialis muscles of a newborn child. However, BRZEZINSKI (1961a, b) showed that the fluid in guinea-pig spindles does not have the histochemical characteristics of lymph, and concluded that it is a highly viscous fluid rich in hyaluronic acids "probably formed by the spindle sheath", i.e. presumably by the capsular sheet cells. The large number of pinocytotic vesicles in these cells supports this view, and suggests that they play a major role in regulating the internal environment of the spindle.

BRZEZINSKI (1961b) suggests that the lymphocytes observed by Jahn may have been introduced into the space as a result of micromanipulation. No lymphocytes have so far been reported in the periaxial space in ultrastructural studies, but GOGLIA (1970) has noted the presence of highly vacuolated cells whose appearance suggests that they may be engaged in secretory activity. One such cell is present in the periaxial space of the spindle illustrated in Plate 2, Fig. 8. Goglia has observed cells of similar appearance in the interlamellar spaces of Pacinian corpuscles. BRZEZINSKI concludes that the capsule and its viscous fluid serve as a kind of shock absorber protecting the sensory endings from being stimulated by pressure. BRIDGMAN and ELDRED (1964), on the contrary, suggest that the spindle may in part act as a pressure-sensitive device, the pressure of extrafusal muscle contraction serving to squeeze fluid out from the ends of the capsule, thereby elongating it so that the nuclear-chain muscle fibres attached to its wall at each end are stretched and activate the secondary endings. However, it seems unlikely that this happens, since, as SHANTHA et al. (1968) point out, the ends of the capsule make a tight collar around each pole which does not allow the intracapsular fluid to escape.

As noted above (p. 14), the periaxial space is continuous with the subperineural space surrounding the incoming spindle nerve trunk. A similar relationship exists between the capsule lumen of the tendon organ and the subperineural space surrounding its afferent supply (SCHOULTZ and SWETT, 1972). All these spaces are full of fluid, and there is some evidence to suggest that the subperineural spaces ultimately communicate with the subpial space surrounding the spinal cord (STEER and HORNEY, 1968). Both central and peripheral nervous systems are thus enclosed within a fluid-filled epithelial envelope, part of which forms the capsule and periaxial space of muscle spindles. That these features should be especially well developed in spindles is presumably because in addition to metabolic insulation there is the specific need to protect the sensory endings from the lateral pressure exerted by muscle contraction. It must be noted, however, that there is no periaxial space in tortoise spindles (CROWE and RAGAB, 1970a).

SHANTHA et al. (1968) found no change in the capsule 14 months after the denervation of spindles in cat hindlimb muscles. There was no increase in thickness, owing to cell proliferation, and no change in enzyme equipment. Thickening of the capsule and enlargement of the periaxial space are, however, characteristic features of a number of muscular and neuromuscular diseases (see CAZZATO and WALTON, 1968).

2. *The Intrafusal Muscle Fibres*

a) Recognition of Two Types. It was long recognized that some intrafusal muscle fibres were of large diameter and others small (see e.g. SHERRINGTON, 1894; CUAJUNCO, 1927, 1940), but the majority of early workers supposed that the small fibres were derived from the large by branching. Thus according to SHERRINGTON (1894) and RUFFINI (1898), large 'parent' fibres in the proximal pole split longitudinally to form smaller 'daughter' fibres, which course through the equatorial region to taper off as long tendinous filaments in the distal pole. BATTEN (1897) maintained that branching in one pole is followed by reuniting of the fibres in the other, and FORSTER (1894) held that branching and reuniting of fibres occurs throughout the length of the spindle. Against this it was urged by KERSCHNER (1888a), BAUM (1900), CUAJUNCO (1927, 1940), and BARKER (1948) that intrafusal fibres remain single, and that the tapering of short fibres alongside longer ones may have been mistaken for instances of branching. The situation became clearer when it was realized that differences in fibre length and diameter were also correlated with differences in equatorial nucleation (COOPER and DANIEL, 1956; BOYD, 1956) so that intrafusal muscle fibres could be classified as long and thick with nuclear bags (nuclear-bag fibres) or short and thin with nuclear chains (nuclear-chain fibres) (BOYD, 1960), which atrophy at different rates after ventral root section (BOYD, 1962a, c).

Once recognized, the presence of bag and chain fibres was reported in the spindles of a variety of mammals, the chains usually outnumbering the bags. Thus the usual complement in rat spindles is 2 bag fibres and 2 chain fibres; in the cat 2 bag and 4 chain fibres; in the dog 2 bag and 6 chain fibres (BANKER and GIRVIN, 1971); in human spindles a range of 1–5 bag fibres and 2–11 chains (CAZZATO and WALTON, 1968), and so on. Mean juxta-equatorial diameters and standard deviations given by BOYD (1962c) for the two types of fibre in cat tenuissimus spindles are 16.86 ± 2.35 μm (range 12–23 μm) for bag fibres and 8.37 ± 1.85 μm (range 4–13 μm) for chains, with mean lengths of 6.83 ± 1.96 mm (range 4.0–11.5 mm) and 3.56 ± 1.08 mm (range 2.0–6.0 mm), respectively (Susa fixation, paraffin sections). Two to three months after cutting the ventral roots the diameter of the chain fibres is reduced from the normal by about 35%, whereas that of the bag fibres is reduced by only about 10% (BOYD, 1962c). After dorsal root section the intrafusal muscle fibres gradually lose their central nuclei and increase slightly in equatorial diameter over a period of several months, so that after about a year their morphology is the same as that in the polar regions, the bags and chains having disappeared and been replaced by myofibrils (TOWER, 1932; BOYD, 1962c). In the atrophy following nerve section (neurogenic atrophy) MAYNARD and TIPTON (1971) found that after eleven weeks the chain fibres of rat spindles had decreased in cross-sectional area by 18.5% as compared with a 15.4% decrease in bag fibres. After the immobilization of a limb, however, (myogenic atrophy) it is the bag fibres that show the greatest decrease in size (MAIER, ELDRED and EDGERTON, 1972).

Both types of fibre become thinner as they pass through the equatorial region, the diameter of the small fibres reducing to little more than that of each nucleus in the nuclear chain. Equatorial nucleation and decrease in diameter account for a sharp drop in myofibril content. Banks, James and Meek (1972) find that in the equatorial regions of rat and cat bag fibres, the myofibrils are reduced in cross-sectional area by at least 80% compared with their polar regions. Each nuclear chain consists of a row of about 20 to 50 nuclei which generally exceeds the length of the nuclear bags and myotube regions of the large fibres. In the cat the nuclear bags are 50–100 μm long flanked on either side by myotube regions, usually of unequal length, 50–150 μm long. They are packed with nuclei 2–3 abreast numbering about 50–100 depending on the length of the bag, the minimum number in transverse section being 3. Bag nuclei are spherical, 6–9 μm in diameter, chain nuclei generally lenticular, 9–12 μm long and not more than 6 μm wide (Winckler and Foroglou, 1972, measurements relating to spindles from sheep extraocular muscles). The maximum width of the bags varies in different species, e.g. in human spindles the nuclei may be packed 5–6 abreast (Cooper and Daniel, 1963; Kennedy, 1970), and there is also some evidence that it varies in different muscles. According to Homma and Seki (1964) the maximum width of the bags, as indicated by the diameter of the widest annulospiral terminal measured in gold chloride preparations of cat and monkey spindles, is greater in the fast tibialis anterior muscle than it is in the slow soleus. The bags lie in parallel with the chains in a localized region of the axial bundle, which may be located at any level within the equatorial region, but usually lies towards one end. The bags themselves do not all lie at the same level, but in such a way that the proximal end of one may begin opposite the distal end of another belonging to a neighbouring bag fibre. Most of the bag fibres extend well beyond the limits of the capsule and finally taper down over a distance of 0.5 mm or so to attach to tendon, aponeurosis, perimysium, or the endomysium of neighbouring extrafusal muscle fibres. The chain fibres taper over shorter lengths; in rat spindles most of them extend beyond the capsule, but in cat and primate spindles the majority attach to its inner wall, or form strong lateral attachments to it before emerging for short distances (Bridgman, Shumpert and Eldred, 1969).

The most obvious differences between intrafusal and extrafusal muscle fibres concern size and innervation, intrafusal fibres being supplied with sensory as well as motor endings. An intrafusal fibre receives the sensory innervation in a region that is both structurally and histochemically different from the rest of the fibre, whereas an extrafusal fibre is both structurally and histochemically uniform along its length (Farrell and Fedde, 1969). The largest extrafusal muscle fibres in the cat's tenuissimus muscle are about 80 mm long and have a diameter around 100 μm. By comparison the largest muscle fibres in tenuissimus spindles, the bag fibres, are about one-tenth as long and one-fifth as thick, their diameter range being similar to that of group I afferent nerve fibres. However, whereas the motor endplate of an extrafusal muscle fibre occupies a length of about 30 μm, the sensory and motor innervation supplied to a bag fibre may occupy a total length twenty to thirty times as great.

b) Are there more than Two Types? In terms of length and diameter intrafusal muscle fibres do not fall cleanly into two classes. In cat spindles there may occasionally be long chain fibres that extend well beyond the limits of the capsule, and, more rarely, short bag fibres that are entirely intracapsular (Bridgman et al., 1969). Histograms of the diameters of intrafusal muscle fibres generally have a bimodal distribution (see e.g. WALKER, 1958; BOYD, 1960, 1962c; SWETT and ELDRED, 1960b; ELDRED, BRIDGMAN, SWETT and ELDRED, 1962), but several investigators recognize a distinct category of medium or intermediate size in addition to large and small fibres (CUAJUNCO, 1927, 1940; BARKER and GIDUMAL, 1961; OGATA and MORI, 1962, 1964; JAMES, 1971a). This could be dismissed as of little significance were it not for the fact that the classification of intrafusal muscle fibres into two types, based on criteria of length, diameter, and equatorial nucleation, is not matched by a comparable duality in terms of enzyme histochemistry. Histochemically three types of intrafusal muscle fibre may be recognized that correspond to glycolytic (type A), oxidative (type B), and oxidative-glycolytic (type C) extrafusal muscle fibres (OGATA and MORI, 1962, 1964, mouse, rat, cat, and human spindles; WIRSEN and LARSSON, 1964, mouse spindles; YELLIN, 1969, rat spindles). OGATA and MORI (1964) correlate oxidative activity with size and distinguish between small 'red' fibres with high activity, 'medium' fibres with moderate activity, and large 'white' fibres with weak activity. Similarly JAMES (1971a) reports that staining for myoglobin in rat and guinea-pig spindles is strong, intermediate, weak or absent according to whether the polar diameters of the intrafusal muscle fibres are respectively large (14–20 μm), intermediate (10–12 μm) or small (7–9 μm). The general conclusion to emerge from YELLIN's (1969) study is that the chain fibres represent a homogeneous category both with respect to their small size and their high succinic dehydrogenase (SD) and phosphorylase (P'ase) activity. Their histochemical profile in these respects generally matches that of the extrafusal type C fibres. The bag fibres, however, display heterogeneity. Some resemble type A (low SD, high P'ase activity) and others type B (moderate SD, low P'ase activity) extrafusal fibres, and the type/size relationship is retained, though on a smaller scale. Another type of bag fibre, usually large, exhibits both low SD and low P'ase activity, and in this respect resembles the multiply-innervated 'type G' fibres present both extra- and intrafusally in sheep extraocular muscles (see BARKER, HARKER, STACEY and SMITH, 1972; BARKER and HARKER, 1972; and pp. 80–83). Each spindle in the muscles examined by Yellin generally contained two or three histochemically distinct fibre types, and there were occasionally some with four.

Against this background it is relevant to recall that BARKER and GIDUMAL (1961) maintained that two types of bag fibre could be distinguished in cat hindlimb spindles, namely, bag fibres and 'intermediate' fibres (see Fig. 14). According to them the bag fibres are the thickest and longest; usually two are present; and their diameter (12–26 μm, mean 16.9 μm) and position in the intrafusal bundle undergo little change. The intermediate fibres are somewhat thinner and shorter and their bags contain fewer nuclei; usually one is present, sometimes two; their position tends to change in the intrafusal bundle, and their diameter (10–16 μm, mean 12.6 μm) undergoes considerable variation. About one in every three spindles examined (chiefly from rectus femoris) possessed fibres of this type. In 10 single rectus femoris spindles the length range of 17 bag fibres was 7.0–8.5 mm as compared with 6.4–7.8 mm for 5 intermediate fibres, and 4.0–6.1 mm for 35 chain fibres. More recently MAYNARD and TIPTON (1971) have shown that the frequency distribution of the cross-sectional area of bag fibres in rat spindles is bimodal with peaks at 7–8 cm^2 and 15–16 cm^2.

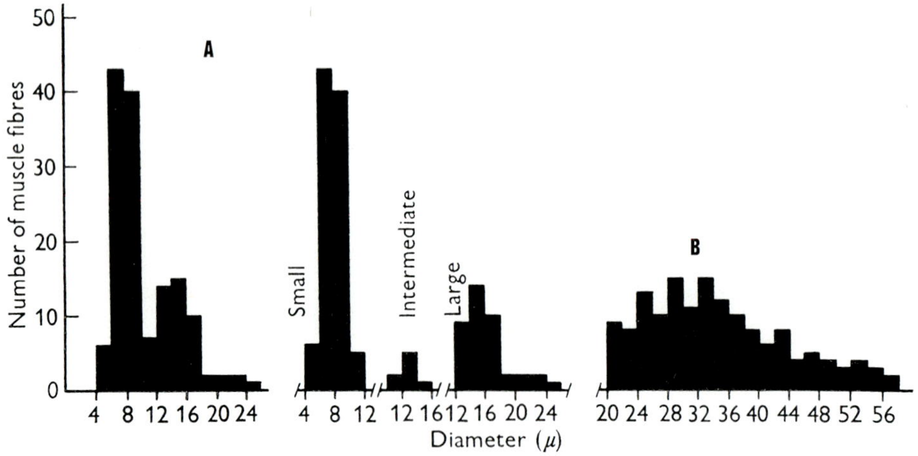

Fig. 14A and B. A Mid-polar diameter histograms of a sample of 142 intrafusal muscle fibres belonging to 18 single and 2 tandem spindles from the rectus femoris muscle of a 2 kg cat. The histogram on the left is broken down on the right in terms of the 94 small, 8 intermediate, and 40 large fibres in the sample. B Diameter histogram of a random sample of 150 extrafusal muscle fibres from the same muscle. Measurements made on Zenker-fixed transverse paraffin sections. (From BARKER and GIDUMAL, 1961)

Further evidence in support of the existence of significantly different types of bag fibre has come from recent studies of rabbit spindles. By combining the use of electron microscopy, histochemical techniques, and teased preparations, BARKER and STACEY (1970) and BARKER, HARKER, STACEY and SMITH (1972) were able to distinguish three types of intrafusal muscle fibre, namely, bag fibres, chain fibres, and intermediate fibres. The differences between them concern length, diameter, equatorial nucleation, ultrastructure and histochemistry. They are summarized in Table 1 and illustrated in Plates 3–5. Morphologically the intermediate fibre is a small bag fibre; it is not quite so thick or so long as the large bag fibres and the nuclear bag itself is smaller, and yet its ultrastructure is similar to that of a chain fibre. Histochemically it has a distinct profile of its own in terms of ATPase and P'ase activity. The ATPase activity, as indicated by the alkali pre-incubated actomyosin ATPase reaction of GUTH and SAMAHA (1969), is either low or medium, and varies inversely with that of the bag fibre(s) in any given spindle. The most usual condition is for the activity to be medium in the intermediate fibre(s) and low in the accompanying bag fibre(s). Various combinations of bag, chain, and intermediate fibres occur, the most frequent being one bag fibre, one intermediate fibre, and two chain fibres. Chain fibres are rarely absent, but 'bag only' spindles consisting of bag and intermediate fibres do occasionally occur (see Plate 3, Fig. 21). Similarly COOPER and DANIEL (1963) found a few small spindles in human neck and lumbrical muscles composed of two bag fibres only (see their Fig. 10A), which may have been combinations of bag and intermediate fibres.

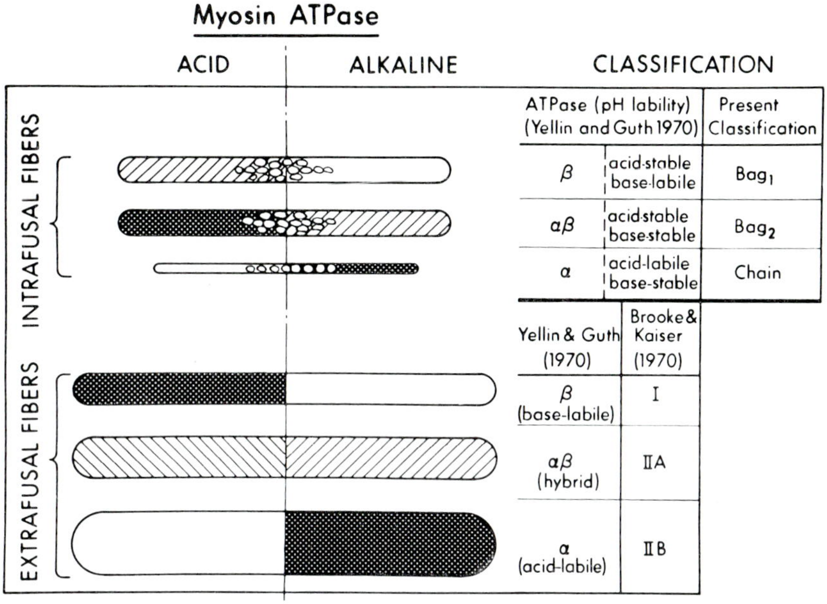

Fig. 15. Schematic representation of the three kinds of intrafusal and extrafusal muscle fibres seen in the monkey and cat after staining for myosin ATPase. The staining reactions after acid pre-incubation (pH 4.35) and alkaline pre-incubation (pH (10.4) are shown on the left and right, respectively. Dark staining indicated by white dots on black, intermediate staining by diagonal lines, and light staining shown as blank. The classifications shown for extrafusal muscle fibres are those of Yellin and Guth (1970) and Brooke and Kaiser (1970). The extrafusal terminology of Yellin and Guth has been extrapolated for the intrafusal fibres, but it must be emphasized that the staining reactions of the extrafusal fibres, although similar, are not identical to those of the intrafusal fibres. (From Ovalle and Smith, 1972)

Table 1. *Rabbit intrafusal muscle fibres*

Feature	Nuclear-bag fibres	Intermediate fibres	Nuclear-chain fibres
Number in spindle	1–2	1–2	2–3
Polar length	2.0–3.0 mm	0.2–0.4 mm shorter than bag fibres	1.0–1.5 mm
Mid-polar diameter	14–18 μm	13–16 μm	7–12 μm
Equatorial nucleation	Large bag: nuclei up to 5 abreast	Small bag: nuclei up to 3 abreast	Single row of nuclei
Ultrastructure			
M line	absent	present	present
Z line	thick	thin	thin
S.R.	poor	rich	rich
mitochondria	small, few	large, many	large, many
Actomyosin ATPase activity	low or medium	low or medium	high
Phosphorylase activity	low	medium	high
Succinic dehydrogenase activity	low	high	high

(From Barker, Harker, Stacey and Smith, 1972)

Earlier work with the light microscope on rabbit spindles maintained that they contained bag fibres only, which differed in length (BARKER, 1948) and were only occasionally accompanied by chain fibres (BARKER and HUNT, 1964). SPIRO and BEILIN (1969) also found only bag fibres in rabbit spindles, but distinguished two kinds histochemically on the basis of the myosin ATPase reaction. In ultrastructural studies, DÜRING and ANDRES (1969) briefly refer to the occurrence of thick and thin muscle fibres in the equatorial region of rabbit spindles, and state that the thick fibres 'often' lack an M line (a significant observation in view of Table 1). CORVAJA and POMPEIANO (1970) describe two types of fibre whose ultrastructure resembles the bag and chain fibres of the cat, but their study was confined to the juxta-equatorial and intracapsular polar regions where it is difficult to distinguish between chain and intermediate fibres.

Plate 3, Figs. 16–21

Figs. 16–20. Toluidine-blue-stained 1 μm transverse sections through the equatorial region of a rabbit tenuissimus spindle showing features of equatorial nucleation. The intrafusal bundle is composed of one typical bag (*typ.b.*) and one intermediate bag (*int.b.*) fibre and two chain fibres (*c.*) surrounded by an axial sheath (*ax.sh.*) and suspended in the periaxial space (*). The typical bag fibre is filled with a maximum of five nuclei abreast, whereas the intermediate bag fibre has a maximum of three abreast (Fig. 19); the myotube nucleation of these fibres is shown in Figs. 20 and 17, respectively. *nf.* nerve fibre. (Figs. 16, 18–20 from BARKER, HARKER, STACEY and SMITH, 1972)

Fig. 21. Transverse section through the equatorial region of a rabbit vastus intermedius spindle composed of four muscle fibres whose equatorial nucleation suggests that they consist of two typical bag (*a, b*) and two intermediate bag (*c, d*) fibres. 10 μm paraffin section stained with haematoxylin and eosin. (From BARKER and HUNT, 1964)

Plate 4, Figs. 22–24

Figs. 22–24. Equatorial nucleation of rabbit intrafusal muscle fibres in a tenuissimus spindle. Electron micrographs of longitudinal sections through a typical bag fibre (Fig. 22), an intermediate bag fibre (Fig. 23), and a chain fibre (Fig. 24). The nuclei in the chain fibre form a single row, in the intermediate bag fibre a double row (transverse sections show three rows; see Plate 3, Fig. 19), and in the typical bag fibre they are packed up to four abreast. Note difference in diameter between the three fibres, and peripheral shell of myofibrils (*myo.*) enclosing their nuclei. Axon terminals of the primary ending (*pr.ax.t.*) wrapping round the fibres are seen cut in transverse section. One terminal (*) is completely enclosed within the intermediate bag fibre

Plate 5, Figs. 25–29

(From BARKER, HARKER, STACEY and SMITH, 1972)

Figs. 25 and 26. Histochemical activity in rabbit intrafusal muscle fibres. Fig. 25. Actomyosin ATPase activity as seen in a transverse section of a flexor hallucis longus spindle is high in the chain fibres (*c.*), medium in the intermediate bag fibre (*int.b.*), and low in the typical bag fibre (*typ.b.*). Fig. 26. Levels of phosphorylase activity in the three types of muscle fibre show a similar pattern in a transverse section of a tenuissimus spindle

Figs. 27–29. Electron micrographs of longitudinal sections of rabbit lumbrical intrafusal muscle fibres show that an M line (*M*), is absent from the H zone (*H*) of the typical bag fibre (Fig. 27), but present in both the intermediate bag fibre (Fig. 28) and the chain fibre (Fig. 29), which also have better developed sarcoplasmic reticulum (*s.r.*). *tr.* triad

Plate 6, Figs. 30–35

Figs. 30–35. Ultrastructural and histochemical differences between intrafusal muscle fibres in spindles from rat peroneus longus muscle. Figs. 30–32. Successive serial transverse sections through the pole of a spindle stained for the alkaline pre-incubated myosin ATPase reaction (Fig. 30), the phosphorylase reaction (Fig. 31), and succinic dehydrogenase (Fig. 32). *typ.b.*, *int.b.*, typical and intermediate bag fibres, *c.* chain fibre. Figs. 33–35. Electron micrographs of longitudinal sections at polar level through typical bag (Fig. 33), intermediate bag (Fig. 34), and chain (Fig. 35) fibres. Note absence of the M line from the H zone (*H*) of the typical bag fibre, and its presence (*M*) in the chain fibre. In the intermediate fibre it is present, but less distinct than in the chain fibre. *tr.* triad; *s.r.* sarcoplasmic reticulum. (Preparations and micrographs by Dr. A. MILBURN)

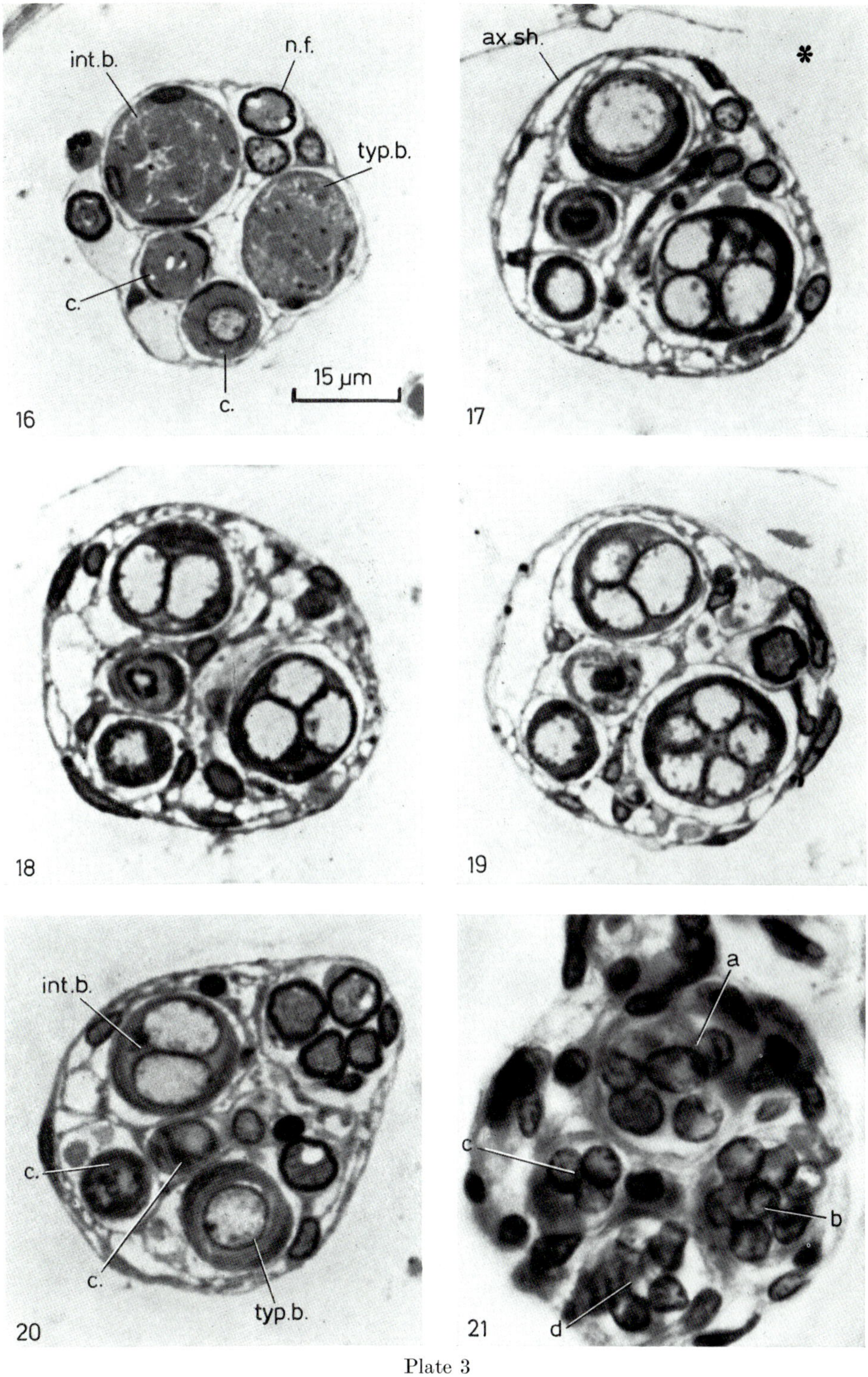

Plate 3

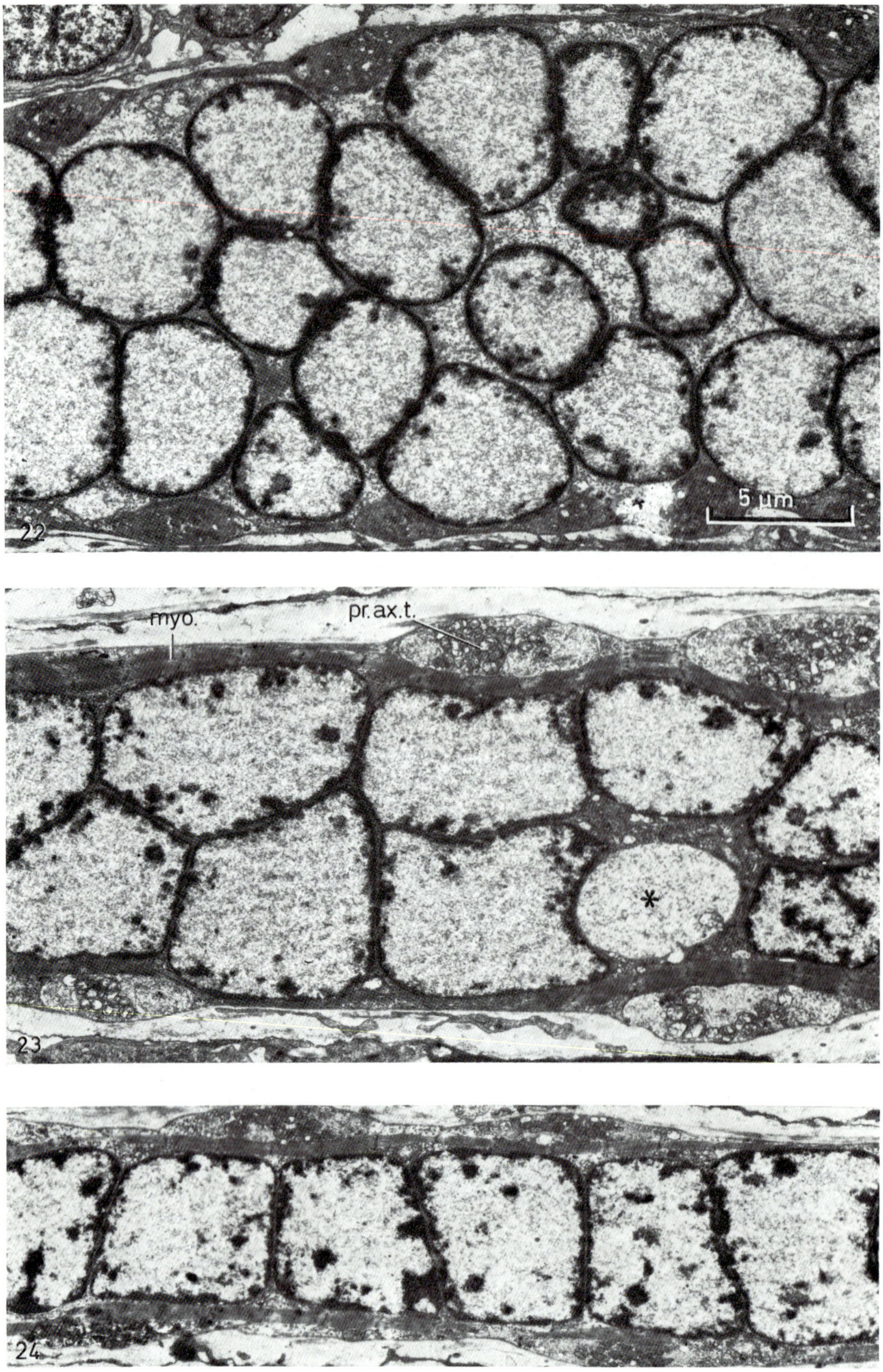

Plate 4

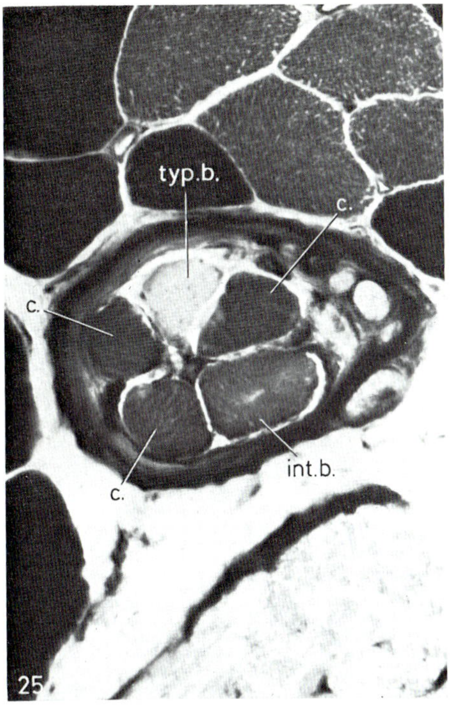

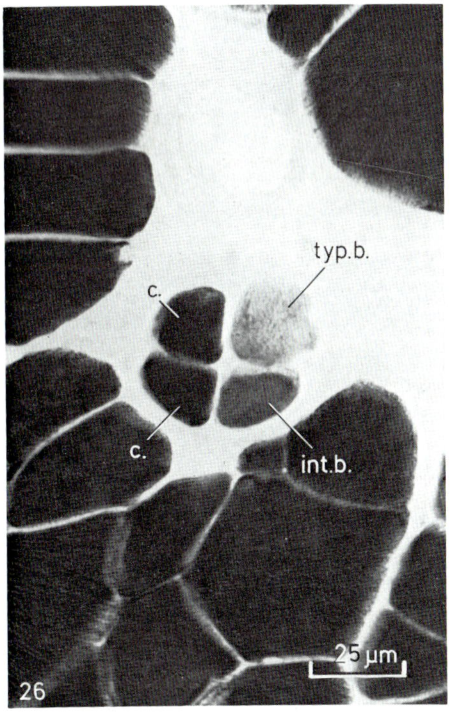

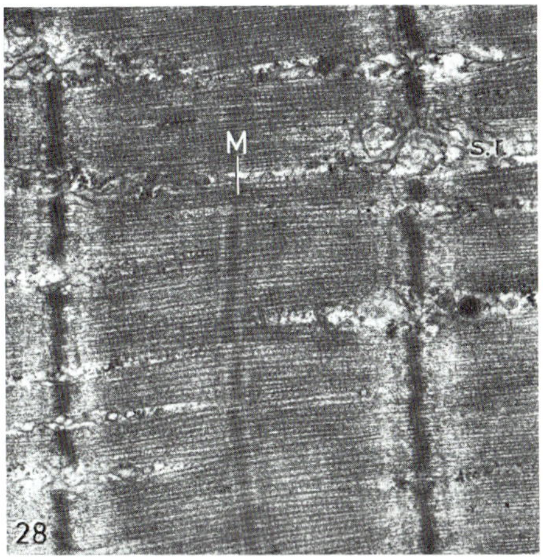

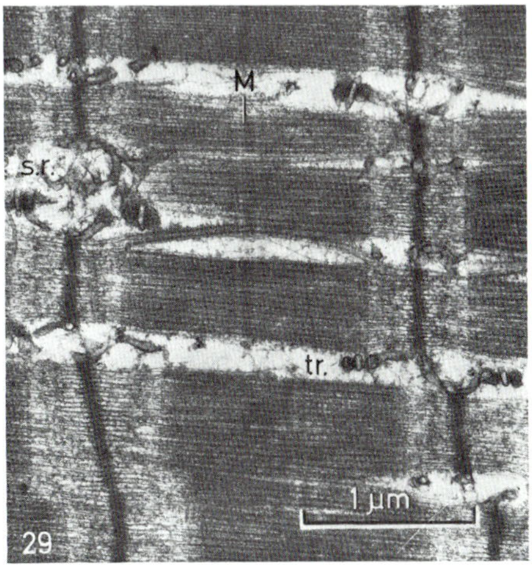

Plate 5

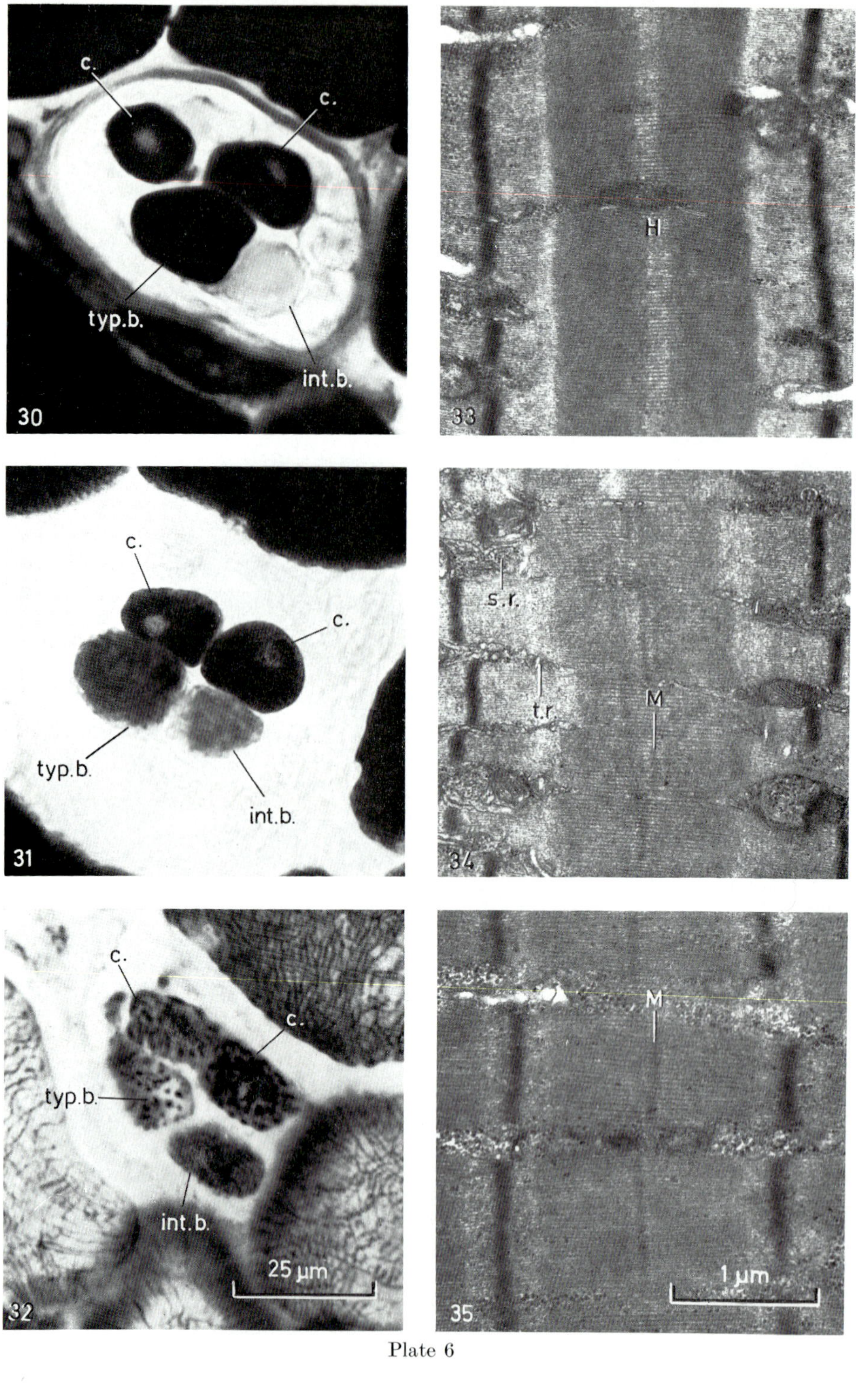
30
c.
c.
typ.b.
int.b.
33
H
31
c.
c.
typ.b.
int.b.
34
s.r.
t.r.
M
32
c.
c.
typ.b.
int.b.
25 µm
35
M
1 µm

Plate 6

That there are different histochemical types of bag fibre is now well established. BANKS (1971) identifies three enzyme profiles in rabbit spindles as belonging to chain fibres and two types of bag fibre. JAMES (1971b) similarly distinguishes three histochemical types of muscle fibre in rat spindles. He designates the chain fibres as Type 1, the intermediate fibres as Type 2, and the bag fibres as Type 3. OVALLE and SMITH (1972) demonstrate three types of fibre in cat and rhesus monkey spindles on the basis of the myosin ATPase reaction. They distinguish between bag fibres that contain the acid-stable form of myosin ATPase only (designated by them as 'bag$_1$ fibres'), and bag fibres that contain ATPase detectable under both acid and alkaline conditions ('bag$_2$ fibres'). The chain fibres contain the alkaline-stable form of ATPase only (see Fig. 15). They consider that the bag$_2$ fibres probably correspond to the intermediate fibres recognized by BARKER and STACEY (1970) in rabbit spindles.

Further indications of heterogeneity among bag fibres have emerged from ultrastructural studies of intrafusal muscle fibres, which will now be considered in detail.

c) Ultrastructure of Intrafusal Muscle Fibres. The following account is based on the descriptions of MERRILLEES (1960, rat), GRUNER (1961, man), LANDON (1966, rat), CORVAJA, MARINOZZI and POMPEIANO (1967, cat), CORVAJA et al. (1969, cat), ADAL (1969, cat), RUMPELT and SCHMALBRUCH (1969, man and rat), BARKER and STACEY (1970, rabbit), GOGLIA (1970, cat, rabbit and rat), BANKER and GIRVIN (1971, dog), SCALZI and PRICE (1971, cat), OVALLE (1971, 1972a, rat), BARKER, HARKER, STACEY and SMITH (1972, rabbit), and unpublished observations made in my laboratory on cat, rat and sheep hindlimb muscles. The most detailed accounts are those of CORVAJA et al. (1969), GOGLIA (1970), BANKER and GIRVIN (1971), and OVALLE (1971, 1972a).

The ultrastructure of nuclear-chain fibres will be described first since their uniform characteristics provide easy reference points for comparison. In these fibres the myofibrils are packed as discrete units in sarcoplasm that is rich in glycogen and contains numerous, large mitochondria and a sarcoplasmic reticulum (SR) that is well developed at the level of the I and Z bands. Transverse sections at this level (Plate 7, Fig. 38) show the myofibrils almost completely encircled by SR elements. Longitudinal sections (Plate 8, Fig. 41) show a distinct M line, usually one and often two triads per sarcomere, and large, long mitochondria that may extend over the length of three or four sarcomeres. The triads are usually located at the junctions of the A and I bands, but also occur well within the A band and are often situated just below the sarcolemma. OVALLE (1971) describes the additional occurrence of dyads, and complex multiple junctional couplings between T tubules and terminal SR cisternae that result in quatrad, pentad and septad configurations. Extensive dilatations of the SR cisternae commonly occur, and are often found in close association with mitochondria or the sarcolemma (OVALLE, 1971). In the polar regions the nuclei are usually peripheral, but equatorially they become centrally placed to form a chain, spaced apart and lenticular in longitudinal outline at each end, and packed closely together and rectangular in the middle (Plate 4, Fig. 24). The chain consists either of a single row of nuclei, or two rows in which the nuclei are so packed as to alternate in zigzag fashion. The spaces in between the nuclei at the ends of the chain are full of sarcoplasm containing mitochondria, ribosomes, Golgi complexes and occasional lipoid droplets. The chain of nuclei is surrounded by a thin peripheral shell of myofibrils (Plate 10, Fig. 45) some of which may be orientated obliquely or at right angles to the axis of the fibre (CORVAJA et al., 1969); occasionally the myofibrils may be reduced to a single strand of myofilaments (see GOGLIA, 1970, Fig. 12A). In the equatorial and juxta-equatorial regions two or occasionally three chain fibres may share the same endomysial envelope and here and there enter into regions of close apposition (CORVAJA et al., 1967; ADAL, 1969). In such regions they become surrounded by a common basement membrane, and two fibres thus paired together present a single oval contour in transverse section. Between their closely apposed surfaces a number of *zonulae adhaerentes* occur so as to form a series of button-like attachments (Plate 9, Fig. 42). As observed with the light microscope these regions of intimate association appear to be places where chain fibres branch or reunite, and were described as such by BARKER and GIDUMAL (1961).

In what may for convenience be designated 'typical' as opposed to 'intermediate' bag fibres there is very little interfibrillar sarcoplasm, little glycogen, poorly developed SR, and mitochondria that are thinner, shorter and fewer than in chain fibres. Transverse sections show the myofibrils to be poorly defined and tightly packed together so as to form a more or less continuous bundle with SR only occasionally encountered at the level of the I and Z bands (Plate 7, Fig. 36). In cat tenuissimus spindles GOGLIA (1970) found the mean diameter of myofibrils in typical bag fibres to be 0.540 μm as compared with a mean diameter of 0.205 μm in chain fibres. The Z lines are thicker than those in chain fibres, but the pseudo-H band is ill-defined, and the M line is either absent (Plate 5, Fig. 27) or present as two parallel thin lines separated by a slightly lighter interspace (Plate 8, Fig. 39). In contrast with chain fibres junctional couplings between T tubules and SR cisternae are rare. Dyads occur, and one triad is found in every 2–4 sarcomeres at the A–I junction, or within the A band, but complex junctional couplings and dilated SR cisternae are absent (OVALLE, 1971). As in chain fibres accumulations of micropinocytotic vesicles are present immediately beneath the sarcolemma. In the myotube region a row of elongated nuclei lies within a core of sarcoplasm surrounded by a peripheral shell of myofibrils. In between adjacent nuclei there are many small mitochondria, clusters of ribosomes, Golgi complexes, and occasional rough endoplasmic reticulum. OVALLE (1972a) describes the occurrence of membrane-bound sarcoplasmic granules (diameter 0.06–0.30 μm) closely associated with each Golgi complex. They are distributed in the sarcoplasmic cores of both bag and chain fibres, as well as under the sarcolemma and within interfibrillar spaces. Ovalle suggests that they may be secretory products. At the level of the nuclear bag the nuclei greatly increase in number so as to fill the muscle fibre up to five abreast, closely packed together and generally polyhedral in outline (Plate 4, Fig. 22; Plate 10, Fig. 44). In the cat a thin continuous peripheral rim of myofibrils encloses the bag, some of which may be orientated obliquely or at right angles to the axis of the fibre (CORVAJA et al., 1969). In the rat LANDON

(1966) describes the myofibrils as splitting into a number of distinct bands, which appear to have longitudinal continuity with the myofibrils of the myotubes on either side. In the dog BANKER and GIRVIN (1971) find the myofilaments "poorly aligned, haphazardly distributed and markedly decreased in number" with Z lines split into "dense bodies" with diameters measuring 700 to 4000 Ångstroms. According to them the ultrastructure of bag fibres undergoes a gradual change between the polar and myotubular regions that involves the loss of the M line and the H and I bands, and a reduction in thickness of the Z lines.

Intermediate forms of bag fibre may vary in ultrastructure in one or several respects from typical bag fibres. In rabbit spindles, as previously mentioned (p.20), the ultrastructure of the intermediate fibre is similar to that of the chain fibre; an M line is present, the SR is well developed, and there are large and numerous mitochondria. Such fibres nevertheless possess small nuclear bags, are intermediate in length and diameter, and have their own histochemical profile. In sheep hindfoot interosseous spindles an intermediate type of bag fibre is present that has an M line, and the mitochondria are larger and the SR is better developed than in typical bag fibres (HARKER, 1974; see Plate 10). The fibre is also slightly shorter and has a smaller nuclear bag. Its histochemical profile, however, in terms of myosin ATPase, P'ase and SD activity, is the same as that of the chain fibres, i.e. high activity of each enzyme, in contrast to typical bag fibres which exhibit low ATPase and P'ase activity and moderate SD activity. In rat peroneal spindles typical bag fibres lack an M line and exhibit high ATPase, moderate P'ase, and low SD activity, in contrast to the intermediate bag fibres which possess an M line, have slightly larger and more numerous mitochondria, and exhibit low ATPase and P'ase activity and moderate SD activity (MILBURN, 1973; see Plate 6). The presence of the M line in the form of two faint parallel lines, as described by OVALLE (1971) in the bag fibres of rat lumbrical muscles, has not so far been observed in rat peroneal spindles. In cat spindles the M line in typical bag fibres is either lacking, or present as a faint double line (HUBBARD and HESS, 1967), whereas in the slightly shorter and thinner intermediate bag fibres the M line is prominent and the mitochondria are larger and more numerous (see Plate 8).

Microladders (also called 'leptofibrils' or 'λ fibrils') occasionally occur in both bag and chain fibres, usually near the surface (GRUNER, 1961; LANDON, 1966; CORVAJA et al., 1969; OVALLE, 1972a). They consist of alternate narrow dark and broad pale bands and resemble miniature myofibrils (see Plate 12, Figs. 51–53). They may be orientated at right angles to the axis of the fibre, or be longitudinally arranged and extend over a length of up to three sarcomeres. They seem to occur most frequently in the sarcoplasm underlying axon terminals. The occurrence of satellite cells is also a feature common to both bag and chain fibres. These are longitudinally orientated fusiform cells generally found in the polar regions. They lie in a trough on the surface of the fibres and are included within its basement membrane (see Plate 12, Figs. 53, 54). Such cells also occur in association with extrafusal muscle fibres. According to RUMPELT and SCHMALBRUCH (1969) they occur more often on bag than on chain fibres, and are more numerous in human than in rat spindles. They are probably stem myoblasts.

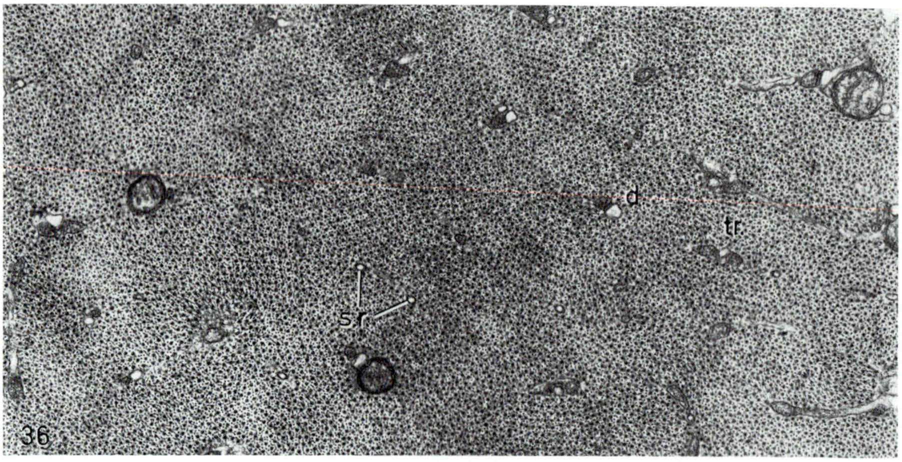

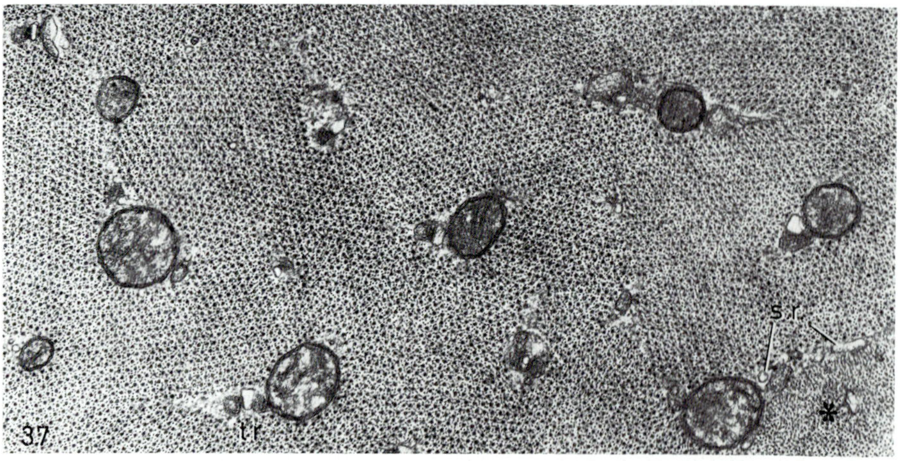

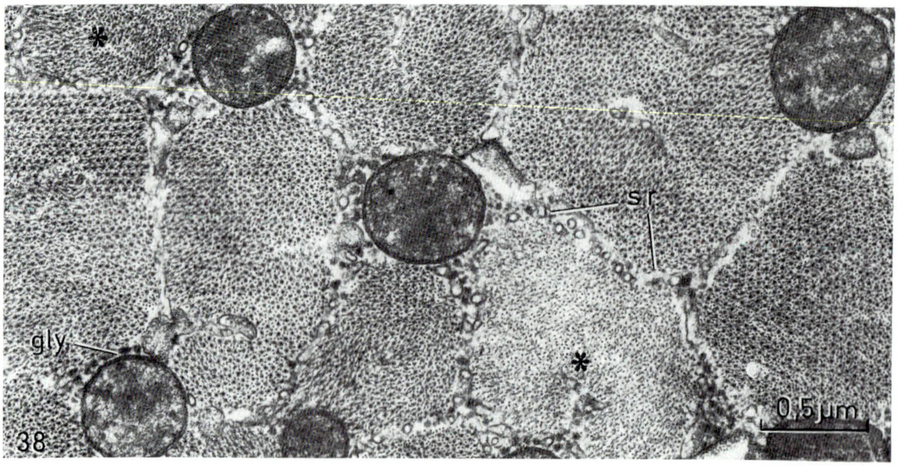

Plate 7

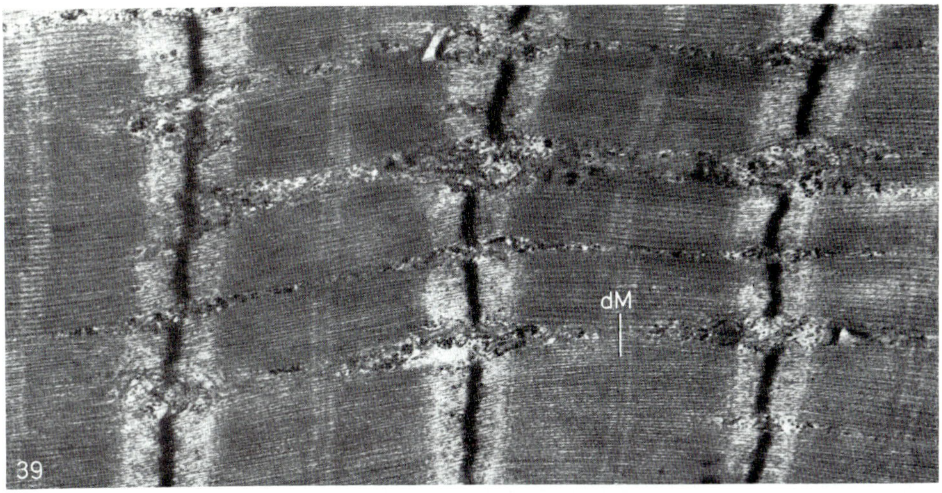

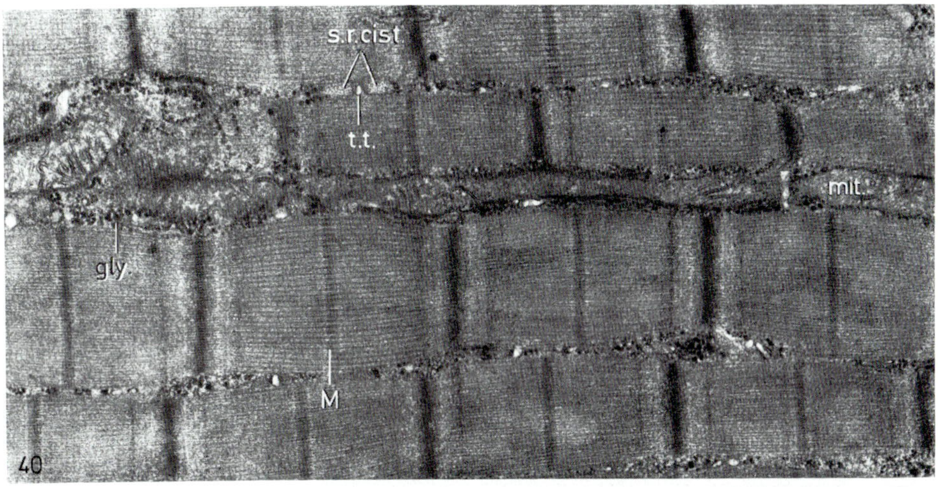

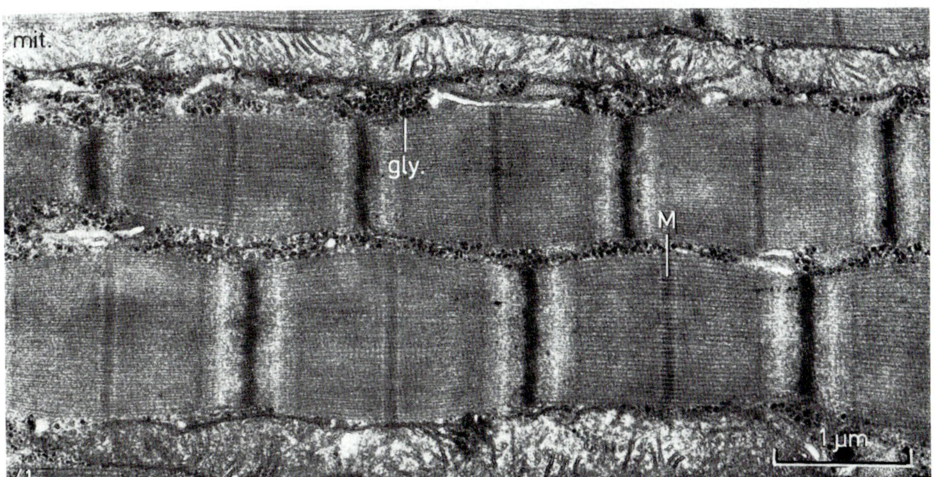

Plate 8

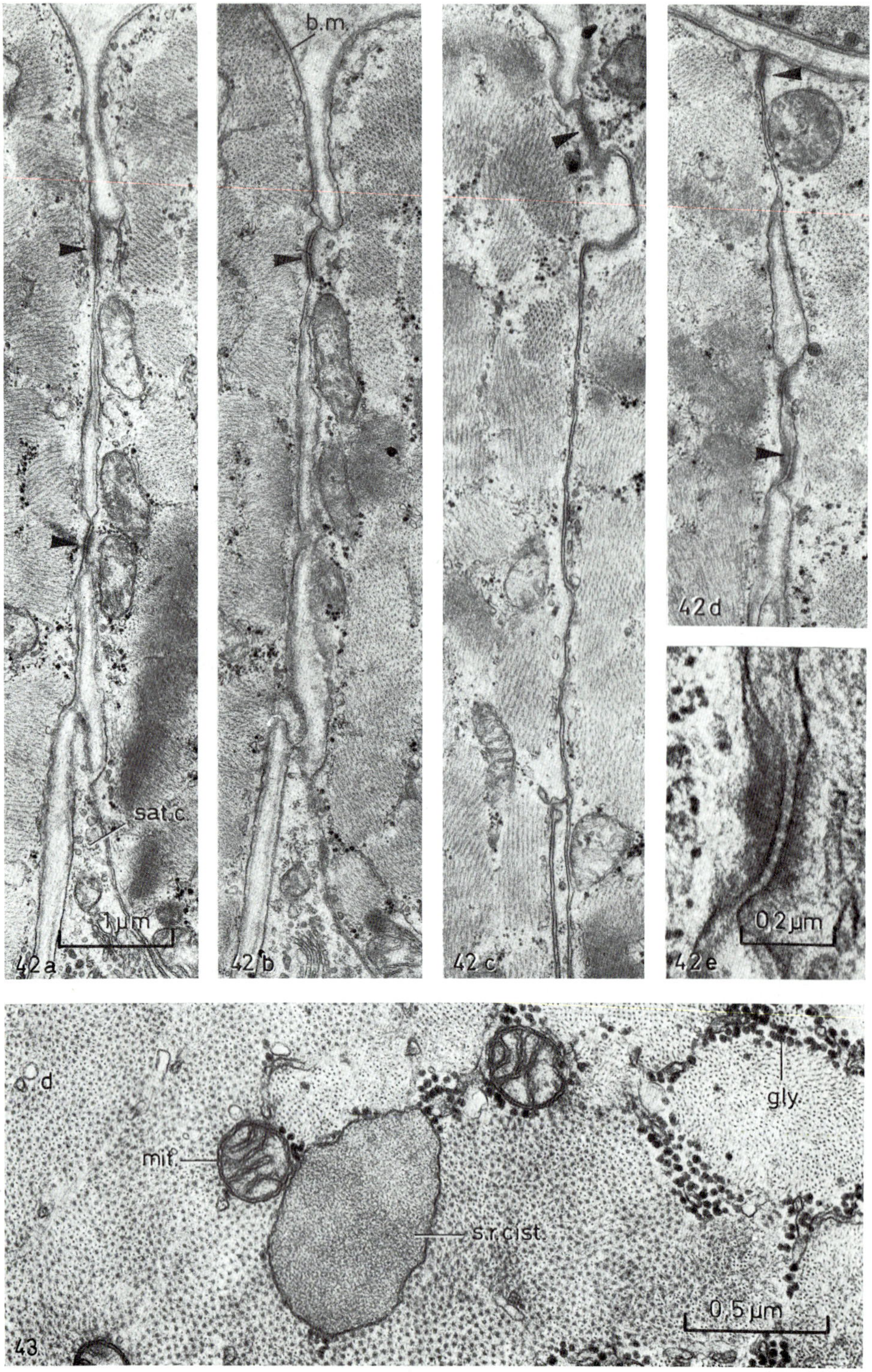

Plate 9

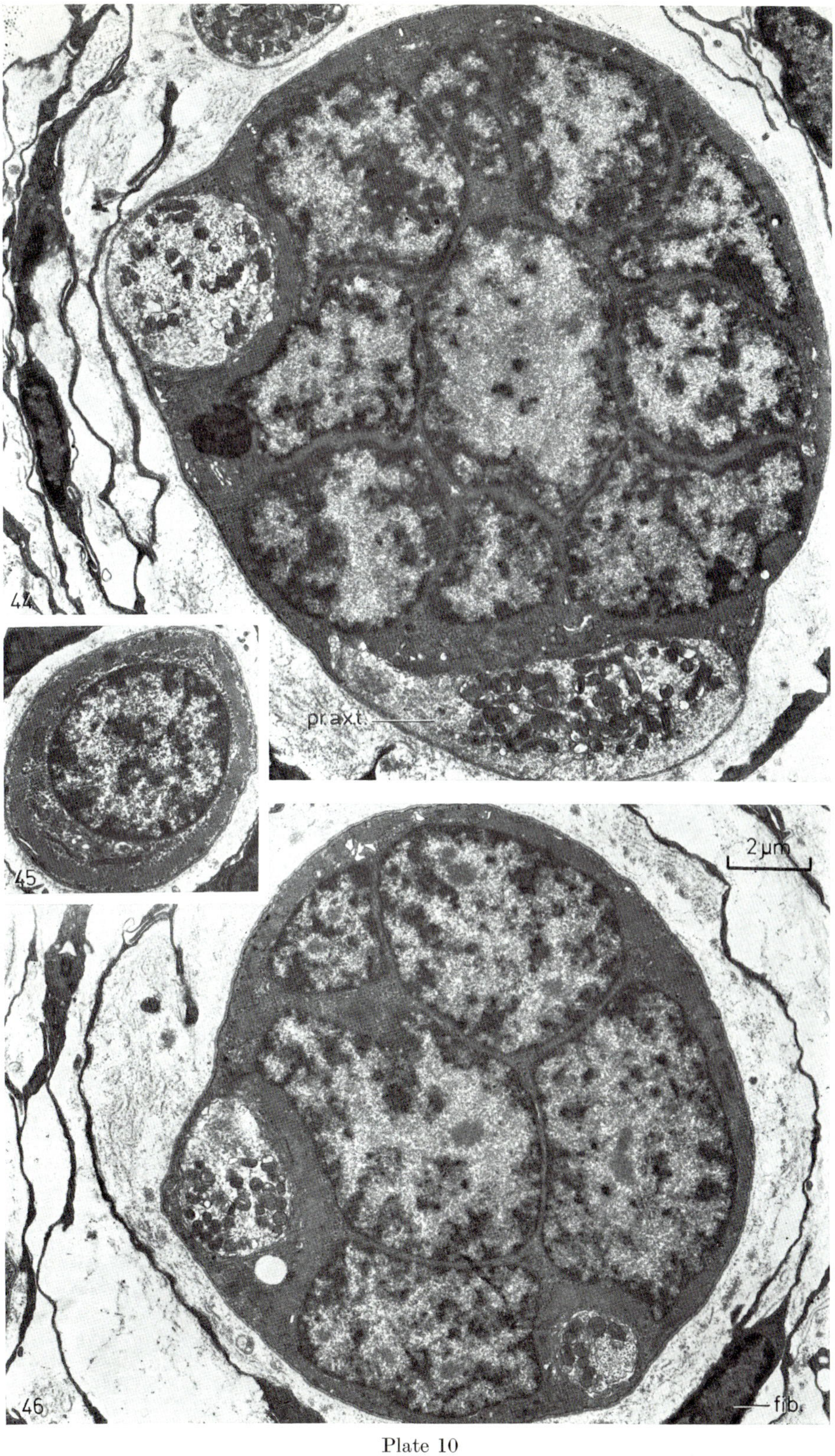

Plate 10

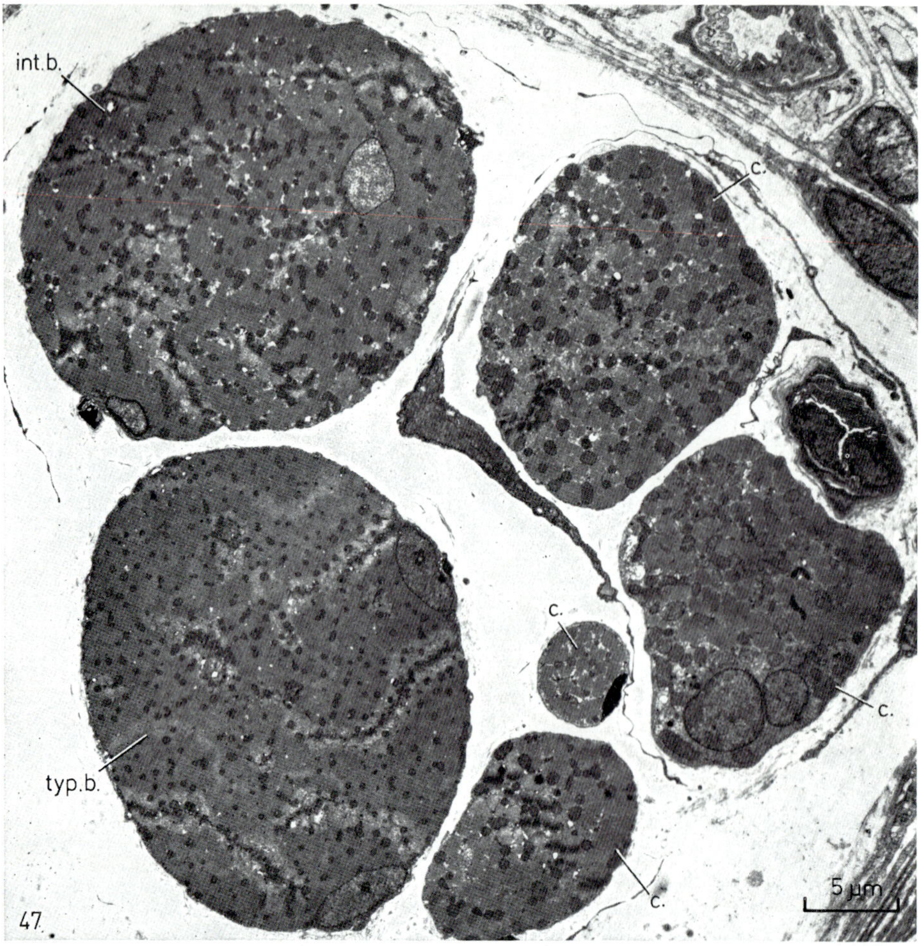

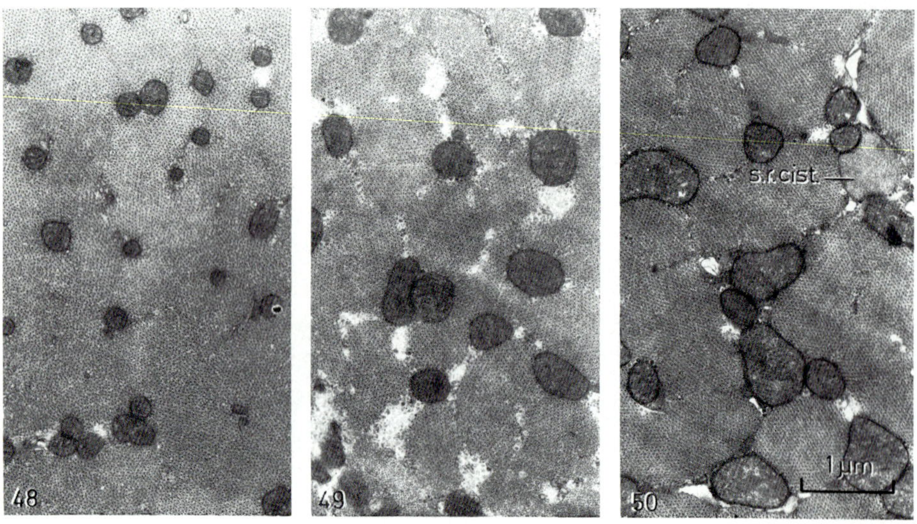

Plate 11

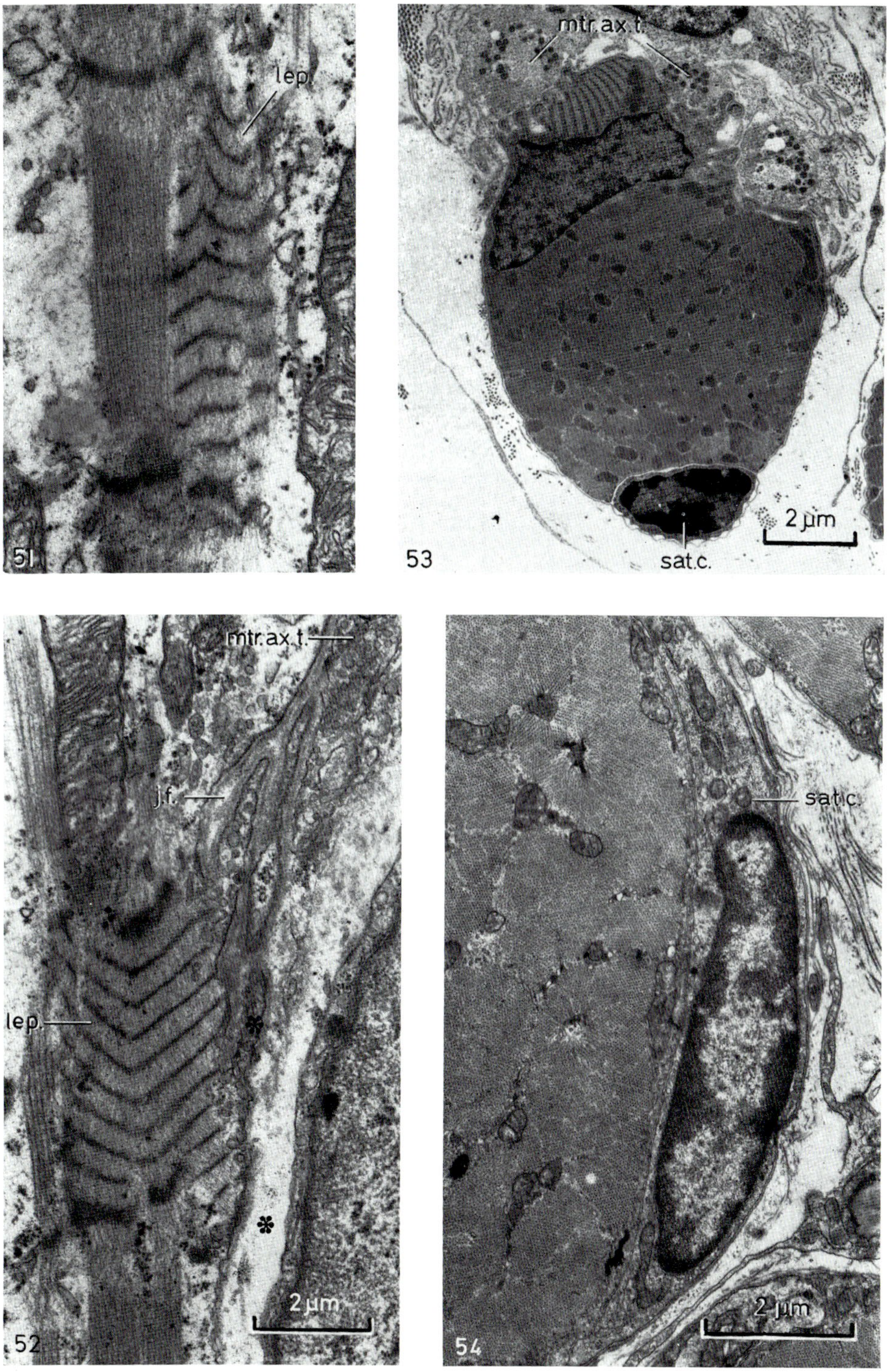

Plate 12

Plate 7, Figs. 36–38

Figs. 36–38. Ultrastructure of cat intrafusal muscle fibres. Electron micrographs illustrate transverse sections at juxta-equatorial level of a typical bag fibre (Fig. 36), an intermediate bag fibre (Fig. 37), and a chain fibre (Fig. 38). The myofibrils are cut through at the level of the A band, except for those asterisked in Figs. 37 and 38, which are cut through at the level of the I band. Note differences in size and number of mitochondria, and development of sarcoplasmic reticulum. A transverse section of the same cat hindfoot lumbrical spindle that contained these muscle fibres is shown in Plate 2; longitudinal sections of the same bag fibres are shown in Plate 8; and transverse sections of the same bag fibres in the region of their equatorial nucleation are shown in Plate 10. *d.* dyad; *gly.* glycogen; *s.r.* sarcoplasmic reticulum; *tr.* triad

Plate 8, Figs. 39–41

Figs. 39–41. Ultrastructure of cat intrafusal muscle fibres. Electron micrographs illustrate longitudinal sections at polar level of a typical bag fibre (Fig. 39), an intermediate bag fibre (Fig. 40), and a chain fibre (Fig. 41). A distinct M line (*M*) is present in the intermediate bag fibre and chain fibre, but in the typical bag fibre it is present in the form of two faint parallel lines (*dM*). A transverse section of the same cat hindfoot lumbrical spindle that contained these muscle fibres is shown in Plate 2, and transverse sections of the same bag fibres in the region of their equatorial nucleation are shown in Plate 10. *gly.* glycogen granules; *mit.* mitochondrion; *s.r.* sarcoplasmic reticulum; *s.r.cist.* terminal cisternae of sarcoplasmic reticulum; *t.t.* transverse tubule

Plate 9, Figs. 42 and 43

Fig. 42. Serial transverse sections through the juxta-equatorial region of a spindle from cat lumbrical muscle showing two nuclear-chain fibres in a region of close apposition. Note the reflection of the basement membrane (*b.m.*) between the adjacent muscle fibres. In Fig. 42c the plasma membranes of the two fibres are in close apposition, and the basement membrane is present only at the top of the figure, where it reflects from one fibre to the other. Arrow heads indicate the *zonulae adhaerentes*; one such junction in Fig. 42d is shown at higher magnification in Fig. 42e. *sat.c.* satellite cell. (From CORVAJA, MARINOZZI and POMPEIANO, 1967)

Fig. 43. Part of a transverse section through an intermediate bag fibre of a spindle from rabbit lumbrical muscle illustrating a dilated terminal cisterna of the sarcoplasmic reticulum (*s.r.cist.*) adjacent to a mitochondrion (*mit.*). *d.* dyad; *gly.* glycogen granules

Plate 10, Figs. 44–46

Figs. 44–46. Equatorial nucleation of cat intrafusal muscle fibres in a hindfoot lumbrical spindle. Electron micrographs illustrate transverse sections of a typical bag fibre (Fig. 44), a chain fibre (Fig. 45) and an intermediate bag fibre (Fig. 46). Note difference in diameter between the three types of fibre (all micrographs at same magnification), and difference in degree of nucleation between the two types of bag fibre. Of the four terminals of the primary ending (*pr.ax.t.*) that have been sectioned, three are of the central rod type lying either in a deep trough on the muscle-fibre surface (as in Fig. 44), or enclosed within the muscle fibre just beneath the surface (as in Fig. 46). A transverse section of the same spindle that contained these muscle fibres is shown in Plate 2, and longitudinal sections of the same bag fibres at polar level are shown in Plate 8. *fib.* fibrocyte forming endomysial envelope

Plate 11, Figs. 47–50

Fig. 47. Low-power electron micrograph of a transverse section through the juxta-equatorial region of a spindle from a sheep's hindfoot interosseous muscle. The intrafusal bundle is composed of one typical bag (*typ.b.*) and one intermediate bag (*int.b.*) fibre and four chain (*c.*) fibres. Note differences in mitochondrial size between the three types of fibre

Figs. 48–50. High-power electron micrographs of transverse sections through the typical bag fibre (Fig. 48), the intermediate bag fibre (Fig. 49), and a chain fibre (Fig. 50) of the spindle shown in Fig. 47. Note differences in mitochondrial size and sarcoplasmic reticulum. *s.r.cist.* dilated terminal cisterna of sarcoplasmic reticulum. (Micrographs by Dr. D.W. HARKER)

d) Conclusions. The classification of mammalian intrafusal muscle fibres into nuclear-bag and nuclear-chain fibres is valid for most species[1], but there is both histochemical and ultrastructural evidence of there being different kinds of bag fibre which may be designated as 'typical' and 'intermediate'. Structurally the features that vary are fibre length and diameter; the size of the bag; the size and number of mitochondria; the degree to which the SR and T systems are developed; and the condition of the M line, which may be a distinct band, a faint double line, or altogether absent. Histochemically bag fibres exhibit a number of different profiles which do not at present appear to correlate consistently with particular structural features. It may well be that such heterogeneity is partly the expression of the variable motor innervation that bag fibres receive from γ fibres (trail endings and p_2 plates) and β fibres (p_1 plates). In cat spindles bag fibres may be supplied with one, two or all three types of motor ending, and the six possible variants may themselves vary in degree of density of innervation, whereas most chain fibres receive trail endings only (BARKER, STACEY and ADAL, 1970). The nature of the 'trophic' influences exercised by motoneurons on bag fibres are therefore likely to be very variable. Such influences, insofar as they determine ultrastructure, evidently continue to operate during the first week or so after birth in the rat, since in 4-day postnatal rat spindles all the intrafusal muscle fibres have the same myofibrillar ultrastructure with M line present (BARKER and MILBURN, 1972), the adult ultrastructural differentiation being achieved 8 days later (MILBURN, 1973). Milburn has shown that intrafusal muscle fibres develop by a sequential process in which the typical bag fibre develops first, followed by the intermediate bag fibre and then the chains (see pp. 89–95). The heterogeneity of bag fibres may thus also be partly due to differences in the duration of their dependence on the morphogenetic influence of the primary afferent.

1 The criterion of differences in equatorial nucleation does not always apply, as R.W. BANKS (personal communication) has recently found in the spindles of guinea-pig lumbrical muscles. Ultrastructurally these have three types of muscle fibre corresponding to chain fibres and typical and intermediate bag fibres. However their equatorial nucleation is sparse, both the intermediate and the chain fibres possessing a single row of nuclei, and the typical bag fibre having only a double row (maximum of two nuclei in cross section). Judged solely in terms of equatorial nucleation all three types of fibre would qualify as chain fibres in cat spindles.

Plate 12, Figs. 51–54

Figs. 51–52. Electron micrographs of longitudinal sections through cat intrafusal muscle fibres (from peroneus digiti quinti) illustrating leptofibrils (*lep.*), which in each case span the length of a sarcomere and part of the width of a myofibril. Each leptofibril lies near the surface of the muscle fibre underneath a motor ending; motor axon terminals (*mtr.ax.t.*) and junctional folds (*j.f.*) are seen in Fig. 52, but in Fig. 51 (same magnification) these structures lie beyond the righthand edge of the photograph. Asterisks in Fig. 52 denote points of attachment of the leptofibril to the sarcolemma

Fig. 53. Electron micrograph of a transverse section through a chain fibre in a cat tenuissimus spindle showing a transversely orientated leptofibril lying underneath motor axon terminals. A satellite cell (*sat.c.*) lies in a trough of the muscle fibre at the bottom of the figure. (This preparation was processed without osmium fixation)

Fig. 54. Electron micrograph of a transverse section through a bag fibre in a spindle from rabbit hindfoot lumbrical muscle showing a satellite cell

Thus the present indications from studies of ultrastructure, histochemistry, and development are that the simple dual classification of mammalian intrafusal muscle fibres is no longer adequate. Indeed histochemical evidence to the contrary dates back to the work of Ogata and Mori (1962, 1964), but it has only recently become clear that it is among the bag fibres that the duality breaks down, as shown by the findings of Yellin (1969), Barker and Stacey (1970), James (1971 a, b), Banks (1971), and Ovalle and Smith (1972). How this may affect terminology remains to be seen. The descriptive terms 'nuclear bag' and 'nuclear chain' have become so well established that a change to Types 1, 2 and 3, as proposed by James (1971a), is unlikely to be acceptable. To retain 'bag' and 'chain', and to distinguish between 'typical' and 'intermediate' bag fibres, or 'bag_1' and 'bag_2' fibres, seems preferable, even in spindles that have a low density of equatorial nucleation.

3. *The Sensory Innervation*

a) The Primary Ending. The afferent fibres that terminate as primary endings are among the thickest found in muscle, and the usual fibre-to-ending ratio is 1:1. Thus von Thiel (1959) found no branching to occur in a primary fibre traced from a spindle in the cat's piriformis muscle to the dorsal root ganglion. Branching to supply primary endings to two spindles lying near together has been observed, but such instances are extremely rare. It may also very occasionally happen that two apparently independent primary fibres enter the same spindle capsule to form a single combined ending (Ruffini, 1898; Barker and Cope, 1962b).

Primary fibres undergo a slight decrease in diameter as they enter the muscle and are distributed in the intramuscular nerve trunks, and a marked decrease as they leave these trunks and approach the spindles. Thus the primary fibres supplying cat rectus femoris measure 12–22 μm in diameter at nerve entry, 9–17 μm in the intramuscular nerve trunks, and 6–15 μm at the sites of the spindles (Adal and Barker, 1962, osmic and gold chloride preparations). The degree to which the diameter decreases between the intramuscular trunk and the spindle depends on the distance involved, so that, for example, a 14 μm fibre drops in diameter by 1.5 μm over 0.5 mm, and by 3.0 μm over 2.0 mm. As the first subdivision prior to termination is approached, however, this reduction in diameter is largely restored. Spindle units that receive one or more secondary endings in addition to the primary generally receive thicker primary afferents than those that do not. The majority of primaries supplying intermediate (one secondary ending) or complex (two to five secondary endings) spindle units are above 10 μm in diameter, whereas the majority of those supplying simple spindle units (primary endings only) are less than 10 μm, as measured in the neighbourhood of the spindles (Adal and Barker, 1962).

The first subdivision of the primary afferent is usually dichotomous and generally takes place within the periaxial space. These first-order branches usually undergo further subdivision to form the terminal branches, which distribute the ending to the nuclear-bag and myotube regions of the bag fibres, and to the adjacent nucleated regions of the chain fibres. A first-order branch may innervate either bag or chain fibres only, or distribute terminals to both types of fibre (Barker and Cope, 1962b). Endings occasionally occur that are restricted entirely to bag fibres (Barker and Cope, 1962b, cat; Jones, 1966, opossum). In human spindles Swash and Fox (1972) describe the primary ending as being predominantly distributed to the bag fibres.

The terminal branches of the ending each give rise to one, or occasionally two systems of annulospiral terminals. These consist of a number of half rings and spirals; complete rings are rare. In rabbit primaries the maximum number of turns made by a spiral terminal is four or five. The rabbit primary ending reconstructed by Barker (1948) consisted of 22 half rings, 3 complete rings, and 12 spiral turns distributed to four muscle fibres[2]. Some terminals are distributed from a central rod and wrap round a muscle fibre to form a configuration that resembles a ribcage. Such terminals may be disposed around two adjacent muscle fibres so as to bind them together. The terminals that wrap around the middle of the nuclear bags are closely set together, but to either side and in the myotube regions they are more widely spaced apart. The few terminals that extend for any distance around the myotube regions are usually of an irregular or clasping form. The terminals on bag fibres are generally thicker than those on the chains (see Plates 13–15).

Primary endings occupy a length of about 300 μm (range 200–400 μm); the longest endings are those formed by afferents that begin to branch before penetrating the capsule. The general appearance of the ending varies in different species. In the cat the terminals on the nuclear bags typically resemble a number of tightly coiled springs and present a strikingly regular appearance. A similar regularity is found in the dog and mouse, and, as Jones (1966) describes, in the opossum, but in man (see Kennedy, 1970), rat, guinea-pig, and rabbit the spiral pattern is less evident (see Barker, 1948). In the cat the ending has an irregular form when it is the only sensory termination present, as in single spindles (Ruffini, 1898; Barker, 1948) or in simple spindle units that form part of tandem spindles (Barker and Ip, 1961).

Electron micrographs of the primary ending (see Plates 10, 16 and 17) show that the axon terminals lie on the surface of the muscle fibres in shallow grooves, the margins of which are prolonged as thin lips of sarcolemma (Landon, 1966). The terminals are not covered by Schwann cells, and are enclosed together with the muscle fibres by a common basement membrane that does not enter the 100–200 Ångstrom wide gap that separates them (Merrillees, 1960). The axon terminals contain numerous small mitochondria, vesicles, some of which contain small granules (Landon, 1966), small tubular systems, neurofilaments, and lipoid droplets. Particles of glycogen also occur either scattered singly or accumulated in dense aggregations, but never in the form of concentric glycogen-membrane complexes such as are found within the cytoplasm of the supplying axons (Corvaja, Magherini and Pompeiano, 1971). The myoneural junction is usually smooth, but some axon terminals have prolongations that interdigitate with the underlying muscle-fibre surface (Corvaja et al., 1969). Here and there the axon

2 These were all regarded as bag fibres, but in the light of subsequent knowledge it is more likely that the intrafusal complement was that which most commonly occurs in rabbit spindles, namely, two bag fibres (one typical, one intermediate) and two chain fibres.

Plate 13, Figs. 55–57

Fig. 55. Equatorial region of a spindle teased from cat peroneus brevis muscle after impregnation with silver. Branches of the primary afferent (*Ia br.*) at top right supply the intrafusal bundle with an annulospiral primary ending. Note how the terminals wrap around the bag fibre (*b.*) nearest the bottom of the figure, consisting of half rings and spirals around the bag and more widely spaced and less regular configurations around the myotube regions. Arrow points to terminals distributed from a central rod. Another bag fibre lies on the upper side of the intrafusal bundle, and the chain fibres lie in the middle (*c.* indicates one of these). *cap.* capsule. (Preparation by Mrs. C. Negus; these, and all subsequent teased, silver preparations illustrated, were processed using the method of Barker and Ip, 1963)

Fig. 56. Part of the equatorial region of a silver preparation of a spindle teased from cat peroneus longus illustrating primary (*pr.e.*) and secondary (*sec.e.*) endings. The secondary ending is of the flower-spray form and is distributed to both bag and chain fibres. II, secondary afferent. (Preparation by Mrs. C. Negus)

Fig. 57. Longitudinal section (20 μm) through equatorial nucleation of two bag fibres in cat hindfoot lumbrical spindle illustrating disposition of primary axon terminals around nuclear bags. Lower bag fibre shows beginning of myotube region on left. Silver method of de Castro (1925). (Preparation by Mrs. M. Cope)

Plate 14, Figs. 58 and 59

Fig. 58. Teased, silver preparation showing a primary ending (*pr.e.*) and three secondary endings (*sec.e.*) in the equatorial region of a cat tenuissimus spindle. The secondary endings are of the annulospiral form and terminate predominantly on the chain fibres. A bag fibre lies on the right of the intrafusal bundle and receives sprays of terminals opposite regions arrowed. *cap.* capsule; *cp.* capillary; I*a br.* branches of primary afferent; II, secondary afferent. (Preparation by Dr. M.C. Ip)

Fig. 59. Teased, silver preparation showing a primary and a secondary ending in the equatorial region of a rabbit peroneus longus spindle. (Preparation by Mr. C.R. Smith)

Fig. 60A–H. Some of Ruffini's drawings of gold chloride preparations of cat spindles that illustrated his classic paper of 1898. A Semi-schematic figure of a spindle with one primary (*pr.e.*) and two secondary (*sec.e.*) endings. The type of secondary shown is the less common flower-spray form. The other endings are motor and were labelled as plate-endings by Ruffini, who regarded them as sensory. They are here identified as trail endings (*tr.e.*) and p_2 plates (p_2 *pl.*). *Cap.* capsule (in gold chloride preparations this is often seen as closely enveloping the intrafusal bundle, the fluid in the periaxial space having been expelled); *int.b.* intrafusal bundle; *n.t.* nerve trunk. B Primary ending of a spindle in which there were also two secondary endings (Ruffini's complex type). Judging by the width of the spirals and the diameters of the muscle fibres, the figure appears to show two chain fibres (*c.*) flanked on either side by two bag fibres (*b.*). C Flower-spray form of secondary ending belonging to a spindle of the complex type. D–H Various types of intrafusal plate-ending selected from ten drawn by Ruffini. D is probably a p_1 plate, E and F p_2 plates, and G and H trail-ending ramifications

Plate 15, Figs. 61–63

Gold chloride preparations (Gairns' [1930] method) of sensory endings in cat spindles

Fig. 61. Equatorial region of a spindle (Ruffini's intermediate type) with one primary (*pr.e.*) and one secondary (*sec.e.*) ending teased from rectus femoris. I*a br.* branches of primary afferent; II, secondary afferent; *fm.ax.* fusimotor axons. (Preparation by Dr. M.C. Ip)

Fig. 62. Equatorial region of a pes interosseous spindle supplied with a primary ending only (Ruffini's simple type). Fig. 62a shows the preparation as demonstrated at the XXI International Congress of Physiological Sciences (Buenos Aires, 1959). Fig. 62b shows the same preparation after it had been squashed by a Congress participant. Pressure applied to the coverslip by the microscope objective has separated the bag (*b.*) and chain (*c.*) muscle fibres to reveal the distribution of the annulospiral terminals. (Preparation initially by Dr. K.K. Yeung)

Fig. 63. Equatorial region of a rectus femoris spindle with a primary ending only. (Preparation by Dr. M.C. Ip)

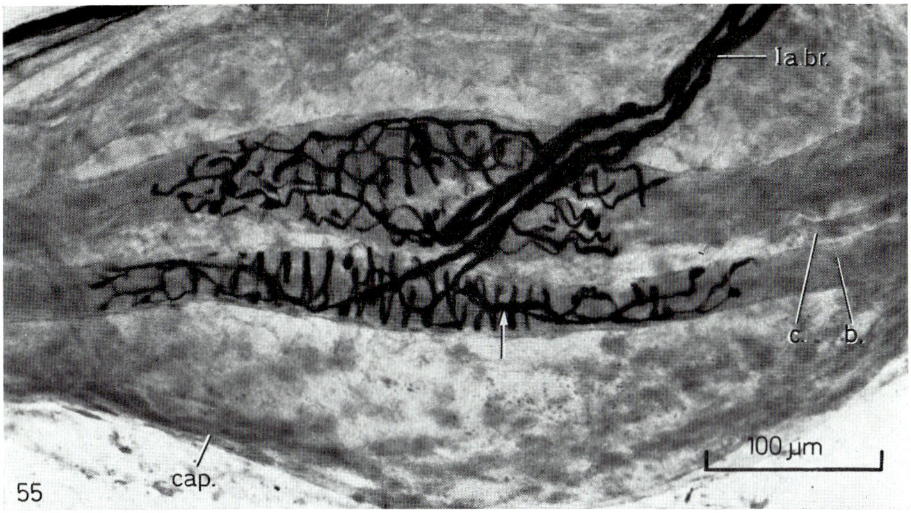

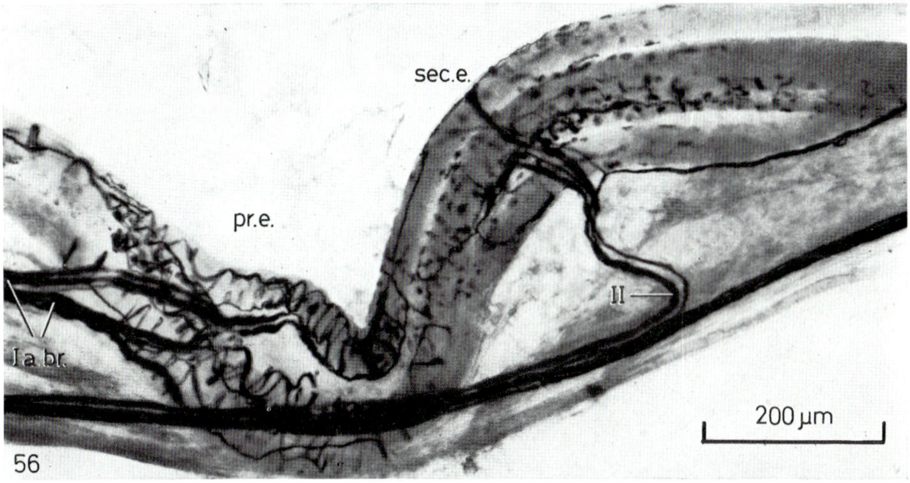

Plate 13

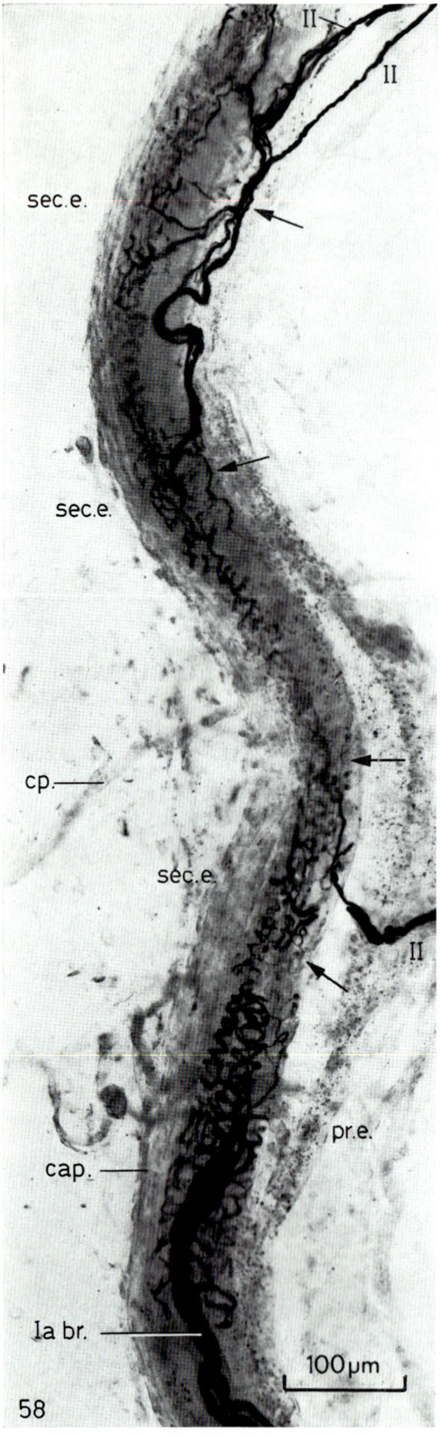
II
II
sec.e.
sec.e.
cp.
sec.e.
II
pr.e.
cap.
Ia br.
100 μm
58

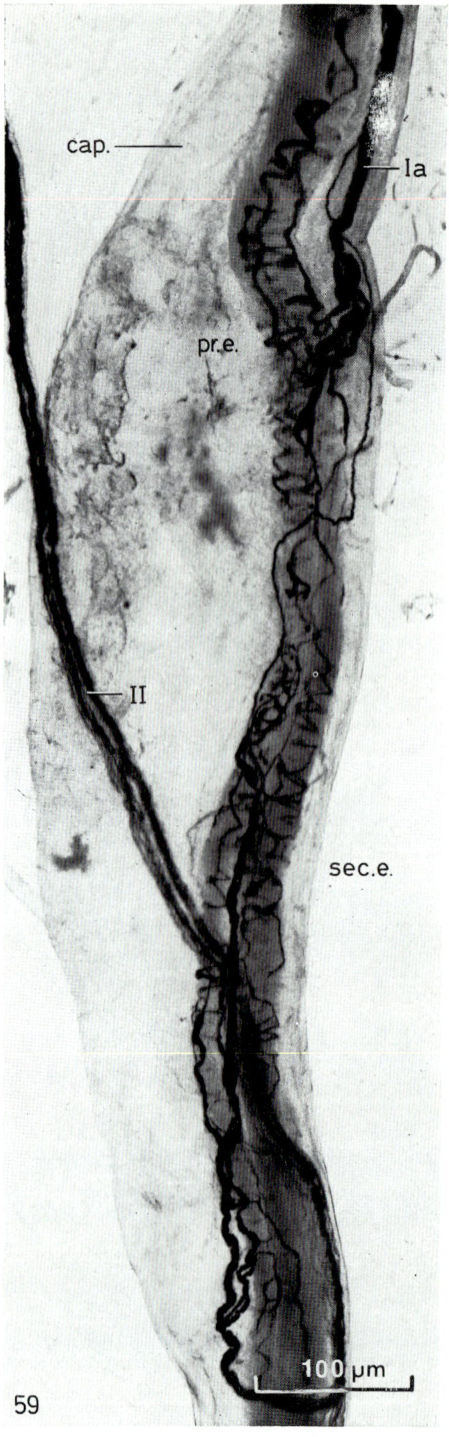
cap.
Ia
pr.e.
II
sec.e.
100 μm
59

Plate 14

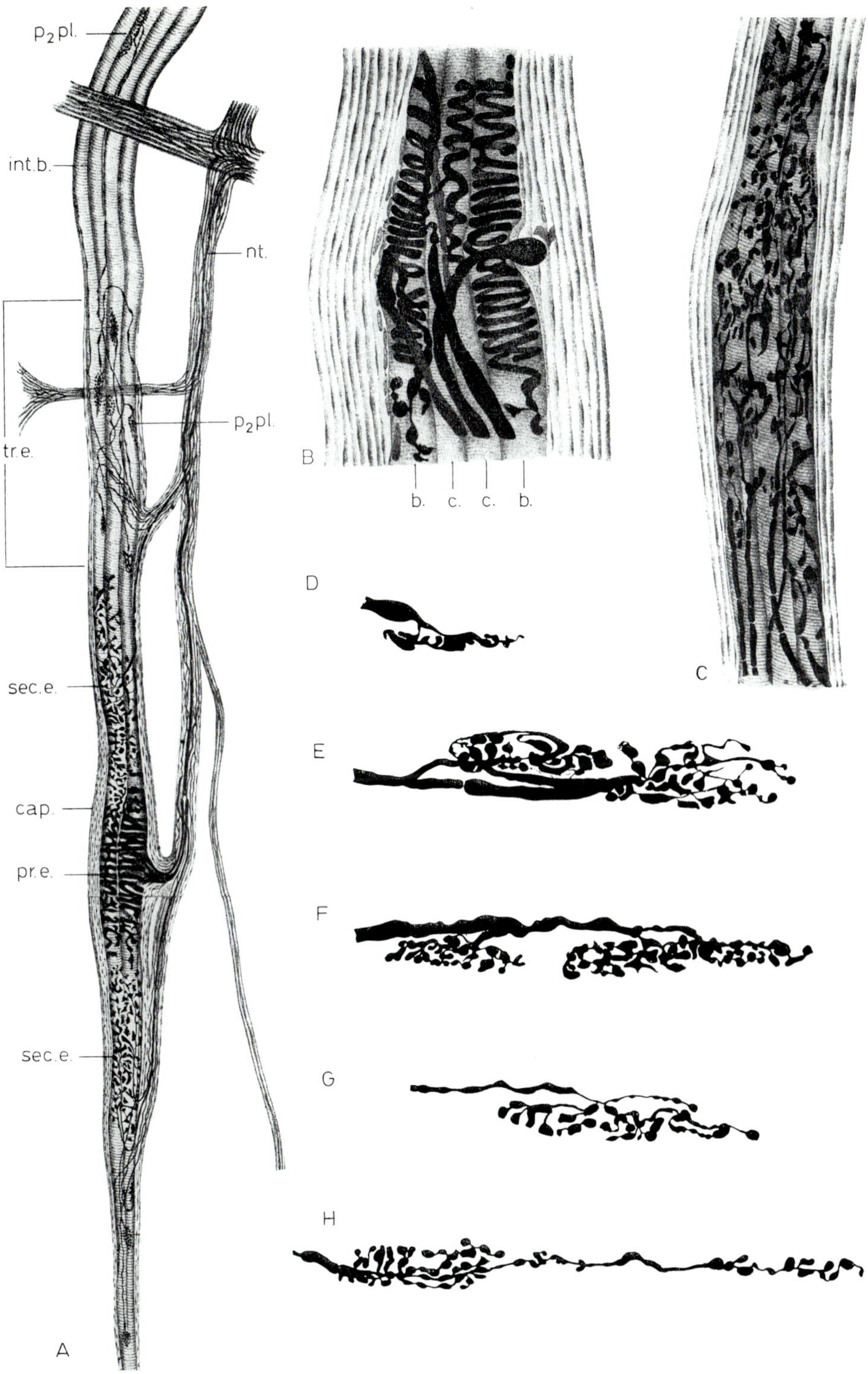

Fig. 60 A–H

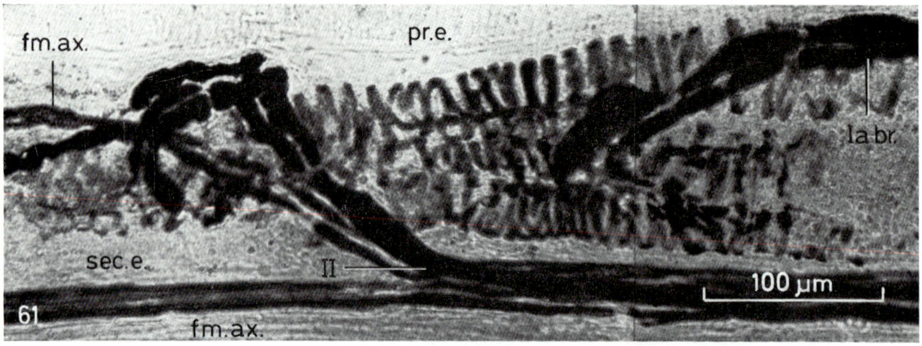
fm.ax.
pr.e.
Ia br.
sec.e.
II
100 µm
61
fm.ax.

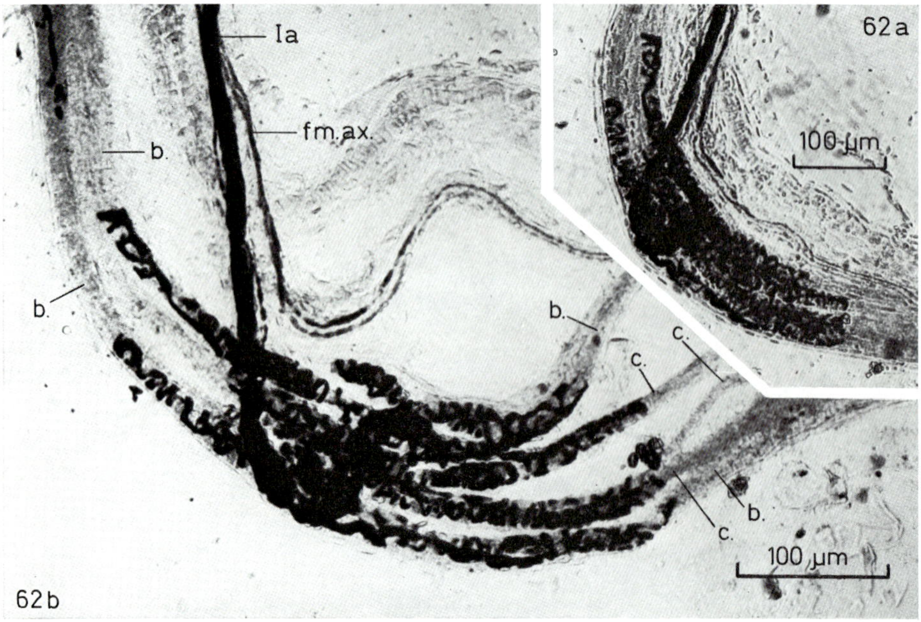
Ia
fm.ax.
b.
b.
b.
c.
c.
c.
b.
100 µm
62b
62a
100 µm

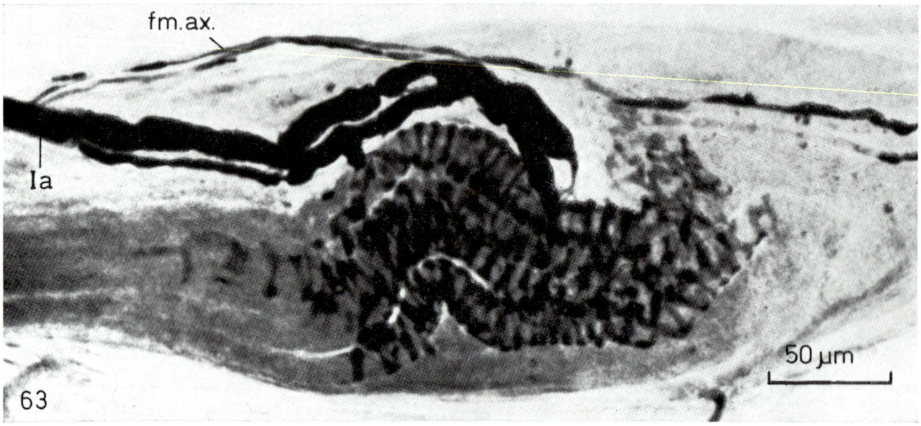
fm.ax.
Ia
50 µm
63

Plate 15

terminal and sarcolemmal membranes are closely apposed to form desmosomes, and in such junctions the sarcolemma may be attached in a trellis-like fashion to an adjacent cisterna of the underlying SR (Düring and Andres, 1969). The terminals on the bag fibres usually bulge outwards from the surface of the muscle fibres, whereas those on the chain fibres are usually low and flattened so that in transverse section nerve and muscle present a circular contour. In such sections the coils of the ending are cut into variously ellipsoid shapes, or may sometimes almost completely encircle a muscle fibre in a horseshoe configuration (Adal, 1969; Corvaja et al., 1969; Hennig, 1969; Banker and Girvin, 1971) (Plate 17, Fig. 67). Where a terminal system springs from a central rod extending along the axis of the muscle fibre, the rod is circular in transverse section, and is embedded in a deep gutter on the surface of the muscle fibre, sometimes being totally enclosed within the sarcoplasm (Plate 10). In some cases an interlocking terminal will make simultaneous synaptic contact with two or three adjacent muscle fibres to form what Adal (1969) has called a "sensory cross-terminal" (Plates 16, 17). Such terminals were also noted by Merrillees (1960) and Gruner (1961); they usually occur between chain fibres, but may also occur between a chain and a bag fibre (Corvaja et al., 1969; Banker and Girvin, 1971). In the one instance so far observed by Stacey and me of a cross-terminal synapsing with a chain and a bag fibre, the bag fibre was of the intermediate type (Plate 16, Fig. 66). This also appears to have been the case in the chain/bag cross-terminal observed by Scalzi and Price (1972) in a monkey spindle.

b) The Secondary Ending. The secondary afferent is smaller than that which supplies the primary ending; from measurements made in gold chloride preparations about 1 mm away from cat tenuissimus spindles Boyd (1962c) obtained a mean diameter of 6.0 μm for secondary afferents as against a mean of 12.4 μm for the primaries (see Fig. 69). When a muscle nerve is severed, the primary

Plate 16, Figs. 64–66

Electron micrographs of sections through the equatorial region of cat spindles illustrating sensory cross-terminals

Fig. 64. Longitudinal section through two chain fibres in a tenuissimus spindle; their nearest polar regions lie beyond the top of the figure. Two cross-terminals are shown. Note elongated nuclei towards the ends of the nuclear chains. Location of the sensory terminals and their low, flat profile in transverse section suggest that they belong to a secondary ending. (This preparation was processed without osmium fixation)

Figs. 65 and 66. Transverse sections through primary cross-terminals linking two chain fibres (Fig. 65), and a chain (*c.*) and an intermediate bag (*int.b.*) fibre (Fig. 66), in a hindfoot lumbrical spindle. A longitudinal polar section of the intermediate bag fibre is shown in Plate 8, Fig. 40, and a transverse section nearer the middle of its nuclear bag is shown in Plate 10, Fig. 46

Plate 17, Figs. 67 and 68

Fig. 67. Electron micrograph of a transverse section through the equatorial region of a rabbit hindfoot lumbrical spindle showing a chain fibre almost completely encircled by a primary axon terminal. Section also shows elastic fibres (*e.*), collagen fibres (*coll.*), and fibrocytes (*fib.*) forming endomysium

Fig. 68. Electron micrograph of a transverse section through the equatorial region of a cat hindfoot lumbrical spindle showing three chain fibres linked together by a primary cross-terminal, and sharing the same endomysial envelope. Note obliquely orientated myofibrils near top of lefthand chain fibre

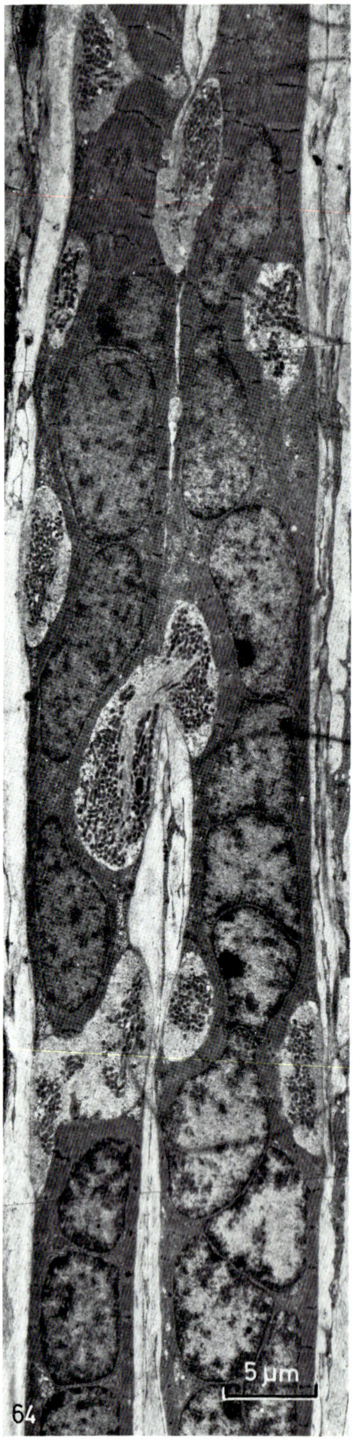

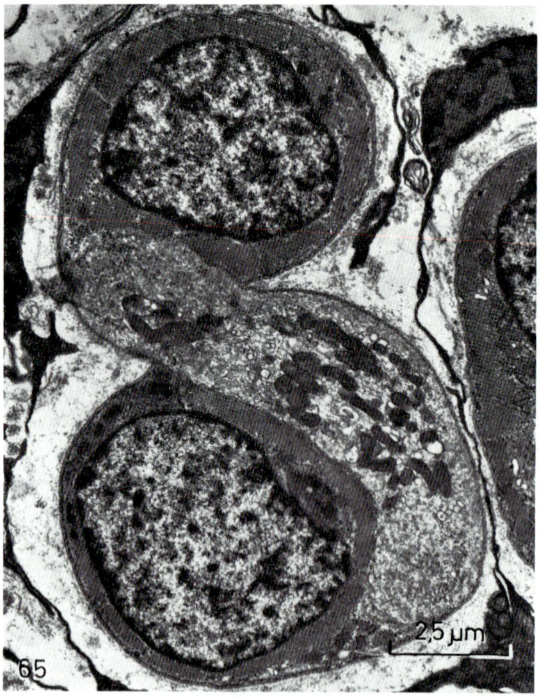

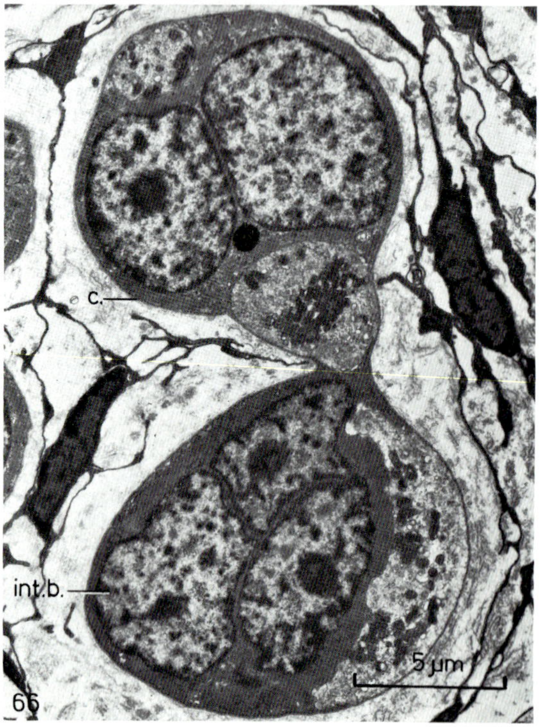

Plate 16

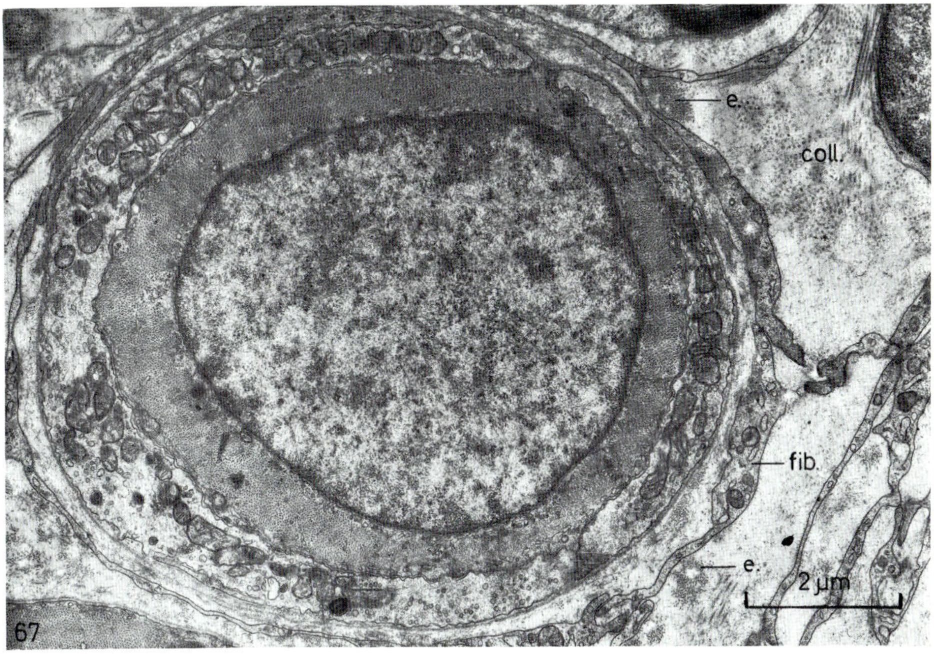

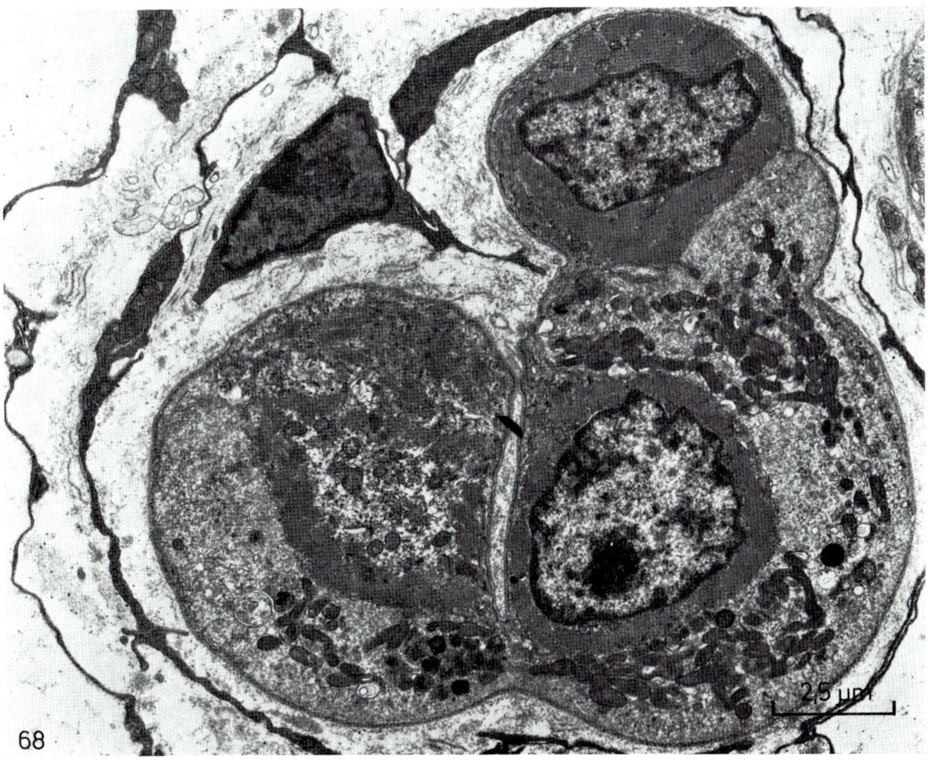

Plate 17

endings, being supplied by the largest afferents, degenerate first; in cat peroneal muscles they disappear 30–48 hours after nerve section, followed by the secondaries at 60–72 hours (BARKER, STACEY and ADAL, 1970). On entering the muscle the secondary afferents decrease in diameter, dropping from a nerve-entry peak of 9.0 μm to one of 6.0 μm near the spindles in cat rectus femoris (ADAL and BARKER, 1962). They may occasionally branch to supply different spindles (WOHLFART and HENRIKSSON, 1960; BOYD, 1962c), but the usual fibre-to-ending ratio is most probably 1:1 as with the primaries. BARKER, IP and ADAL (1962) found branching to occur in only 8 of 279 secondary fibres traced 0.5–4.0 mm from their spindles in cat soleus muscles, and STACEY (1969) found no branching in 40 secondary fibres traced 0.8–9.5 mm back to nerve entry in various cat hindlimb muscles. Similarly JONES (1966) reports a 1:1 ratio in his study of opossum hindlimb spindles.

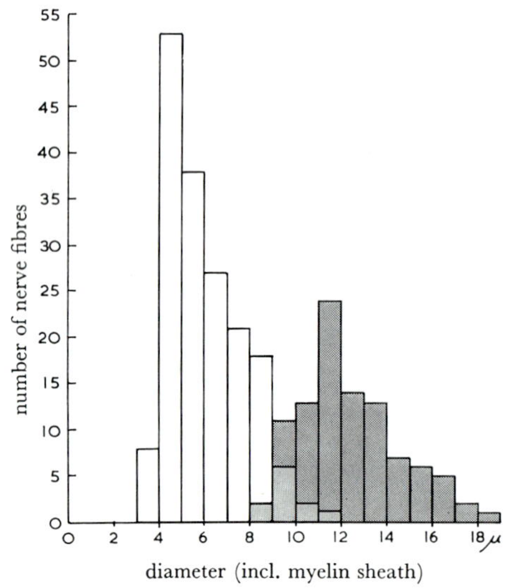

Fig. 69. Fibre-size histogram of the total diameters of cat tenuissimus spindle afferent fibres. Group Ia fibres from primary endings, stippled shading (mean diameter 12.4 μm); group II fibres from secondary endings, uniform dark shading (mean diameter 6.0 μm). Fibre diameters measured in teased, gold chloride preparations about 1 mm from the spindles (11 de-efferented, 91 normal). (From BOYD, 1962c)

Secondary endings are located in the equatorial region on one or both sides of the primary ending (Plate 13, Fig. 56; Plate 14). In rare cases the endings on each side of the primary may be supplied by the same afferent (BARKER, 1948; JONES, 1966). In most cat and rabbit hindlimb muscles the usual condition is for there to be one secondary ending lying adjacent to the primary (RUFFINI's (1898) intermediate type of spindle). From the results of teasing seven entire cat rectus femoris muscles treated by the gold chloride method, BARKER and IP (1961) found the intermediate type of single spindle to occur, on average, with almost double the frequency of either the simple or complex types. Teasing other entire cat muscles indicated that this is also true of soleus, semitendinosus and flexor digitorum longus. In the fifth interosseous muscle of the foot, however, the simple type predominates, and in the fourth internal intercostal muscle, the complex type (BARKER, 1962b).

Secondary endings occupy a length of about 400 μm (range 250–700 μm), those lying adjacent to the primary ending generally being the least extensive. BOYD (1962c) divides the regions on either side of the primary into a series of 400 μm zones, and designates the secondary endings occupying them as S_1, S_2, S_3, and so on, according to their position relative to the primary ending region, designated as P. In 311 spindles teased from cat tenuissimus, soleus, and interosseous muscles, the PS_1 combination accounted for 41% of the spindles, and P spindles and $S_1\ P\ S_1$ spindles each accounted for 16%. The $P\ S_1\ S_2$ and $S_1\ P\ S_1\ S_2$ arrangements were quite common, but spindles with more than three secondaries were rare. The overall ratio of secondary to primary endings in the cat hindlimb muscles studied by BOYD (1962c) and BARKER (1962b) works out at 1.3:1 in a range of 0.85–1.7:1.

The ramifications of the secondary ending are generally more dispersed than those of the primary. In silver preparations (Plate 14) they appear as fine tendrils rather than wide strands of neurofibrils, and wrap around the muscle fibres so as to form a delicate tracery rather than a set of coarse bands. The most common form of the ending is annulospiral (BARKER, 1948), and the terminals are mainly distributed to the chain fibres (BOYD, 1959), which they enlace with coils and spirals, but rarely with complete rings. A few terminals are supplied to the bag fibres, usually in the form of small sprays. A less regular 'flower-spray' form of the ending occurs in cat spindles in which the scattered varicose terminals appear to be distributed more or less equally to both bag and chain fibres (Plate 13, Fig. 56). Of 98 secondary endings in cat rectus femoris spindles, BARKER and IP (1960) found that 36% were of this type. KENNEDY (1970) describes the secondary ending in human spindles as flower-spray, but SWASH and FOX (1972) report the occurrence of both annulospiral and flower-spray types. The secondary endings are stylized in the flower-spray form by RUFFINI (1898) in his schematic Fig. 1 (see Fig. 60A). His anxiety to establish the existence of the secondary ending, then in dispute, may have led him to emphasize this form in cat spindles rather than the more common annulospiral type.

The ultrastructure of the secondary ending is very similar to that of the primary ending, the main criterion for recognition being the position of the terminals relative to the area of equatorial nucleation, and their location mainly on chain fibres. Longitudinal sections through the ending (Plate 16, Fig. 64) show that the terminals lie in shallow grooves, and present low flattened profiles (HENNIG, 1969; CORVAJA et al., 1969). Sometimes the margins of the groove close over the surface of a terminal so that it is completely enclosed within the sarcoplasm. MAYR (1970) has published micrographs showing one chain fibre encircled by an annulospiral terminal, and another supplied with a number of separate knob-like terminals, which are evidently part of a flower-spray configuration. Both types of terminal were connected to the same secondary afferent.

c) Free Endings. Free-ending axon terminals have been observed in association with all types of tissue in muscle except the capillary network (STACEY, 1969; see Fig. 1). The non-myelinated group IV sensory fibres are exclusively free-ending in nature, and myelinated sensory fibres belonging to groups II and III, and possibly group I, also participate in free-ending innervation (STACEY, 1969). STACEY (1967, 1969) has observed group III fibres supplying extensive ramifying terminals within the capsule wall of spindles and among the connective tissue between the intrafusal muscle fibres in the polar regions. Some branches may also terminate in the intramuscular connective tissue outside the spindle. Only six examples of such sensory endings were found by STACEY (1967) in a total of 318 spindles examined in teased, silver preparations of de-efferented and sympathectomized cat hindlimb muscles.

4. The Motor Innervation

The motor innervation of the mammalian spindle has been the subject of controversy for a number of years. The first detailed analysis was made by BARKER (1948, rabbit), who described each intrafusal muscle fibre as being supplied with one or two motor end-plates, similar to extrafusal end-plates, in each pole. Differences of opinion arose when it was realized that in addition to the intrafusal motor end-plate, the spindle also receives another more diffuse type of motor ending. This was first detected in cholinesterase preparations, which revealed the presence of diffuse enzymic activity near the equatorial region, in contrast to that shown by the discrete sub-neural apparatuses of plates located further along the poles (COËRS and DURAND, 1956; KUPFER, 1960; HESS, 1961a; COËRS, 1962). BOYD (1962c) maintained that the end-plates are supplied by thick 'γ_1 fibres' and are always located on bag fibres, whereas the diffuse endings, supplied by thin 'γ_2 fibres', terminate exclusively on the chain fibres as a 'γ_2 network'. The contrary view, that there is not an exclusively selective innervation of this kind, was stated by BARKER and IP (1965), who named the diffuse ending the 'trail ending'. It then transpired that there are, in fact, three types of motor ending present, the plates being of two kinds that degenerate at different rates after nerve section and have different sources of supply, namely, from β fibres (p_1 plates) and from γ fibres (p_2 plates) (BARKER, 1966, 1967). The terms 'γ_1' and 'γ_2' were formally abandoned at the Nobel Symposium in 1965 (see GRANIT, 1966, p. 119) since it had by then become clear that there is no correlation between diameters of γ-stem fibres in the muscle nerve and those of their terminal branches (ADAL and BARKER, 1965a); and that β fibres also contribute to fusimotor innervation (BESSOU, EMONET-DÉNAND and LAPORTE, 1963, 1965; ADAL and BARKER, 1965a).

There remains the possibility of there being a fast and direct fusimotor α pathway to the spindle in addition to the fusimotor γ and skeleto-fusimotor β innervation. The case for this, initially inconclusive (see MATTHEWS, 1964), has been consistently urged by HAASE and his colleagues (see e.g. BARRIOS, HAASE and HEINRICH, 1967), and is also supported by CARLI, DIETE-SPIFF and POMPEIANO (1966, 1967a, b). The matter cannot be settled histologically, since if the collateral origin of a p_1 fibre cannot be located (as is usually the case), there is no criterion by which one can determine whether the fibre is connected to a β fibre, or to an

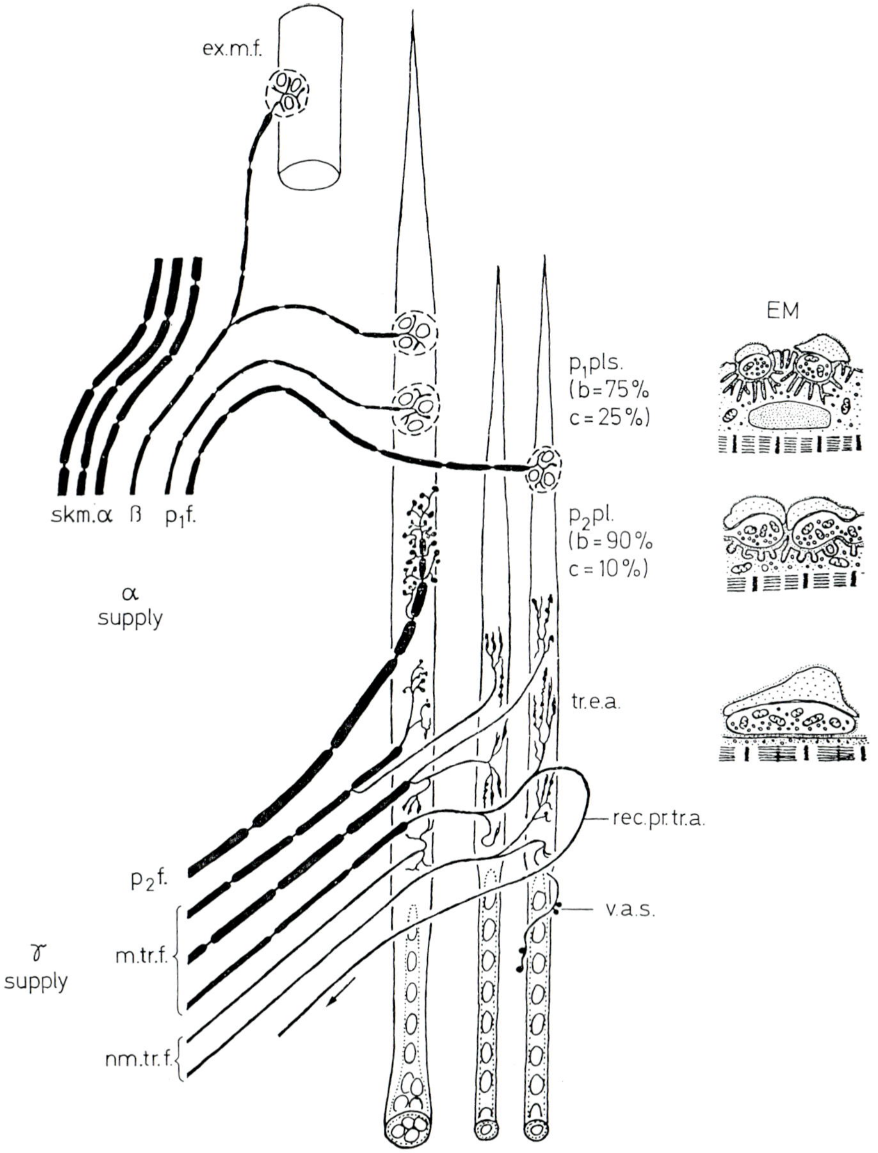

Fig. 70. Schematic diagram of the motor innervation of nuclear-bag (*b*) and nuclear-chain (*c*) muscle fibres in the cat, based on observations made on spindles from the peroneal, flexor hallucis longus, and soleus muscles. Non-myelinated trail fibres (*nm.tr.f.*) do not always participate in the innervation: in the peroneals they are present in addition to myelinated trail fibres (*m.tr.f.*) in just over one third of the spindles. Less than 5% of spindles have recurrent preterminal trail axons (*rec.pr.tr.a.*) leaving them. Some p_1 fibres ($p_1f.$) can be histologically identified as collaterals of fibres (β) that also innervate extrafusal muscle fibres (*ex.m.f.*); this is not possible for the majority, but there are nevertheless good reasons for assuming that they have the same derivation. Sketches of the ultrastructure of the myoneural junctions of the three types of fusimotor ending are shown on the right of the figure under *EM*. Other lettering: $p_1pls.$ p_1 plates; $p_2pl.$ p_2 plate; $p_2f.$ p_2 fibres; *skm.a* skeletomotor *a* fibres; *tr.e.a.* trail-ending area; *v.a.s.* vesicular axonic swelling. (From Barker, Stacey and Adal, 1970)

α fibre that has an exclusively fusimotor distribution. However, the results recently obtained by EMONET-DÉNAND, JOFFROY and LAPORTE (1971) and ELLAWAY, EMONET-DÉNAND, JOFFROY and LAPORTE (1972) are not compatible with the existence of exclusively fusimotor α fibres. Stimulation of 1813 individual α fibres supplying cat tibialis anterior (689), extensor digitorum longus (429) and flexor hallucis longus (695) muscles resulted in 1793 of the fibres eliciting contraction of extrafusal muscle fibres. The remaining 20 axons were probably damaged since their repetitive stimulation failed to elicit either a muscle action potential or an increase in the rate of spindle discharges. The fusimotor innervation will therefore

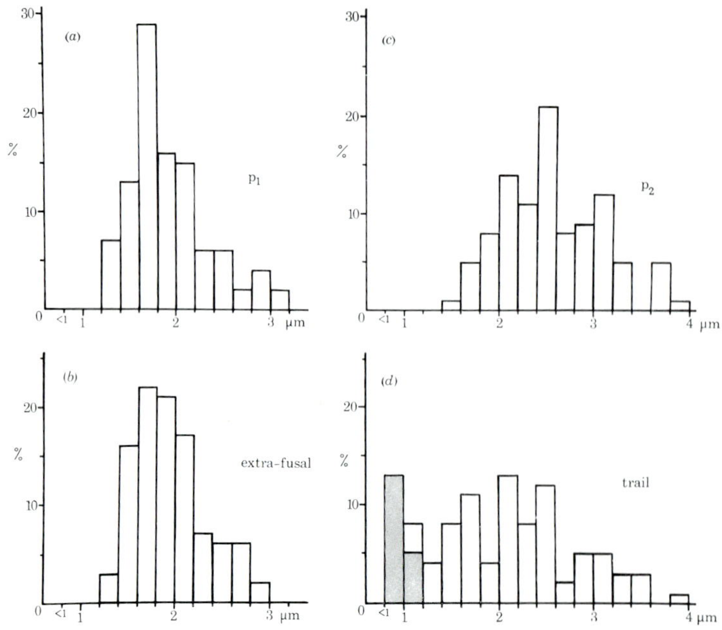

Fig. 71a–d. a, c, d Histograms of the total diameters of 100 fusimotor fibres, or fibre branches, supplying each type of motor ending in cat peroneal spindles as measured in teased, silver preparations up to 0.5 mm from spindle entry; b similar histogram of 100 terminal branches of extrafusal motor fibres. Stippled columns in d indicate non-myelinated trail fibres. (From BARKER, STACEY and ADAL, 1970)

be described in terms of the β innervation (p_1 plates) and the γ innervation (p_2 plates and trail endings), but before doing so it is useful to examine the diameters of the motor fibres supplying them near spindle entry. The account that follows is based mainly on the detailed analysis of cat fusimotor innervation by BARKER, STACEY and ADAL (1970); their schematic diagram of the innervation is reproduced in Fig. 70.

a) Diameters of Fusimotor Fibres Close to the Spindle. Histograms of the total diameters of fusimotor fibres or fibre branches supplying endings to cat peroneal spindles are shown in Fig. 71a, c, d, and are compared with a similar histogram of the diameters of the terminal branches of extrafusal motor fibres. The diameters of the p_1 fibres (range 1.2 to 3.0 μm, peak 1.6 μm) are seen to be very similar to those of the terminal branches of extrafusal motor fibres (range 1.2 to 2.8 μm, peak 1.6 μm), and to be generally smaller than those of the p_2 fibres (range 1.4 to 3.8 μm, peak 2.4 μm). The trail fibres are both myelinated (diameter range 1.0 to 3.8 μm) and non-myelinated (diameters of 1.0 μm or less). The latter are branches of myelinated trail fibres and in electron micrographs measure 0.2 to 1.6 μm in diameter with a peak at 0.8 μm (BARKER and STACEY, 1968). The histograms show that the diameter of a myelinated fusimotor fibre near spindle entry is no guide as to the type of motor ending it is connected to, though it may be said that the probability of p_1 plates being formed by fibres measuring 3.0 to 4.0 μm thick in teased, silver preparations may virtually be excluded. In tracing the course of 22 γ-stem fibres from muscle nerve to spindle entry in cat deep lumbrical muscles, ADAL and BARKER (1965a) found that seven entered the receptors without prior branching. About one third of the myelinated fibres in the p_2 and trail histograms may therefore probably be regarded as γ-stem fibres rather than fibre branches.

When a muscle nerve is severed, the nerve endings degenerate at rates that differ according to the diameter of the supplying nerve fibre, those supplied by the largest fibres being the first to disappear. After cutting the common peroneal nerve in the cat, fragmentation and disappearance of the extrafusal motor end-plates and their branches occurs between 48 and 60h later. The p_1 plates and their fibres follow the same time course. The γ innervation, however, persists intact throughout this period. The p_2 plates disappear between 72 and 78h, followed by the trail endings at 78 to 84h. By 96h all trace of fusimotor innervation has disappeared. The different rates of degeneration of the p_1 and p_2 plates thus confirm their derivation from stem motor fibres that are respectively large and small in diameter, as would be expected in terms of the β and γ sources of supply. If γ fibres participate in p_1 plate innervation we should expect to find some p_1 plates intact 60 to 84h after nerve section, but none is present. The persistence of trail endings for the longest period after nerve section indicates that the thinnest γ fibres are included in providing the trail innervation.

b) The β Innervation. In its general appearance the p_1 plate resembles an extrafusal motor end-plate consisting of axon terminals in the form of tapers and, less frequently, knobs or rings, ramifying upon a nucleated sole plate that has a Doyère eminence (Fig. 72a; Plate 19). The plates are often longer than their extrafusal counterparts, probably because of the considerably smaller diameters of intrafusal as compared with extrafusal muscle fibres. The majority of p_1 fibres terminate in one plate only, but branching may occur to give rise to two or three separate but adjacent plates on the same muscle fibre. In some cases two p_1 fibres of apparently separate origin will form adjacent plates on the same muscle fibre, and on occasion may contribute to one and the same plate. A feature of the p_1 innervation is its frequent derivation from relatively large intramuscular nerve trunks passing close by the spindle, rather than from spindle nerve trunks, though this also occurs. Typically, single p_1 fibres will leave such trunks to innervate a spindle pole, while the rest of the nerve fibres continue on their course, many of

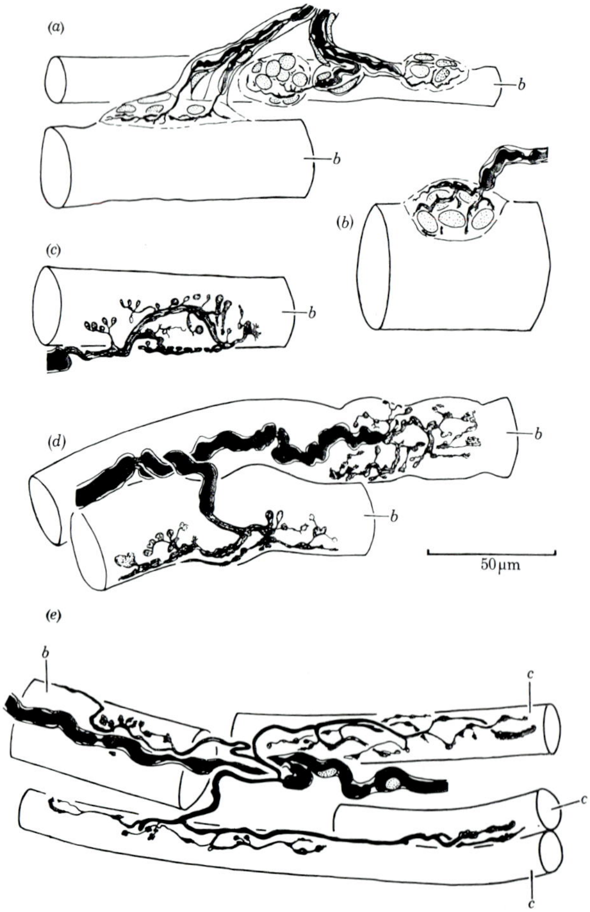

Fig. 72a–e. Drawings of motor endings traced from photographs of teased, silver preparations of de-afferented cat spindles. a Three p_1 plates supplying a pes interosseous spindle compared with b an extrafusal motor end-plate from the same muscle. c A p_2 plate supplying a peroneus digiti quinti spindle. Note sideview of knob-like terminals applied to surface of muscle fibre. d Two p_2 plates supplying a pes interosseous spindle. e Three trail-ending ramifications from a trail-ending area in one pole of a soleus spindle. *b* bag fibre, *c* chain fibre. Nuclei have been omitted from (*c*) and (*d*), and, except for two Schwann nuclei, from *e*. (From BARKER, STACEY and ADAL, 1970)

them destined to innervate extrafusal muscle fibres. It is sometimes possible to locate the origin of p_1 fibres as branches of fibres within these trunks, but the hazards of teasing, together, perhaps, with the nature of β fibre branching, are such that it is only rarely possible to obtain a preparation in which the other branch can be seen to terminate extrafusally, as in Plate 19, Figs. 78–80.

In cat spindles the p_1 innervation is mainly confined to the nuclear-bag fibres. In a sample of 100 p_1 plates in cat peroneal and flexor hallucis longus (FHL) spindles, BARKER, STACEY and ADAL (1970) found that 75 were located on bag fibres and 25 on chain fibres. Two instances were observed in which a p_1 fibre branched within the spindle to innervate both bag and chain fibres. The plates are usually located in the polar regions either just before or just after the muscle fibres pass out of the capsule (see Fig. 3), but they may also occur towards the extreme ends of the pole. In rabbit spindles, which have fewer intrafusal muscle fibres, it is usually possible to distinguish between typical and intermediate bag fibres in teased, silver preparations, and to ascertain the distribution of the p_1 innervation between the two types (see Plate 21). In a sample of 63 p_1 plates supplied to rabbit spindles in various hindlimb muscles, BARKER and STACEY (1970) found that 33 (52.4%) were distributed to typical bag fibres, 27 (42.9%) to intermediate bag fibres, and 3 (4.7%) to chain fibres. In one case a p_1 fibre branched to supply a plate to each of the two types of bag fibre.

In rat spindles two types of intrafusal motor end-plate have been described which differ in size, form of terminal, cholinesterase staining, and ultrastructure (GLADDEN, 1969; HENNIG, 1969; MAYR, 1969; OVALLE, 1972b). Gladden (teased, silver preparations, tail muscles) identifies the smaller of the two as corresponding with the cat p_1 plate. Mayr (cholinesterase and Sudan black B staining, lumbrical muscles) reports that such plates are distributed to the chain fibres, whereas according to Ovalle (silver sections and electron microscopy, lumbrical muscles) they are restricted to the bag fibres. Hennig (phase and electron microscopy, lumbrical muscles) shows them distributed to both types of fibre in her Fig. 1. Other evidence suggests that the distribution is non-selective, but predominantly to bag fibres as in cat and rabbit spindles. Thus KARLSEN (1965, silver sections, jaw muscles) observes that extrafusal-type (i.e. p_1) plates are most frequently located on bag fibres, and PORAYKO and SMITH (1968, teased, silver preparations, lumbrical muscles) report six instances of the innervation of bag fibres by β collaterals. Beta fibres are almost solely responsible for the excitation of the primary endings of spindles located in the distal muscles of the rat's tail (KIDD, 1964), and ANDREW, PART and WAIT (1971) have shown that the motor innervation of some of these muscles completely lacks a γ component. The β collaterals that innervate the spindles in these muscles presumably distribute p_1 plates to both bag and chain fibres.

Plates of the extrafusal type, equivalent to cat p_1 plates, have been observed on bag fibres in human spindles by COOPER and DANIEL (1963), KENNEDY (1970), and SWASH and FOX (1972), and in spindles of the Australian opossum by JONES (1966), who calls them 'nuclear bag end plates'. SWASH and FOX also observed that p_1 plates were only rarely located on the chain fibres of human spindles. An instance of a plate supplied to a bag fibre by a β collateral was seen by BERNDT, OSWALDO-CRUZ and ROCHA-MIRANDA (1969) in the tenuissimus muscle of the American opossum.

BARKER, STACEY and ADAL (1970) maintain that p_1 innervation is a widespread and integral feature of fusimotor innervation in cat peroneal, FHL, and soleus muscles. They found it to be more prevalent in fast muscles (FHL, peroneus digiti quinti) than slow (soleus). In a sample of 23 spindle poles from FHL, p_1 fibres (i.e. presumed collaterals of β fibres) accounted for 28.8% of the total fusimotor supply as compared with 14.2% in a similar sample from soleus. In order to account for the high frequency of the p_1 innervation in FHL it is necessary to assume that both fast and slow conducting β fibres are involved in the supply.

This would mean that the spindles are associated with both slow and fast contracting motor units composed of type B and type C muscle fibres, respectively (see Discussion in BARKER, STACEY and ADAL). ANDREW and PART (1973) have in fact recently found such an association to obtain in the β-innervated spindles in rat segmental tail muscles. They have established that stimulation of some of the β fibres produces a dynamic effect on the primary response, and that stimulation of other β fibres produces a static effect. They found that extrafusally the dynamic β fibres supplied slow motor units, and that static β fibres supplied fast motor units (the units have not as yet been typed histochemically).

Conduction velocities so far determined for β fibres supplying cat and rabbit muscles range from 26 to 90.5 m/sec (see BESSOU et al., 1965; EMONET-DÉNAND, JANKOWSKA and LAPORTE, 1970; BARKER, EMONET-DÉNAND, LAPORTE, PROSKE and STACEY, 1971; ELLAWAY, EMONET-DÉNAND and JOFFROY, 1971). The maximum conduction velocities of all types of motor fibre are slower in the rat; the range of conduction velocity of the β fibres innervating the segmental tail muscles is 20–50 m/sec (ANDREW and PART, 1972). The results of ADAL and BARKER's (1965a) histological study of cat deep lumbrical muscles suggest that the distribution of the slowest β fibres is mainly intrafusal. The stem diameter of five β fibres whose intramuscular course was traced ranged from 6.0 to 12.5 μm, and at these two extremes their distribution varied from being predominantly fusimotor to predominantly skeletomotor, respectively.

Electron micrographs of the cat p_1 plate published by BARKER, STACEY and ADAL (1970) and GOGLIA (1970) show an ultrastructure closely resembling that of extrafusal motor end-plates (Plate 24). The axon terminals lie in synaptic gutters on a substantial sole plate. The postsynaptic membrane is folded, the spacing, dimensions, and branching of the junctional folds being very similar to those of extrafusal plates. The two plates sectioned by BARKER, STACEY and ADAL are both located on intrafusal muscle fibres whose ultrastructure includes the presence of an M-line, indicating that they are either chain fibres or intermediate bag fibres. At present it is not clear whether the ultrastructure of a cat p_1 plate located on a typical bag fibre differs in any way. Goglia's micrograph does not include myofibrils and he does not state what type of muscle fibre the plate was located on. CORVAJA et al. (1969) encountered a plate on a typical bag fibre whose ultrastructure led them to doubt that it was either of the p_1 or p_2 type. However, if the oblique plane of sectioning and presence of preterminal axons are taken into account, the micrographs may most probably be interpreted as illustrating a p_1 plate, though no definite conclusions can be drawn about the folding of the postsynaptic membrane. The rat p_1 plate has a myoneural junction in which junctional folds are present according to HENNIG (1969), and absent according to OVALLE (1972b, his 'nuclear-bag plate'). Each publishes micrographs that include both smooth and folded regions of the postsynaptic membrane!

c) The γ Innervation: p_2 Plates. The location of p_2 plates in cat spindles is the same as that of p_1 plates (see Fig. 3). Their most characteristic features are the knob-like axon terminals; the lack of any obvious sole plate; the large size of many of the plates; and the relatively large diameter of the supplying fibre (Fig. 72c, d; Plates 18 and 20). BARKER, STACEY and ADAL (1970) found the average length of 45 p_2 plates in peroneal spindles to be 72.9 μm (range 27–120 μm),

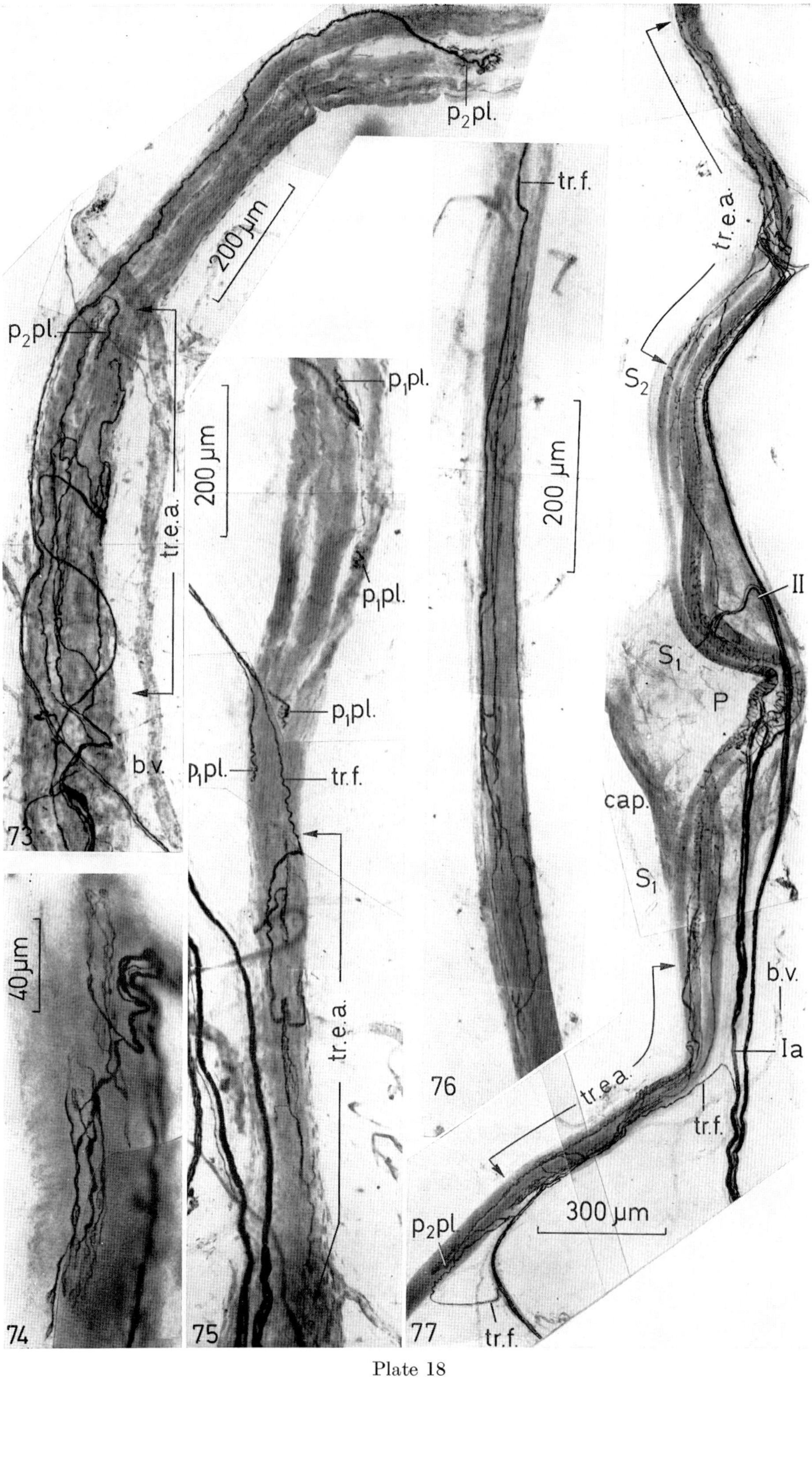

p2pl.
200 µm
p2pl.
tr.e.a.
b.v.
73
40µm
74
p1pl.
200 µm
p1pl.
p1pl.
p1pl.
tr.f.
tr.e.a.
75
tr.f.
200 µm
76
tr.e.a.
S2
II
S1
P
cap.
S1
b.v.
Ia
tr.e.a.
tr.f.
300 µm
p2pl.
tr.f.
77

Plate 18

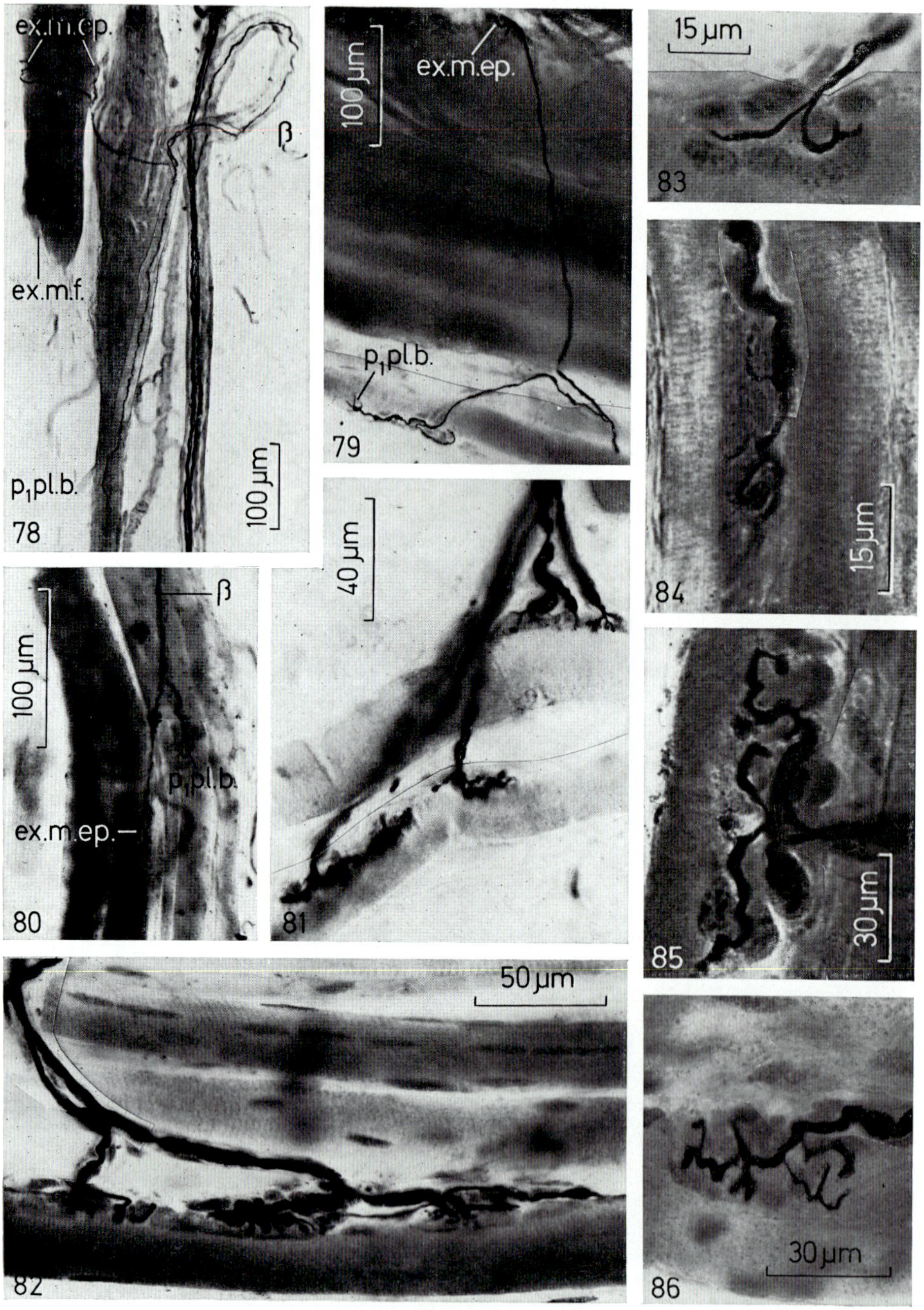
ex.m.ep.
β
ex.m.f.
$p_1$pl.b.
100 μm
78
100 μm
ex.m.ep.
$p_1$pl.b.
79
15 μm
83
84
15 μm
β
100 μm
$p_1$pl.b.
ex.m.ep.
80
40 μm
81
85
30 μm
50 μm
82
30 μm
86

Plate 19

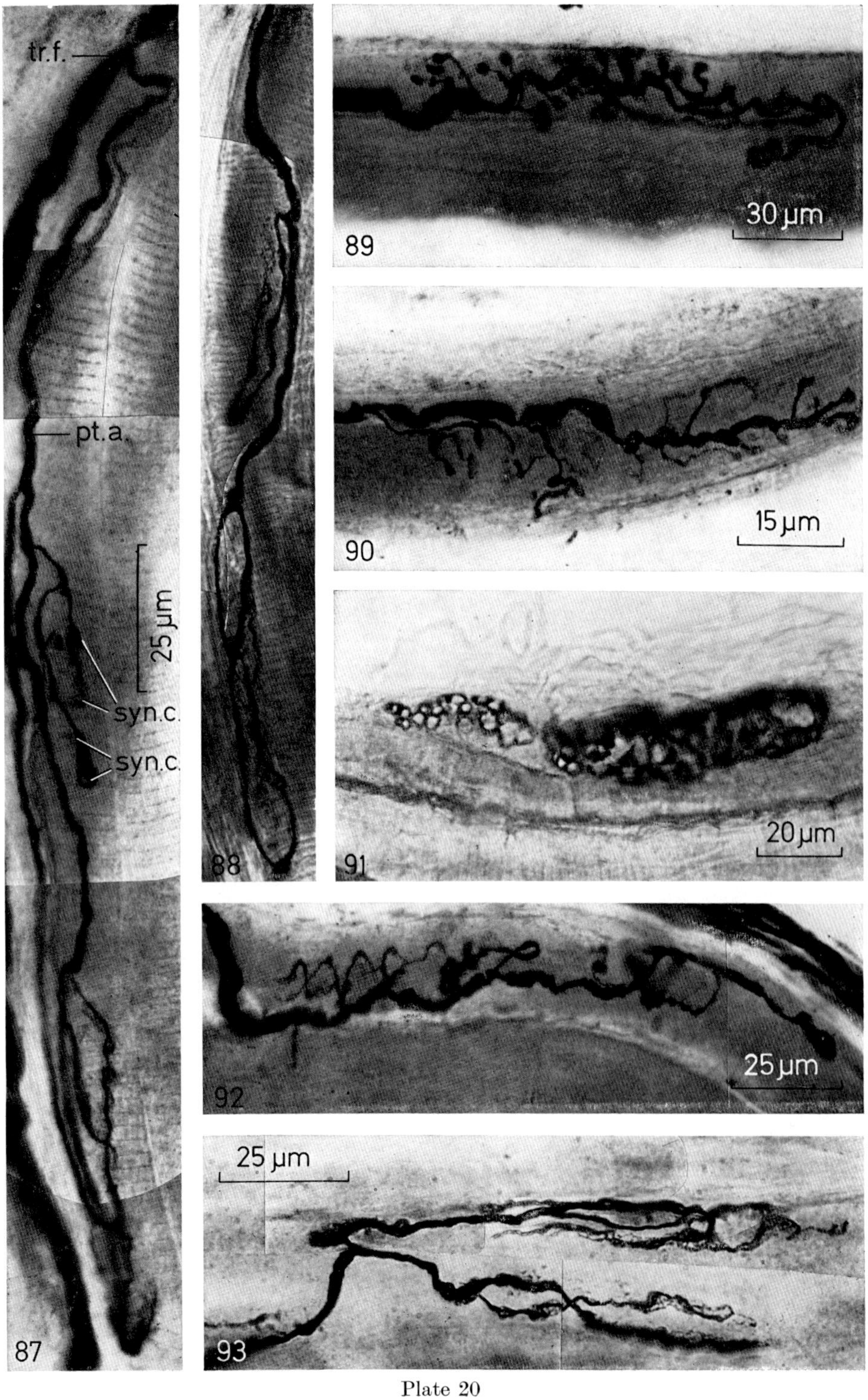
tr.f.
pt.a.
25 µm
syn.c.
syn.c.
87
88
30 µm
89
15 µm
90
20 µm
91
25 µm
92
25 µm
93

Plate 20

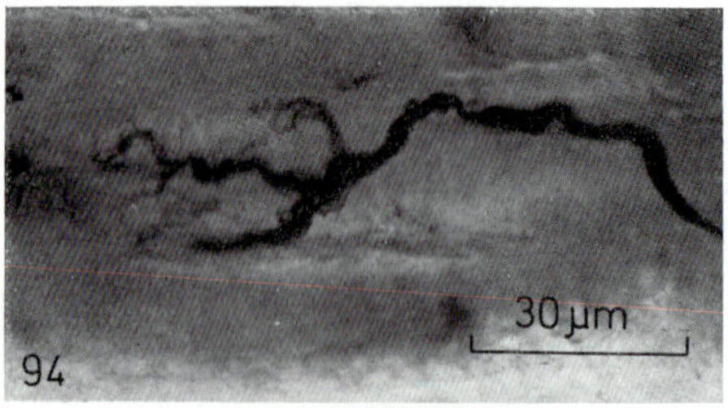
30 μm
94

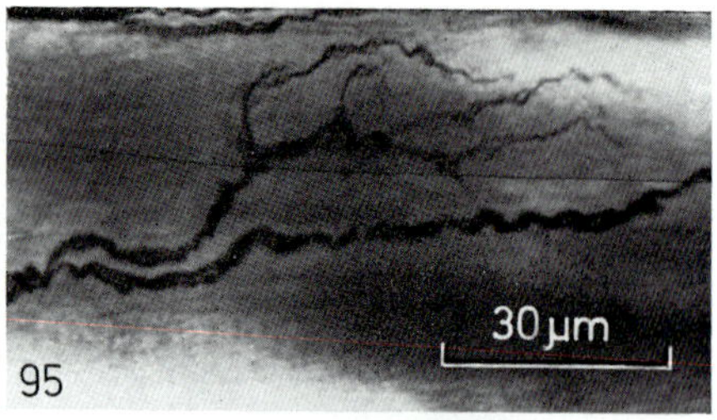
30 μm
95

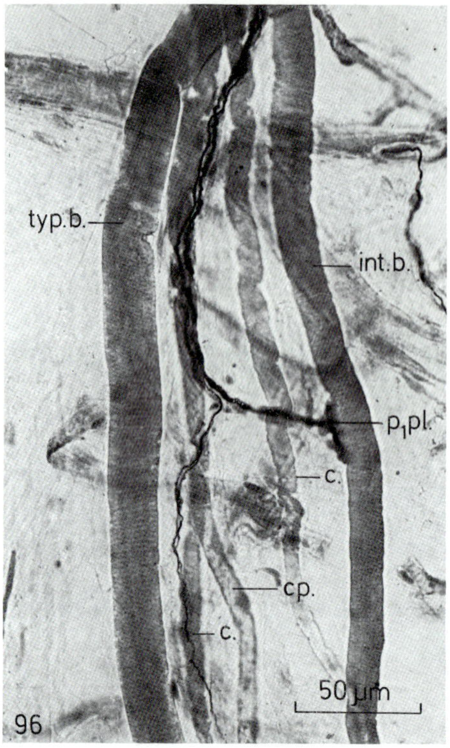
typ.b.
int.b.
$p_1$pl.
c.
cp.
c.
50 μm
96

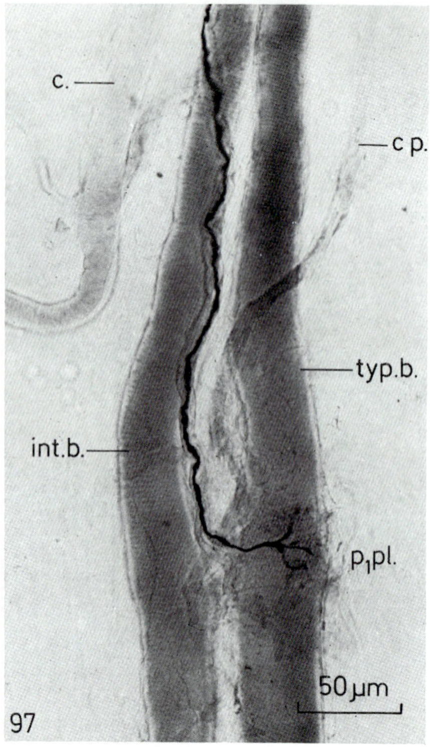
c.
c p.
typ.b.
int.b.
$p_1$pl.
50 μm
97

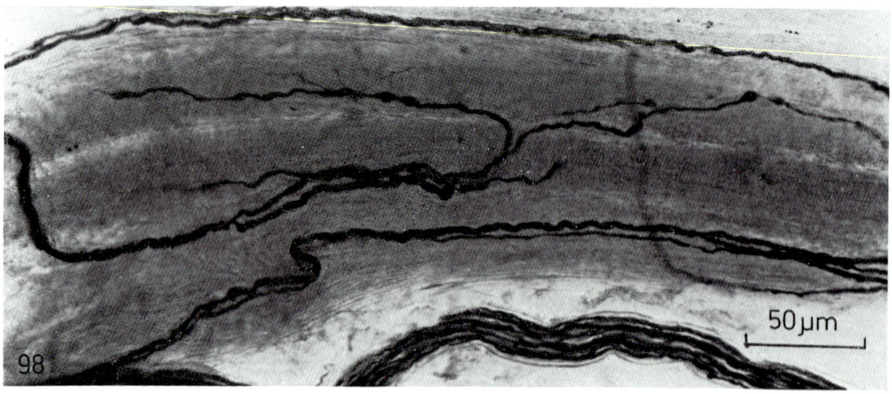
50 μm
98

Plate 21

Plate 18, Figs. 73–77

Photographs of teased, silver preparations of cat spindles. *b.v.* blood vessel; *cap.* capsule; *p_1 pl.* p_1 plate; *p_2 pl.* p_2 plate; *tr.e.a.* trail-ending area; *tr.f.* myelinated trail fibre; I*a* primary afferent; II secondary afferent; *P* primary ending; S_1, S_2 secondary endings. Fig. 76 from Barker (1967); other figures from Barker, Stacey and Adal (1970)

Fig. 73. Pole of a spindle supplied with trail endings and p_2 plates. From peroneus digiti quinti

Fig. 74. Two complex trail ramifications. Part of a trail ending innervating a peroneus brevis spindle

Fig. 75. Pole of a spindle supplied with trail endings and p_1 plates. From peroneus digiti quinti. (Preparations in Figs. 73–75 by Dr. M.J. Stacey)

Fig. 76. One pole of a de-afferented peroneal spindle innervated by a single myelinated trail fibre which enters at the top of the figure from the equatorial region. (Preparation by Dr. M.C. Ip)

Fig. 77. Spindle teased from peroneus brevis and photographed at low power to give general impression of sensory and motor innervation. The lefthand wall of the capsule has been torn in teasing. Only the innervated part of each pole, amounting to about one third of the total polar length in each case, is included in the figure. The sensory innervation consists of one primary and three secondary endings, i.e. from bottom upwards, S_1 P S_1 S_2. The upper S_1 ending is of the flower-spray type, the other two are annulospiral. Each pole receives trail endings; the lower pole is in addition supplied with a p_2 plate. (Preparation by Mrs. C. Negus)

Plate 19, Figs. 78–86

Photographs of teased, silver preparations illustrating β innervation and p_1 plates in cat (Figs. 79–86) and rabbit (Fig. 78) hindlimb muscles. β beta fibre; *ex.m.ep.* extrafusal motor end-plate; *ex.m.f.* extrafusal muscle fibres; *p_1 pl.b.* p_1 plate on a bag fibre. Figs. 78, 80 from Barker (1968); Figs. 79, 81–83, 86 from Barker, Stacey and Adal (1970); Figs. 84, 85 from Barker (1967). (Preparations in Figs. 79, 81–86 by Dr. M.J. Stacey)

Fig. 78. A β fibre divides into two branches one of which supplies a p_1 plate to a bag fibre in one pole of a rabbit interosseous spindle, while the other terminates in extrafusal motor end-plates nearby. (Preparation by Dr. M.C. Ip)

Fig. 79. A β fibre in cat soleus muscle branches to supply two p_1 plates to a bag fibre in a spindle, and also terminates in an extrafusal motor end-plate nearby

Fig. 80. Similar preparation to that shown in Fig. 79. Cat interosseous muscle. (Preparation by Dr. M.C. Ip)

Fig. 81. Two bag fibres receive p_1 plates in a cat FHL spindle

Fig. 82. A p_1 fibre branches to supply three adjacent plates on a bag fibre in a cat FDL spindle

Fig. 83. Simplest type of p_1 plate; note sole-plate nuclei. Cat FHL spindle. Plate is located on a bag fibre

Fig. 84. Side view of p_1 plate in a de-afferented cat interosseous spindle. Note nucleated sole plate and Doyère eminence

Fig. 85. Surface view of p_1 plate in a de-afferented cat interosseous spindle

Fig. 86. Surface view of p_1 plate located on a bag fibre in a cat FHL spindle

Plate 20, Figs. 87–92

Photographs of teased preparations of cat spindles illustrating trail innervation and p_2 plates. All are silver preparations except Fig. 91, which is a cholinesterase preparation (modified Karnovsky and Roots [1964] technique). All the p_2 plates shown are located on bag fibres. (From Barker, Stacey and Adal, 1970)

Figs. 87, 88 and 93. Details of trail-ending ramifications supplied to de-afferented soleus (Fig. 87) and pes interosseous (Fig. 88) spindles, and to a normal spindle teased from peroneus brevis (Fig. 93). Fig. 87 shows the nodal origin of a preterminal axon (*pt.a.*) from a myelinated trail fibre (*tr.f.*), and two presumed regions of synaptic contact (*syn.c.*). Similar regions of contact are evident in the ramifications formed by the preterminal axons in Figs. 88 and 93. Preparations by Dr. M.C. Ip (Figs. 87, 88) and Mrs. C. Negus (Fig. 93).

more than twice the average lengths of extrafusal and p_1 plates. The axon terminals consist of knobs, rings, and, less frequently, tapers. Ruffini (1898, p. 203) describes them as 'a number of most delicate and elegant knobbed aborescences' and was so impressed by the difference between these plates and extrafusal ones that he concluded they were sensory (see Fig. 60D, E).

The majority of cat p_2 fibres enter a pole to terminate in one plate only; occasionally a p_2 fibre may be seen branching outside the spindle so as to supply both poles. The majority of p_2 plates innervate bag fibres; of 50 peroneal p_2 plates examined by Barker, Stacey and Adal (1970), 90% were located on bag fibres, 10% on chain. In two instances a p_2 fibre was seen to branch so as to supply one plate to a bag fibre and one to a chain fibre. One p_2 plate was seen to terminate on a bag fibre as well as on an adjacent chain fibre; another spread its terminals over two bag fibres.

In rabbit spindles the p_2 plate is claw-like with well-defined and loosely arranged neurofibrils in the axon terminals, which gives them a somewhat straggling appearance in teased, silver preparations (Barker, Harker, Stacey and Smith, 1972; see Plate 21, Figs. 94, 95). They are distributed to both typical and intermediate bag fibres, but have not been observed on chain fibres (Barker and Stacey, 1970). A similar type of motor ending has been identified by Gladden (1969) as the p_2 plate in rat spindles, and this appears to correspond with the larger of the two types of plate distinguished by Ovalle (1972b). Cholinesterase preparations of the large plates (mean length 26.5 μm) in rat spindles reveal a subneural apparatus consisting of a loose arrangement of circular subunits, whereas in the small plates (mean length 18.5 μm) it has a more compact structure with some resemblances to that of extrafusal plates (Mayr, 1969). These differ-

Fig. 89. A p_2 plate located in a peroneus longus spindle. Note ultraterminal fibre with terminal bulb; the apparent bulb immediately preceding this is a kink in the ultraterminal fibre. (Preparation by Dr. M.J. Stacey)

Fig. 90. A p_2 plate supplying a de-afferented peroneus longus spindle. Note contributions of axon terminals from the two nodes that precede the preterminal one. (Preparation by Dr. M.C. Ip)

Fig. 91. Cholinesterase preparation of a p_2 plate supplying a peroneus longus spindle. Enzymic activity reveals a subneural apparatus consisting of a cluster of subsynaptic rings. (Preparation by Miss G.M. Geller)

Fig. 92. A p_2 plate supplying a de-afferented peroneus longus spindle. Note ultraterminal fibre with terminal bulb. (Preparation by Dr. M.C. Ip)

Plate 21, Figs. 94–98

Photographs of teased, silver preparations of motor endings in rabbit peroneal spindles. *c.*chain fibre; *cp.* capillary; *int.b.* intermediate bag fibre; p_1 *pl.* p_1 plate; *typ.b.* typical bag fibre. Preparations by Mr. C.R. Smith. (From Barker, Harker, Stacey and Smith, 1972)

Figs. 94 and 95. Examples of p_2 plates on bag fibres

Fig. 96. A p_1 plate seen in side view on an intermediate bag fibre. In this extracapsular part of the polar region the other members of the intrafusal bundle (a typical bag fibre and two chain fibres) are clearly seen

Fig. 97. A p_1 plate seen in surface view on a typical bag fibre

Fig. 98. A trail ending. The ramifications and axon terminals are less complex than those in cat trail endings

ences are similar to those reported by BARKER, STACEY and ADAL (1970) for cat p_1 and p_2 plates. According to MAYR (1969) the large plates are distributed to the bag fibres in rat spindles, but OVALLE (1972b) maintains that they are restricted to the chains (his 'nuclear chain plate').

In human spindles KENNEDY (1970) reports the presence of large-diameter 'multibranched' γ fibres that supply pairs of plates to that part of the polar region which typically receives the p_2 innervation in cat spindles. The fibres may distribute the plates to bag fibres only, to chain fibres only, or to both bag and chain fibres. SWASH and FOX (1972) state that the p_2 plates are a prominent and easily recognizable feature of human fusimotor innervation. They found them located on both bag and chain fibres, their length ranging from 47 to 108 μm. JONES (1966) describes the occurrence of irregular and diffuse plates on the bag fibres of opossum spindles, which may be of the p_2 type.

The main ultrastructural features of the p_2 plate in the cat are the 2 to 3 μm wide knob-like axon terminals, which typically lie in shallow depressions of an extensive, thinly spread sole plate, and are apposed to a folded postsynaptic membrane (ADAL and BARKER, 1967; BARKER, STACEY and ADAL, 1970; see Plate 24). The folds are wider and shallower than those of extrafusal and p_1 plates, and the basement membrane usually lines the sides of the folds without fusing in the middle. HENNIG (1969) has described the ultrastructure of an ending tentatively identified as a p_2 plate in a rat spindle, which consists of a row of end-bulbs enclosed in deep inlets of sarcoplasm and covered by tongues of sarcoplasm and Schwann cytoplasm. The ending appears to be located on a typical bag fibre and lacks junctional folds. OVALLE (1972b) describes the ultrastructure of his rat 'nuclear-chain plates' (? p_2 plates) as myoneural junctions with a moderate number of unbranched junctional folds.

d) The γ Innervation: Trail Endings. The main characteristics of the trail ending are its almost invariable presence; its juxta-equatorial location; the extensive area occupied by its ramifications; and the frequent occurrence of preterminal axons that course for long distances before terminating. A number of trail fibres enter a spindle pole to contribute to the trail-ending area. If intramuscular branching is taken into account (see ADAL and BARKER, 1965a), it is evident that there is a considerable overlap in trail supply, so that the trail-ending area in most spindles is supplied with ramifications ultimately connected to several different trail stem fibres in the muscle nerve. Non-myelinated trail fibres innervate just over one-third of the spindles in cat peroneal muscles, and comprise between one quarter and one third of the total γ supply to these spindles (BARKER, STACEY and ADAL, 1970). Most of these fibres probably proliferate from the nodes of myelinated trail fibres some distance from the spindle. A few may be recurrent preterminal trail axons that leave the site of a trail-ending area in one spindle and ultimately reach other spindles nearby to contribute to their trail innervation, appearing as non-myelinated axons in their nerve supply (BARKER, 1968b). Such inter-connexions are not common; in a sample of 227 cat peroneal spindles, only 10 (4.4%) were detected by BARKER, STACEY and ADAL (1970) as having such axons leaving them. A more frequent occurrence is for preterminal axons to leave one pole and pass through the equatorial region to terminate in the other. Some

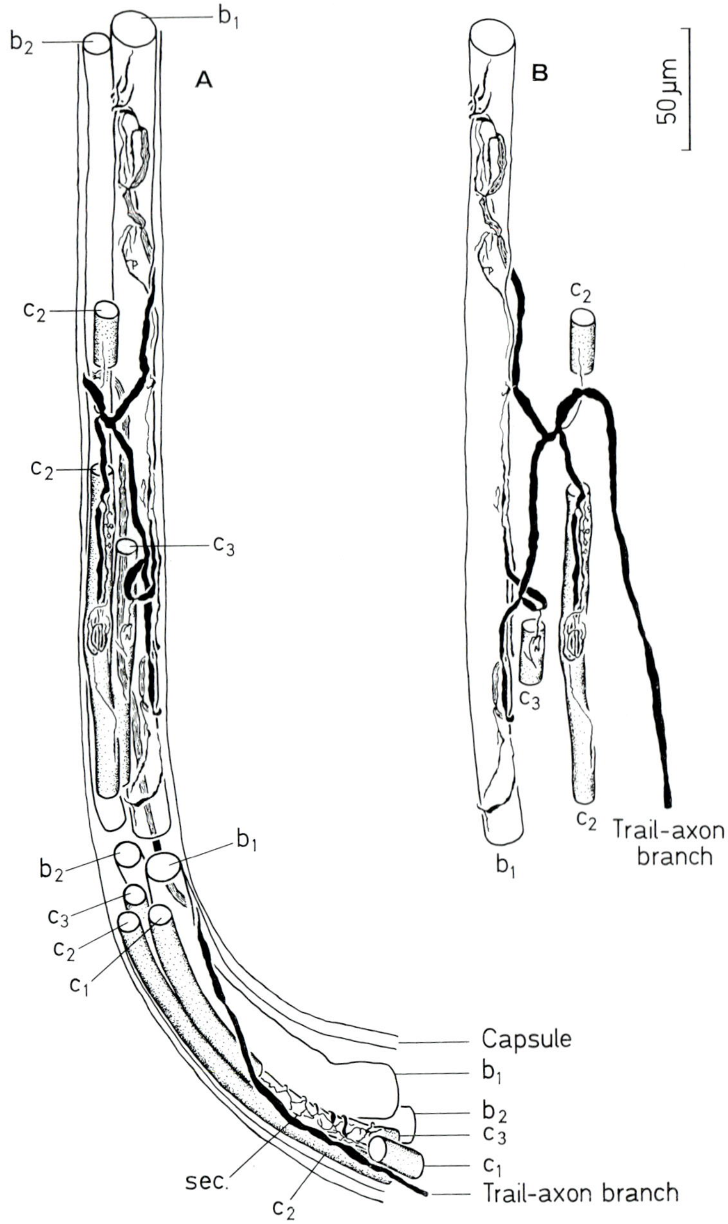

Fig. 99. Trail ending distributed to one pole of a cat tenuissimus spindle innervated by a static γ axon. All other motor fibres supplying the muscle were cut and 220 hr allowed for their degeneration. The pole consists of two bag fibres (b_1,b_2) and and three chain fibres (c_1,c_2,c_3 shown dotted). Trail-ending ramifications are distributed to one of the bag fibres (b_1) and two of the chain fibres (c_2 and c_3). A Distribution of the innervation shown in the form of a schematic optical dissection based on tracings of the photographs of the teased, silver preparation (compare with Plate 22). *sec.* part of secondary ending on c_3. B Distribution shown in the form of a plan of the optical dissection. (From BARKER, EMONET-DÉNAND, LAPORTE, PROSKE and STACEY, 1973)

Plate 22, Fig. 100

Photographs of a teased, silver preparation of a trail ending distributed to one pole of a cat tenuissimus spindle innervated by a static γ axon. All other motor axons were cut and 220 hr allowed for their degeneration. The pole has been photographed in four planes of focus in order to demonstrate the distribution of trail ramifications to the bag fibre b_1 and the chain fibre c_2. The photographs should be compared with the schematic optical dissection of this pole and its innervation shown in Fig. 99

Fig. 100a–d. a Bottom plane of focus. The branch of the static γ trail axon (*tr.ax.br.*) that supplies the innervation is seen entering the pole at the bottom of the figure. The most extensive trail ramification supplied to b_1 (*tr.rm.*b_1) lies further up the pole on the bottom surface of the fibre (beneath this is the lowest plane of focus, which reveals the capsule). This fibre shows bulbing, and one such bulb (*bb*) is seen near the top of the figure. b Lower middle plane of focus. This level reveals the course of the bag fibres b_1 and b_2 through the pole. Note the wavy pattern of their Z lines. c Upper middle plane of focus. The bag fibre b_1 is still in evidence together with that part of its extensive trail ramification which lies on its upper surface. The chain fibre c_2 is now in focus for part of its course; note its smaller diameter and straight Z lines as compared with b_1. *cp.* capsule. d Top plane of focus. This level reveals the most extensive trail ramification, received by c_2 (*tr.rm.*c_2), which lies on the top surface of the fibre (above this is the highest plane of focus, which reveals the capsule). (From Barker, Emonet-Dénand, Laporte, Proske and Stacey, 1973)

Plate 23, Figs. 101–102

Fig. 101a–c illustrate part of a trail ending distributed to one pole of a cat tenuissimus spindle innervated by a static γ axon. All other motor axons were cut and 168 hr allowed for their degeneration. Teased, silver preparation. The pole has been photographed in two planes of focus, which demonstrate the presence of one bag fibre (*b*) and two chain fibres (c_1,c_2) and the distribution of trail ramifications to *b* and c_2

Fig. 101a–c. a Bottom plane of focus. The branch of the static γ trail axon (*tr.ax.br.*) enters the field of view at the top of the figure from the direction of the equatorial region. Having supplied a short branch to innervate c_2 with a trail ramification (*tr.rm.*c_2; the terminals visible lie on the bottom surface of the fibre), it passes out of the field of view from whence a recurrent branch enters the bottom of the figure to supply the trail ramification lying on the bottom surface of *b* (*tr.rm.b*). Beneath this is the lowest plane of focus, which reveals the capsule. A small bundle of free-ending afferent axons (*f.-e.ax.*) is seen outside the pole; it passes beneath it to terminate among fat cells and connective tissue in the neighbourhood of the spindle. b Lower middle plane of focus. This level shows the trail terminals lying on the top surface of c_2, and the muscle fibres *b*, c_1 and c_2 can be distinguished. c Same plane of focus as in b as photographed with Zeiss Nomarski optics, which make the three muscle fibres and their cross striations more distinct. *cp.* capsule

Fig. 102a–d illustrate the trail innervation of the pole of another spindle in the same, teased, silver preparation. The branch of the static γ trail axon (*tr.ax.br.*) enters the field of view at the bottom of the figures from the direction of the equatorial region and innervates one of the two bag fibres shown. This is the only trail ramification supplied to this spindle. a Bottom plane of focus. The main part of the ramification and terminals lie on the bottom surface of the bag fibres. Beneath this is the lowest plane of focus, which reveals the capsule. Some elastic fibres (*e.*) are visible. b Middle plane of focus. c Middle plane of focus as photographed with Zeiss Nomarski optics. Note wavy pattern of the Z lines in the cross striations of both bag fibres. d Top plane of focus. This shows how part of the ramification curls round the bag fibre to terminate on its top surface, and is a good example of how such dispositions serve to individuate the muscle fibre that is innervated. (From Barker, Emonet-Dénand, Laporte, Proske and Stacey, 1973)

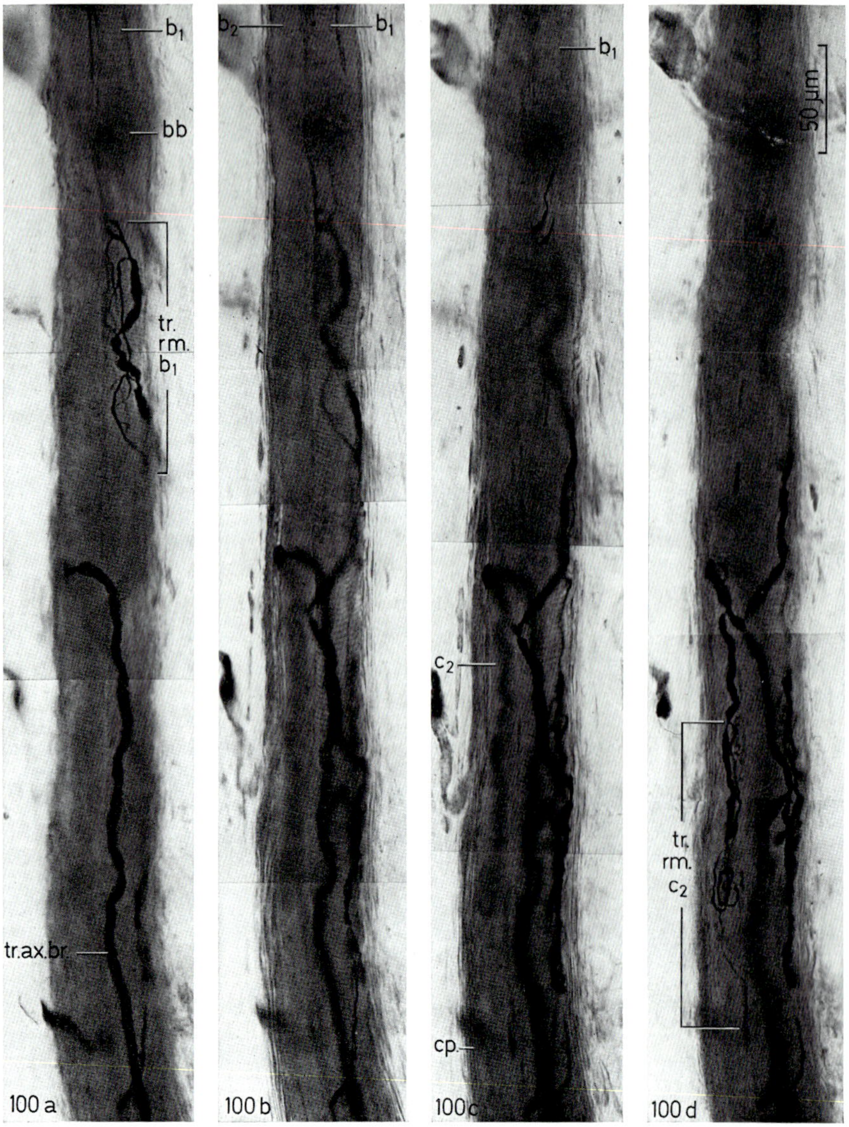

Plate 22

may fail to reach the other pole, and end among the sensory terminals in a single vesicular swelling.

The branching of myelinated trail fibres within the spindle is sometimes very profuse. Up to five myelinated branches, or myelinated branches and preterminal axons, may arise from a single node. The ramifications that ultimately form the area of synaptic contact vary from a simple brush-like taper 10 μm long to complex configurations that may extend up to 210 μm long and contain a variety of neurofibrillar contacts in the shape of brushes, hooks, knobs and rings (see Fig. 72e; Plates 18 and 20). In the detailed quantitative study made by Barker,

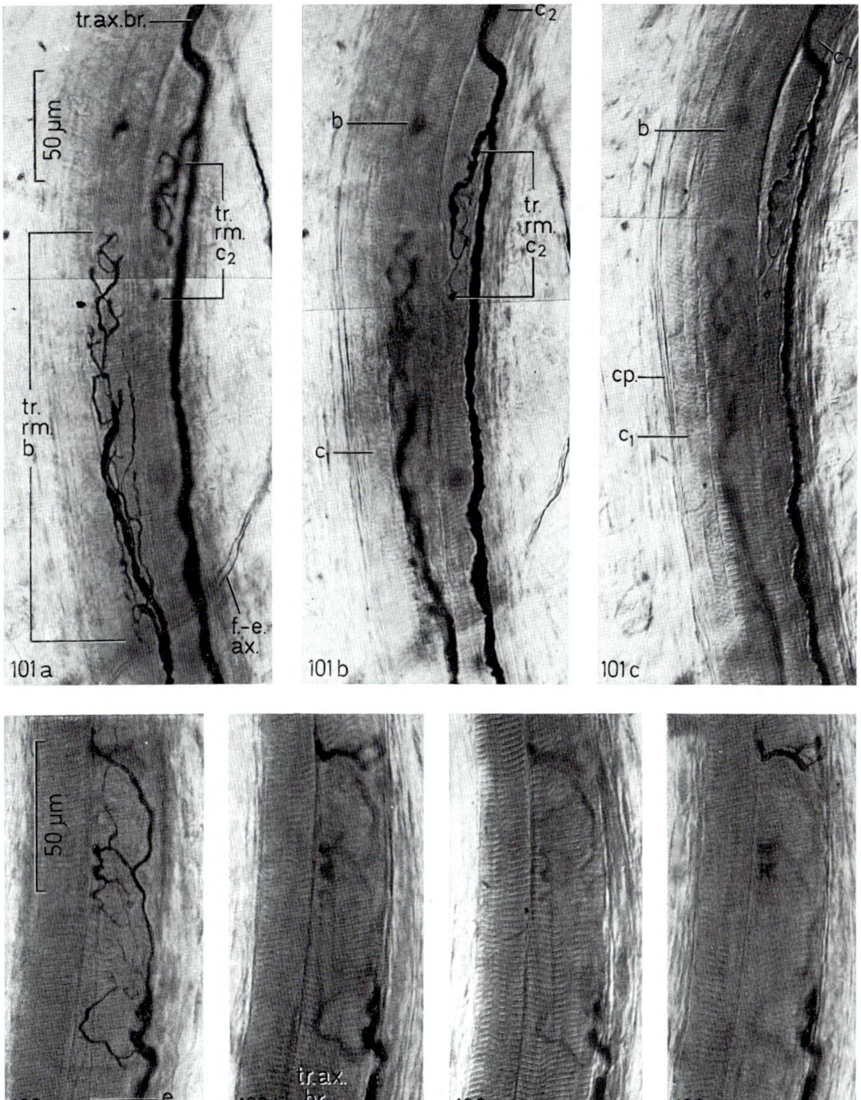

Plate 23

Stacey and Adal (1970) the mean number of ramifications per trail fibre was 3.7, and the mean number of ramifications in trail-ending areas in a number of spindle samples varied from 10 to 18.5. In a sample of 89 peroneal spindle poles, the mean number of myelinated trail fibres per pole was 2.6, and non-myelinated fibres, when present, contributed a mean number of 1.6 fibres per pole. The length of the trail-ending areas in a sample of 30 poles ranged from 0.44 to 1.58 mm, mean 0.84 mm. Trail endings may overlap with secondary endings, and both trail and sensory axon terminals may occasionally occur side by side on the same muscle fibre (Plate 25, Fig. 107).

The distribution of trail endings to both bag and chain muscle fibres, first reported by BARKER and IP (1965) in cat spindles, has since been confirmed in spindles of the cat (ADAL and BARKER, 1967; CORVAJA et al., 1969; BARKER, STACEY and ADAL, 1970), rabbit (BARKER and STACEY, 1970), rat (HENNIG, 1969; OVALLE, 1972b), dog (BANKER and GIRVIN, 1971), and man (KENNEDY, 1970; SWASH and FOX, 1972). In recent collaborative experiments carried out by LAPORTE and BARKER and their colleagues (BARKER, EMONET-DÉNAND, LAPORTE, PROSKE and STACEY, 1970, 1971, 1973), it has been possible to assess the degree to which trail endings are distributed to the two types of muscle fibre in cat tenuissimus spindles. Tenuissimus muscles were prepared in which the motor innervation was reduced to a single γ axon by cutting all the other motor axons and allowing them to degenerate during a period of 7–12 days. The function of the surviving γ axon was then determined, and the distribution of its endings ascertained in teased, silver preparations. In the ten muscles successfully prepared the function of the surviving γ axon was static and the motor innervation distributed to the spindles consisted of trail endings. The conduction velocities of the axons ranged from 33 to 48 m/sec.

A detailed histological analysis of thirty spindles innervated by six of the surviving static axons showed that the trail endings supplying the spindle poles were distributed to both bag and chain fibres with about twice the frequency of being supplied exclusively to one type of muscle fibre or the other (see Fig. 99 and Plates 22, 23). The density of trail innervation supplied by the static axons to the bag fibres, in terms of the mean number of terminals per fibre, was typically from one and a half to twice that supplied to the chain fibres. On the other hand, whereas the number of bag fibres supplied with trail endings in a spindle pole was seldom more than one, the number of chain fibres innervated was usually two in a range of 1–4.

The following examples from Table 1 in BARKER et al. (1973) give some idea of how trail endings supplied by a static γ axon vary in their distribution to cat intrafusal muscle fibres. In the annotation used b represents a bag fibre; c a chain fibre; a colon indicates the equatorial region; and each fibre carries a number indicating the number of trail terminals received.

(i) $b^{22}\ c^{4}\ c^{10}$
(ii) $b^{3}\ c^{12}\ c^{7}\ c^{3}\ c^{11}$
(iii) $c^{11}\ c^{8}\ c^{4} : b^{13}\ c^{9}\ c^{2}\ c^{7}$
(iv) $c^{18}\ c^{7}\ c^{1}\ c^{12}$
(v) b^{12}

Unfortunately it is not possible to ascertain the distribution of trail endings to typical and intermediate types of bag fibre in silver preparations of cat spindles. However, electron micrographs show that both types receive them and this is also true of rabbit spindles (BARKER and STACEY, 1970) (Plate 25).

Further evidence showing that static γ axons innervate both bag and chain fibres has come from another histophysiological investigation currently in progress. In experiments on cat tenuissimus spindles Procion Yellow is injected electrophoretically inside intrafusal muscle fibres impaled by microelectrodes and activated by single static γ axons. The injected muscle fibre is then located by fluorescence microscopy and its type confirmed by examining its ultrastructure and equatorial nucleation with electron microscopy. Of ten muscle fibres so far impaled, six proved to be bag fibres and four chain fibres. In one experiment two muscle fibres activated by the same static γ axon were impaled, and the subsequent microscopy showed that one was a bag fibre and the other a chain (Barker, Bessou, Jankowska, Pagès and Stacey, 1972) (see Fig. 103).

The common innervation of bag and chain fibres by static γ axons has also recently been demonstrated by Brown and Butler (1973) using the glycogen depletion technique of Edström and Kugelberg (1968). They found that prolonged stimulation of single static axons innervating cat tenuissimus spindles depletes glycogen in both bag and chain fibres. In a sample of fifteen spindles from five separate experiments in which static axons were stimulated there was only one in which the depletion was restricted to chain fibres.

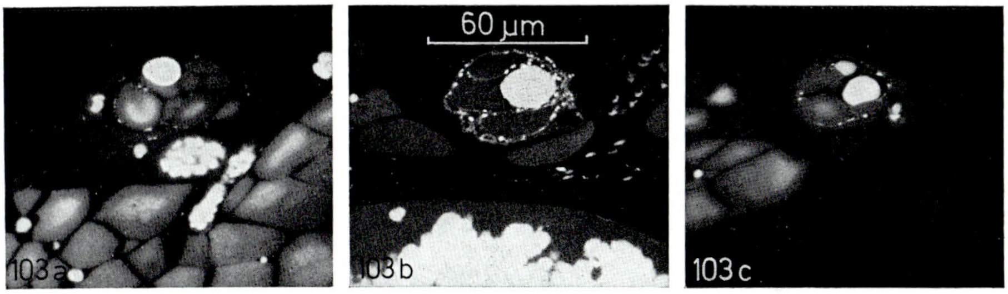

Fig. 103a–c. Transverse sections (5 μm thick) of cat tenuissimus spindles showing fluorescence of muscle fibres impaled and injected with Procion Yellow following recording of their membrane potentials on activation by static γ axons. The fluorescent muscle fibre in a is a chain fibre from which an action potential was recorded; in b it is a bag fibre from which a junctional potential was recorded. In c two muscle fibres were impaled and injected that were activated by the same static γ axon. The large fluorescent muscle fibre is a bag fibre from which a junctional potential was recorded; the small fluorescent fibre is a chain fibre from which an action potential was recorded. (From Barker, Bessou, Jankowska, Pagès and Stacey, 1972)

In contrast to these findings, Boyd (1971b), reporting on the functional distribution of 139 fusimotor axons to cat tenuissimus spindles, observed that only 11% elicited the contraction of both bag and chain fibres on being repetitively stimulated in the muscle nerve. The remainder elicited the contraction of either bag fibres or chain fibres. These results are regarded as "strongly supporting the functional concept of separate motor control of the two intrafusal systems", and it is clear that Boyd envisages this control as being implemented by some fusimotor axons terminating as plates on bag fibres, and others terminating as trail endings on chain fibres, in accordance with his original thesis (1962c). An alternative interpretation would make allowance for some of the fusimotor axons being trail fibres that on stimulation elicited the contraction of bag fibres only, and others, much less frequently, being plate axons (β p_1 axons or γ p_2 axons) that elicited the contraction of chain fibres only. There is also the possibility that in many instances it may be extremely difficult, if not impossible, to observe the simultaneous contraction of both bag and chain fibres, e.g. as in the distribution of trail terminals to the two types of muscle fibre that exists in example (ii) above. Judging from BOYD's (1971a) film, which illustrates a weak, 'focal' type of contraction that occurs in bag fibres as compared to stronger local contractions exhibited by chain fibres, one has the impression that in such instances of dual innervation the activation of the bag fibre might well be overlooked. The small proportion of axons that Boyd observed to activate both types of muscle fibre may represent only those instances in which the bag fibre was extensively innervated and contracted strongly, e.g. a condition such as obtains in example (i) above. There is also the added complication of having to interpret contractions that involve the participation of intermediate bag fibres. More recently Boyd has developed a preparation in which intrafusal contractions are observed on stimulating single γ axons of known function innervating spindles in the abductor digiti quinti medius muscle of the cat (BOYD, GLADDEN and MCWILLIAM, 1973). Of eight static γ axons so far studied, four activated chain fibres only, and four activated a bag fibre in addition to chain fibres.

The simple model of the spindle originally proposed by BOYD (1962c), in which two types of intrafusal muscle fibre are separately innervated by two types of γ axon, continues to find favour with BOYD (1971b) and MATTHEWS (1972) as the most satisfactory basis for explaining the actions of static and dynamic γ axons. However, the most recent evidence, as we have seen, does not fit in with this scheme. In supplying trail endings to spindles, static γ axons distribute the terminals either to bag fibres or to chain fibres, or, more frequently, to both[3].

Electron micrographs of trail endings (Plates 25, 26) show the axon terminals applied to the surface contour of the muscle fibre overlying a thinly spread sole plate that usually lacks depressions or guttering. In some junctions there is no trace of any sole plate. Junctional folds are usually absent, but wide and shallow folds are sometimes present spaced apart at irregular intervals. Occasionally, as reported by DÜRING and ANDRES (1969), there may be a break in continuity of the basement membrane in the synaptic cleft resulting in a closer apposition of axon terminal and muscle fibre. Preterminal axons that run in between the muscle fibres and are located near the terminals are a characteristic feature of most sections. According to OVALLE (1972b) trail axon terminals in rat spindles contain large numbers of flattened synaptic vesicles and very few of the round synaptic vesicles that predominate in the intrafusal plate endings.

Electron micrographs frequently show trail terminals located on bag fibres. Some idea of the involvement of both types of muscle fibre in the innervation may be gained by obtaining a single sample of transverse sections of the juxta-equatorial region of a number of spindles, and noting the disposition of the terminals in the first micrograph in which any appear. In 17 cat and 4 rabbit spindles

3 MATTHEWS (1972) describes the condition where both types of muscle fibre are innervated as 'cross-innervation', a term that has hitherto been used in connexion with experiments in which cross unions are made between the severed ends of nerves supplying fast and slow muscles in order to ascertain the effects of the subsequent reinnervation (see e.g. BULLER and LEWIS, 1965). It is to be hoped that few will follow him in extending the use of the term from this experimental context to describe a normal condition. The same objection applies to his use of 'cross-innervation' to describe skeleto-fusimotor innervation by β axons.

sectioned in this way Stacey and I found terminals located on both bag and chain fibres in 6 of the samples; on bag fibres only in 10; and on chain fibres only in 5.

e) Types of Motor Innervation Received by Spindle Poles. The most common type of fusimotor fibre supplying cat and rabbit hindlimb spindles is the γ trail fibre. Analyses of the fusimotor supply near the level of spindle entry to samples of cat soleus, peroneal, and FHL spindle poles have shown that trail fibres account for approximately two-thirds (64.6–74.8%) of the total input in each sample (BARKER, STACEY and ADAL, 1970). The remaining third consists of β and γ fibres that supply p_1 and p_2 plates, respectively, in proportions that vary from muscle to muscle. A low frequency of p_1 innervation is offset by a high frequency of p_2 (as in peroneus longus), or vice versa (as in FHL). In peroneus longus the high provision of p_2 plates is reflected in the γ supply to the muscle by the provision of a relatively high number of γ-stem fibres per spindle. In rabbit peroneal spindles the trail fibre input exceeds that in cat, comparable figures given by BARKER, HARKER, STACEY and SMITH (1972) being 80.7% trail, 11.7% p_1 and 7.6% p_2. Only one of 345 cat spindle poles examined by BARKER, STACEY and ADAL (1970) lacked a trail innervation. In that analysis 15.6% of the poles were supplied with trail endings only; most of these inserted into tendon, usually as very short poles. On the other hand, the 'trail only' pole is much more common in rabbit peroneal spindles, 48.8% of 125 poles receiving an innervation of this type (BARKER, HARKER, STACEY and SMITH, 1972). In a few cat spindles (less than 10%) the motor innervation consists entirely of trail endings. On rare occasions cat spindles may possess one pole devoid of any motor innervation, as JONES (1966) has observed in opossum spindles.

f) Identity of Static and Dynamic Fusimotor Fibres. The circumstantial evidence in favour of correlating γ static and dynamic fibres respectively with trail fibres and p_2 fibres may be summarized as follows. (i) The high proportion of trail fibres participating in spindle innervation, as compared with p_2 fibres, corresponds with the numerical preponderance of static as opposed to dynamic γ fibres isolated in single-fibre experiments in both cat and rabbit, e.g. 47 static fibres as opposed to 17 (26.6%) dynamic fibres isolated in a sample of cat tibialis posterior γ fibres (BROWN, CROWE and MATTHEWS, 1965); 25 static fibres as opposed to 5 (16.7%) dynamic in a rabbit tenuissimus sample (EMONET-DÉNAND and LAPORTE, 1969). The latter proportion compares with 35 (9.7%) p_2 fibres in a sample of 406 γ fibres, or fibre branches, supplying a sample of rabbit peroneal spindle poles (BARKER, HARKER, STACEY and SMITH, 1972). (ii) The fact that trail endings are the most resistant to degeneration after nerve section suggests that the γ-stem fibres supplying them include those of the smallest diameter in the γ range, and this agrees with the finding that the slowest γ fibres are invariably static (BROWN et al., 1965). (iii) Static fibres have a strong excitatory effect upon secondary endings (APPELBERG, BESSOU and LAPORTE, 1966; BESSOU and PAGÈS, 1969), the majority of which terminate mainly on chain fibres, and most chain fibres receive a motor innervation that is exclusively trail. In addition to these points there is now the direct evidence identifying static γ fibres as trail fibres provided by the experiments of BARKER, EMONET-DÉNAND, LAPORTE, PROSKE and STACEY (1970, 1971, 1973; see p. 68). Similar direct evidence identifying dynamic γ fibres as p_2

fibres awaits future experiment. Meanwhile it must be noted that cat hindlimb muscles assessed histologically as having a high or low γ p_2 supply do not show comparable fluctuation in their supply of dynamic γ fibres. Thus whereas Barker, Stacey and Adal (1970) found seven times more p_2 fibres supplying peroneus longus spindles than FHL spindles, Ellaway, Emonet-Dénand, Jami and Joffroy (1972) found that in samples of γ fibres tested for function the proportion of dynamic γ fibres was the same (40%) for both muscles.

Plate 24, Figs. 104–106

Electron micrographs illustrating the ultrastructure of intrafusal motor end-plates (Fig. 104 p_2, Fig. 105 p_1) in the cat. A micrograph of part of an extrafusal end-plate is included for comparison (Fig. 106). (From Barker, Stacey and Adal, 1970)

Fig. 104. Longitudinal section through the mid-polar region of a peroneus longus spindle (teased out from the muscle prior to preparation for sectioning) showing part of a p_2 plate located on a chain fibre. A thin sole plate underlies knob-like axon terminals. The junctional folds are wide, shallow and unbranched and are occasionally confluent at their bases. *S.c.* Schwann cell. (Micrograph by Dr. M. N. Adal)

Fig. 105. Longitudinal section of part of a p_1 plate located on a chain fibre in the polar region of a tenuissimus spindle that was teased out of the muscle prior to preparation for sectioning. The ultrastructure is very similar to that of extrafusal plates. *s.p.n.* sole-plate nucleus

Fig. 106. Longitudinal section through part of an extrafusal end-plate supplying a deep lumbrical muscle fibre showing an axon terminal and junctional folds. (Micrograph by Dr. M. N. Adal)

Plate 25, Figs. 107–109

Electron micrographs illustrating the ultrastructure of trail endings in a rabbit hindfoot lumbrical spindle. All show absence of sole plate

Fig. 107. Transverse section through the equatorial region of a chain fibre (note part of chain nucleus) showing a trail axon terminal (*tr.ax.t.*) synapsing alongside a sensory axon terminal (*s.ax.t.*), which is partially covered by a tongue of sarcoplasm. *e.* elastic fibre; *fib.* fibrocyte with part of nucleus shown; *pr.ax.* preterminal axons (either sensory or trail); *S.c.* Schwann cell

Fig. 108. Transverse section through the juxta-equatorial region of an intermediate bag fibre showing the presence of two trail axon terminals. *lip.* lipoid droplet

Fig. 109. Transverse section through a trail axon terminal located on a typical bag fibre in the juxta-equatorial region. *s.r.cist.* dilated terminal cisterna of sarcoplasmic reticulum

Plate 26, Figs. 110–113

Ultrastructure of the axon terminals of trail endings in cat spindles. *ax.t.* axon terminal; *j.f.* junctional fold; *pt.tr.ax.* preterminal trail axon; *S.c.* Schwann cell; *s.p.* sole plate. (Figs. 110 and 111 from Barker, Stacey and Adal, 1970)

Fig. 110. Transverse section through the mid-polar region of a de-afferented and sympathectomized peroneus longus spindle showing the smooth junctions of two trail axon terminals on a chain fibre. Note relatively thick sole plate, and presence of two preterminal trail axons. (Micrograph by Dr. M.N. Adal)

Fig. 111. Longitudinal section through the juxta-equatorial region of a tenuissimus spindle showing a trail axon terminal on a chain fibre. At the level of the asterisk there is a break in the continuity of the basement membrane and a decrease in the width of the synaptic gap. Note thin sole plate, and presence of occasional shallow junctional folds

Fig. 112. Transverse section through the juxta-equatorial region of a peroneus longus spindle showing two trail axon terminals on a chain fibre. Note occasional shallow junctional folds. (Micrograph by Dr. M.N. Adal)

Fig. 113. Transverse section through the juxta-equatorial region of a lumbrical spindle showing trail axon terminals with smooth myoneural junctions on a chain fibre. The terminals lie in shallow depressions of a thin sole plate

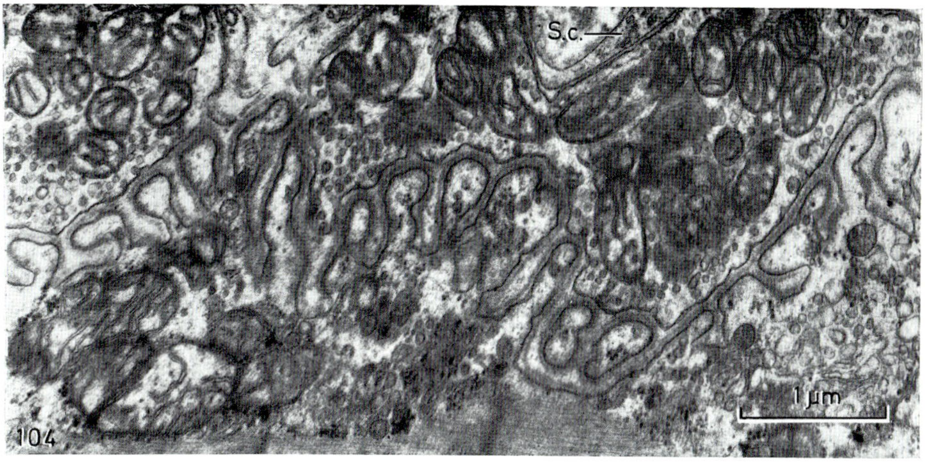

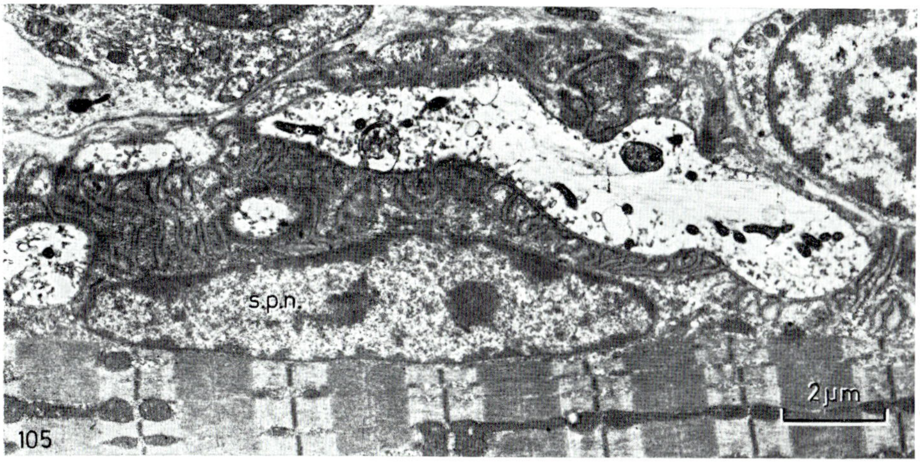

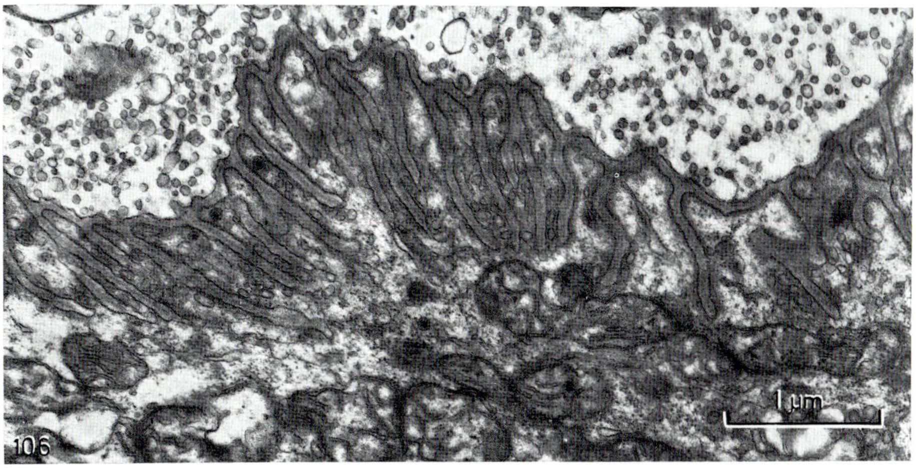

Plate 24

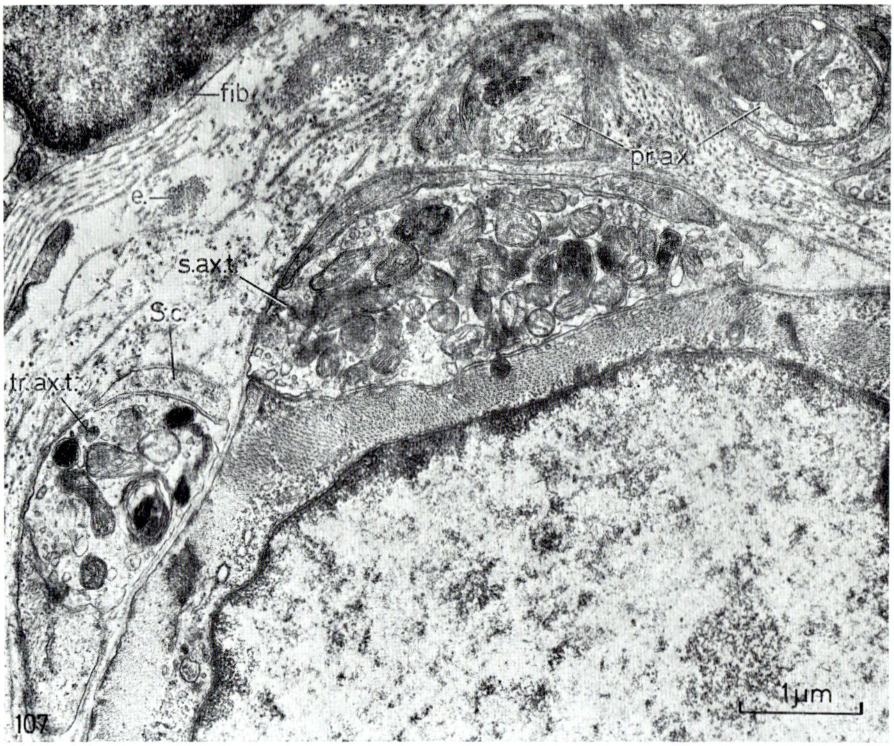

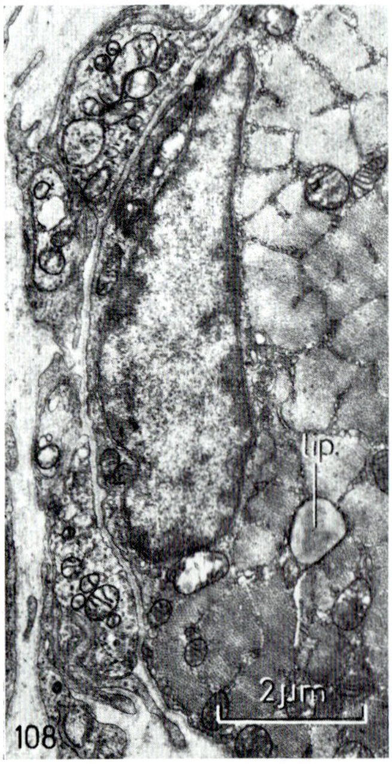

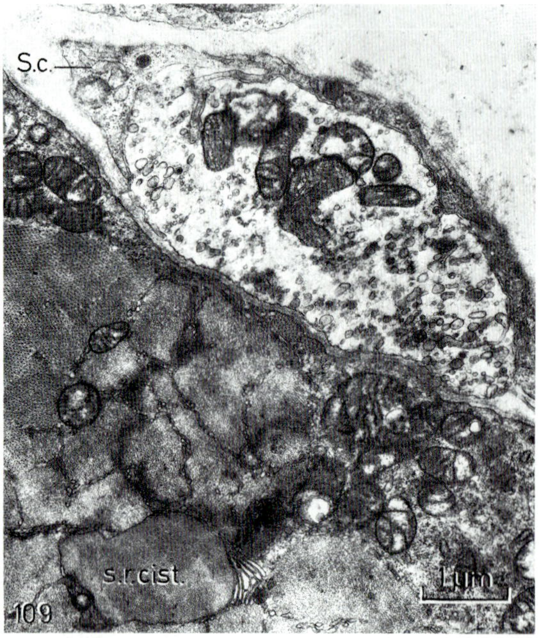

Plate 25

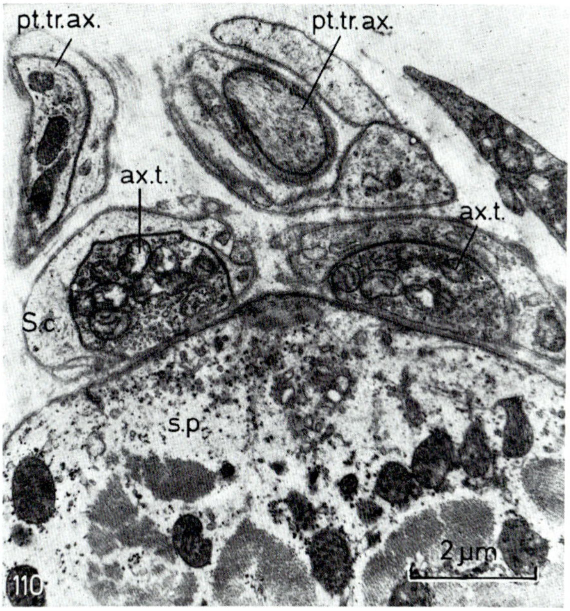
pt.tr.ax.
pt.tr.ax.
ax.t.
ax.t.
S.c.
s.p.
2 μm
110

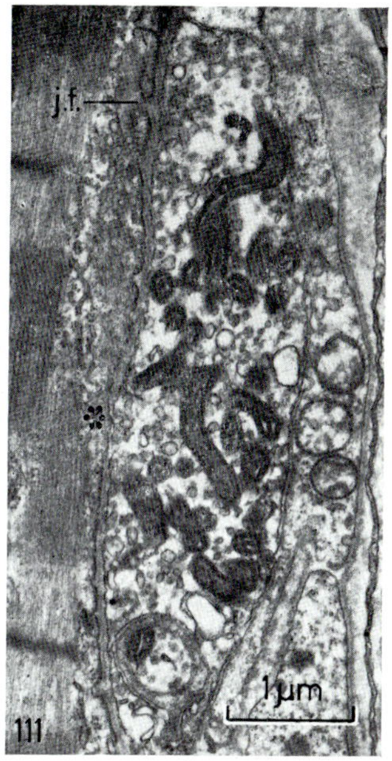
j.f.
1 μm
111

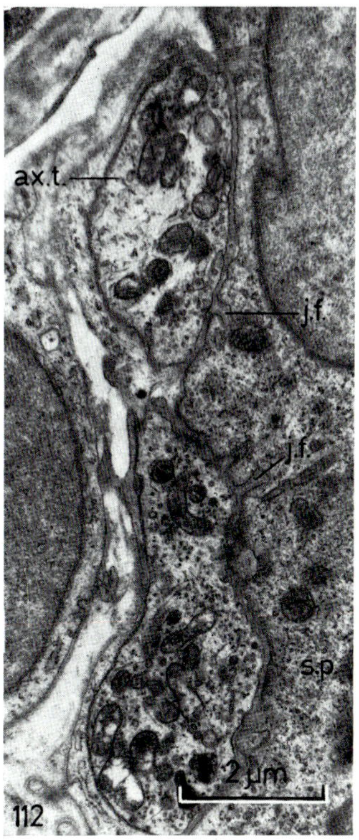
ax.t.
j.f.
j.f.
s.p.
2 μm
112

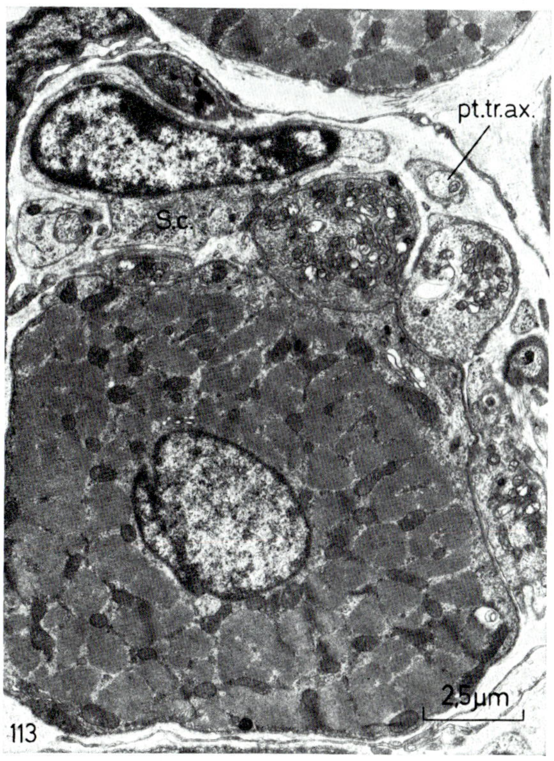
pt.tr.ax.
S.c.
2,5 μm
113

Plate 26

Some of the β motor fibres that collaterally innervate rabbit lumbrical spindles exert a dynamic effect on the responses of the primary ending, others exert a static effect (EMONET-DÉNAND, JANKOWSKA and LAPORTE, 1970). In the cat, the fusimotor branches of β fibres have been traced to p_1 plates (BARKER, STACEY and ADAL, 1970; BARKER, EMONET-DÉNAND, LAPORTE, PROSKE and STACEY, 1971); the majority of p_1 plates are located on bag fibres (see p. 54); and the stimulation of β fibres so far isolated has produced a dynamic effect (BESSOU et al., 1965). These facts, taken in conjunction with the disposition of rabbit p_1 innervation (see p. 55), suggest that rabbit dynamic β fibres are those that supply p_1 plates to the typical bag fibres, whereas the static β fibres account for the supply of p_1 plates to the intermediate bag fibres, and, much less frequently, to the chain fibres[4]. It follows that future experiment should succeed in revealing the existence of static β fibres in the cat, since 25% of p_1 plates in cat spindles are located on chain fibres. In this context the incidence of p_1 plates on cat intermediate bag fibres, at present unknown, has to be taken into account. As already noted (p. 56), the presence of static and dynamic β fibres in rat segmental tail muscles has recently been demonstrated by ANDREW and PART (1973).

The general conclusion to be drawn from the dual functional role played by β fibres in rabbit spindles is that it is the nature of the muscle fibre and its contractile properties that are important in determining static and dynamic effects, rather than the type of motor ending initiating the contraction. It remains to be seen how this may be reconciled with the fact that the most frequent disposition of the static γ trail innervation is for it to be supplied to both bag and chain fibres. EMONET-DÉNAND, JOFFROY and LAPORTE (1972) suggest that such common innervation could explain their observations on the frequency-depending action of some γ fibres. Stimulation of these fibres at 10–30/sec increases the phasic sensitivity to sinusoidal stretch, whereas higher rates of stimulation (60–150/sec) decrease it, i.e. a dynamic response is succeeded by a static one. The present indications thus suggest that it is an over-simplification to correlate functionally distinct fusimotor fibres with fibres terminating in morphologically distinct endings.

5. *The Sympathetic Innervation*

Recent ultrastructural studies by BANKER and GIRVIN (1971) and SANTINI and IBATA (1971) claim that in addition to the somatic innervation intrafusal muscle fibres are innervated by postganglionic sympathetic nerve fibres. Before examining this claim we may note that non-myelinated fibres present in spindles may be of four different kinds. (i) Sympathetic vasomotor fibres engaged in innervating the extensive vascular supply to the spindle. (ii) Somatic group IV afferents (diameter range 0.15–1.2 μm, peak 0.35 μm), and the non-myelinated preterminal axons derived from myelinated free-ending stem fibres belonging to afferents in groups II and III (see STACEY, 1969). (iii) Preterminal axons derived from γ trail fibres (diameter range 0.2–1.6 μm, peak 0.8 μm; see p.53). (iv) Occasional growing sprouts from β and γ motor axons or endings (see BARKER and

4 MATTHEWS (1972) ignores the existence of the intermediate bag fibres and their p_1 innervation, and suggests that the static β fibres may terminate in trail endings.

IP, 1966; TUFFERY, 1971). The existence of a sympathetic fusimotor innervation would therefore add a fifth category of non-myelinated fibre.

In early work with the light microscope HINSEY (1927) reported that sympathetic vasomotor fibres were the only ones to remain intact after degeneration of the somatic innervation of the spindle, and HINES and TOWER (1928) found that these were the only fibres that degenerated in sympathectomized material. Other workers, however, reported the presence of non-myelinated fibres that bore no relation to spindle capillaries, but appeared to take part in the innervation of the spindle itself (TELLO, 1922, cat, man; GARVEN, 1925, hedgehog; BARKER, 1948, rabbit). BARKER (1948) describes them as running within the walls of the capsule and dividing extensively in the polar regions to form a fine anastomosing network. The fibres are less than 0.5 μm thick, are not always present, and appear to terminate either within the axial sheath or on the intrafusal muscle fibres, though never among the sensory endings. He suggested that they might be "somatic sensory fibres subserving the sensation of pain", i.e. group IV afferents. In a later paper BARKER and COPE (1962b) reported the presence of non-myelinated fibres in cat deafferented spindles that innervated the muscle fibres in the equatorial region. They suggested that these were most probably of sympathetic origin. However this was before trail endings had been identified and described, and it was later acknowledged by BARKER (1967) that the fibres in question were more likely to be preterminal trail axons.

Recent evidence in favour of an intrafusal sympathetic innervation is based on the characteristics of the vesicle population in sympathetic terminals. A feature of this is the presence of small granular vesicles that contain norepinephrine (BLOOM and BARRNETT, 1966), and this affords an opportunity for the detection of the terminals by the use of fluorescence methods for the histochemical demonstration of catecholamines. Employing such methods BANKER and GIRVIN (1971) observed fluorescent nodules (? autofluorescent fibrocytes) running parallel and close to the intrafusal muscle fibres in the equatorial region of dog gastrocnemius spindles. In electron micrographs they also observed small (0.4–0.5 μm diameter) non-myelinated axons in the equatorial region that contained both large and small granular vesicles. Some of these were located near to small arterioles within the capsule wall, while others were located outside the axial sheath in the periaxial space. They were therefore identified as sympathetic axons. Other non-myelinated axons (0.6–0.7 μm diameter), devoid of granular vesicles, were observed underneath the axial sheath close to the intrafusal muscle fibres. The most obvious interpretation of these observations is to identify the former axons as sympathetic vasomotor fibres, and the latter as, most probably, preterminal axons of γ trail fibres. However, Banker and Girvin also observed that small granular vesicles are numerous in the terminals of secondary sensory endings. This leads them to the remarkable conclusion that "the prominence of granular vesicles in secondary sensory endings and the pattern of catecholamine fluorescence in this zone suggest a sympathetic innervation", vasomotor fibres and group II afferents apparently being all one and the same since "the small non-myelinated nerves and the 'sensory' endings which were conspicuous in the myotubular zones contained the same granular vesicles which are prevalent in the sympathetic nervous system" (p. 190). Such confusion is inevitable if it is believed that granular vesicles are a diagnostic feature of sympathetic axons and terminals.

In the ultrastructural study of SANTINI and IBATA (1971, cat tenuissimus and lumbrical spindles) use is made of TRANZER and THOENEN's (1967) finding that 5-hydroxydopamine (5-HDA) specifically accumulates as a highly osmiophilic substance in the small and large granular vesicles of postganglionic sympathetic terminals. In the spindles of normal cats Santini and Ibata observed small non-myelinated axons (diameter not less than 0.5 μm), devoid of granular vesicles, within the capsule and in the periaxial space. They were not associated with blood vessels. In the spindles of cats injected with 5-HDA they observed similar non-myelinated axons, some with large granular vesicles and others without. They advance various arguments in favour of identifying the former axons as sympathetic, a conclusion that is no doubt correct since BANKER and GIRVIN (1971) observed such axons in their normal material anyway. There is no good reason why such axons should not be vasomotor. Indeed, in accomplishing the innervation of the spindle's extensive vascular supply it seems obvious that vasomotor fibres should course apart from the blood vessels from time to time. SANTINI and IBATA's (1971) suggestion that sympathetic axons thus located are not vasomotor but fusimotor, engaged in making a "significant, direct, i.e. non-humoral, contribution .... to motor integration via 'sympathetic bias' on the muscle spindle" can only be regarded as an interesting speculation. For such an idea to gain credence it would be necessary to demonstrate the existence of sympathetic fusimotor terminals in an ultrastructural study made on spindles deprived of their somatic innervation by degeneration after cutting the dorsal and ventral roots. Meanwhile the sceptic is unlikely to be convinced by evidence from normal material that rests ultimately on the size and density of vesicles, or on the assumption that the vasomotor identity of a sympathetic axon is necessarily excluded if it is not found in contact with a blood vessel.

6. *Tandem Spindles*

The tandem type of spindle has two or more encapsulated sensory regions spaced along the length of the intrafusal bundle, in contrast to the single spindle, which has one such region situated equatorially between two poles. The term 'tandem' was introduced by COOPER and DANIEL (1956) to describe such spindles occurring in human hand and neck muscles, though they had been observed before in the muscles of various mammals by BAUM (1900), CUAJUNCO (1927), and BARKER (1948). The successive capsular regions are linked together by one or two bag fibres, while the chain fibres are restricted to each capsular unit in the series.

The most common type of tandem spindle is that in which two encapsulated sensory regions occur in linear succession. BARKER and IP (1961) found that of 64 tandem spindles in cat rectus femoris, 88% were double, 9% triple and 3% quintuple. The length of these tandem spindles ranged from 8.43 to 22.30 mm (mean 13.0 mm); the length of single spindles in the same muscle ranged from 2.40 to 13.68 mm (mean 7.02 mm). Some of the tandem spindles extend for the entire length of the extrafusal muscle bundles in this bipennate muscle. The typical condition is for a large capsular unit receiving a primary and one or more secondary endings to be linked with one or two small capsules that receive a primary ending only. The tandem fraction of the total spindle population was estimated by BARKER and IP (1961) to be 16–20% for cat rectus femoris; about the same for soleus; slightly higher for semitendinosus and flexor digitorum longus; and considerably lower for the pes interossei. SWETT and ELDRED's (1960a) estimations are also of this order, being 11% for cat soleus and 24% for medial gastrocnemius. Voss (1962b) found only 6% in human lumbrical muscles.

7. *Spindles in Extraocular Muscles*

The eye muscles of mammals are of particular interest because they are composed of two separate motor systems, some of the muscle fibres being focally innervated by end-plates, and others being supplied with a distributed innervation that consists of multiterminal grape endings. This is a common feature of the muscles of lower vertebrates, but is restricted in mammals to the rectus and oblique extraocular muscles (HESS, 1961b); the tensor tympani (ERULKAR, SHELANSKI, WHITSEL and OGLE, 1964) and stapedius (FERNAND and HESS, 1969) muscles of the ear; and the striated muscles of the oesophagus (CSILLIK, 1965; FLOYD, 1971). The extraocular rectractor bulbi (BACH-Y-RITA and ITO, 1965) and levator palpebrae (DIETERT, 1965) are composed of plate-innervated muscle fibres only, like the skeletal muscles.

Muscle spindles are present in the eye muscles of man and the chimpanzee (COOPER and DANIEL, 1949); the macaque monkey (GREENE and JAMPEL, 1966a); the albino mouse (MAHRAN and SAKLA, 1965); the horse (BONAVOLONTA, 1956); and the following artiodactyls: wild and domestic sheep, goat, wild boar and pig, cow, ox, deer, giraffe, and gnu (CILIMBARIS, 1910; COOPER and DANIEL, 1949). Spindles have been reported absent from the eye muscles of the rabbit, hare, rat dog, cat and fox (CILIMBARIS, 1910), and from those of the cheetah and bear (COOPER and DANIEL, 1949). In this curious distribution primates thus share possession with the horse and the artiodactyls, and the carnivores lack them in common with the rodents (except the mouse). No information is available with regard to the occurrence of spindles in the eye muscles of other placental mammals, or of marsupials or monotremes. In compiling this tally it should be noted that some negative reports later proved incorrect. For example, CILIMBARIS (1910) failed to find spindles in pig or horse eye muscles, but their presence was later reported in the pig by SUTTON (1915), COOPER and DANIEL (1949) and BONAVOLONTA (1958) and in the horse by BONAVOLONTA (1956a). Also the spindles in the eye muscles of the macaque monkey were finally detected by GREENE and JAMPEL (1966), having previously eluded SHERRINGTON (1897) and COOPER and DANIEL (1949). They proved to be sparsely distributed in the rectus muscles and absent from the oblique muscles. HOSOKAWA (1961) draws attention to papers published in Japanese (HAMADA, 1928; YOKOMATSU, 1932) that report the presence of spindles in human eye muscles prior to Cooper and Daniel's work. According to SAS and APPELTAUER (1963) their presence in these muscles was first pointed out by SIEMMERLING (1888).

A spindle in a mammalian rectus or oblique eye muscle is thus surrounded by extrafusal muscle fibres composing two motor systems, a situation comparable with that of a spindle located in a frog toe muscle. Such frog spindles are collaterally innervated by branches from the two motor systems (GRAY, 1957). Is a similar motor innervation supplied to spindles in mammalian eye muscles? This and related problems have recently been investigated by HARKER (1972a, b) in a detailed histochemical, light and electron microscopic study of the sheep superior rectus muscle (see also BARKER, 1968; BARKER, HARKER, STACEY and SMITH, 1972; BARKER and HARKER, 1972). Histochemical preparations of this muscle show that it is composed of types A and C twitch fibres, and a third type that has a low glycolytic and oxidative activity (as in frog non-twitch fibres; see ENGEL and IRWIN, 1967), and a distributed grape motor innervation. These 'type G' fibres (BARKER, HARKER, STACEY and SMITH, 1972) possess characteristics that differ according to whether they are located within the central core of thick muscle fibres that lies adjacent to the eyeball, or within the outer zone of thin muscle fibres that occupies the orbital rim. The orbital rim G fibres (diameter 10–25 μm) occur in high peripheral concentration in the proximal and distal thirds of the muscle, and receive a grape innervation in the form of extensive droplet chains on average 200 μm long spaced 610 μm apart. Their myofibrils are closely packed without delineation, and there is no M line. The sarcotubular system is poorly developed, the mitochondria are very small, and the myoneural junctions are smooth with occasional wide, shallow junctional folds (Plate 27, Fig. 114). The central core G fibres are scattered singly throughout the core and comprise about 7% of its total fibre population. With an average diameter of 30 μm they are the largest fibres in the muscle, and their grape innervation consists of small clusters of terminals on average 60 μm long spaced 850 μm apart. Their myofibrils are delineated into large irregular groups; an M line is present; and the mitochondria are small and very scarce. Other ultrastructural features are similar to those of the orbital rim G fibres.

The muscle contains approximately 150 spindles peripherally distributed in the orbital rim zone, and transverse sections show a typical mammalian equatorial nucleation of bags (average 2) and chains (average 5). The bag fibres are derived from the orbital rim G fibres. These increase in number and decrease in diameter as they proceed towards the belly of the muscle from each end owing to subdivision to form daughter fibres, or short branches that form tendinous attachments with other muscle fibres. (ZENKER and KRAMMER (1967) have

---

Plate 27, Figs. 114–119

Figs. 114–119. Ultrastructure of intrafusal muscle fibres and their extrafusal counterparts in sheep superior rectus muscle as seen in longitudinal section. The ultrastructure of an orbital rim G fibre (Fig. 114) compares closely with that of a nuclear-bag fibre (Fig. 115); the M line is absent, the sarcoplasmic reticulum is poorly developed, and the mitochondria are very small. Similarly, the ultrastructure of an orbital rim small C fibre (Fig. 117) compares closely with that of a nuclear-chain fibre (Fig. 118); the M line is present, the sarcoplasmic reticulum is well developed, and the mitochondria are large. The ultrastructure of the motor endings is also very similar. Fig. 116 shows a grape axon terminal lying on the surface of a bag fibre cut in transverse section. The myoneural junction is smooth and could equally well be that of a grape axon terminal on an orbital rim G fibre. Fig. 119 shows an axon terminal of a plate ending on a chain fibre (longitudinal section). It overlies a thin sole plate with regular junctional folds that are shallow and wide. The ultrastructure of the plate endings on orbital rim small C fibres is identical. (From HARKER, 1972a, b)

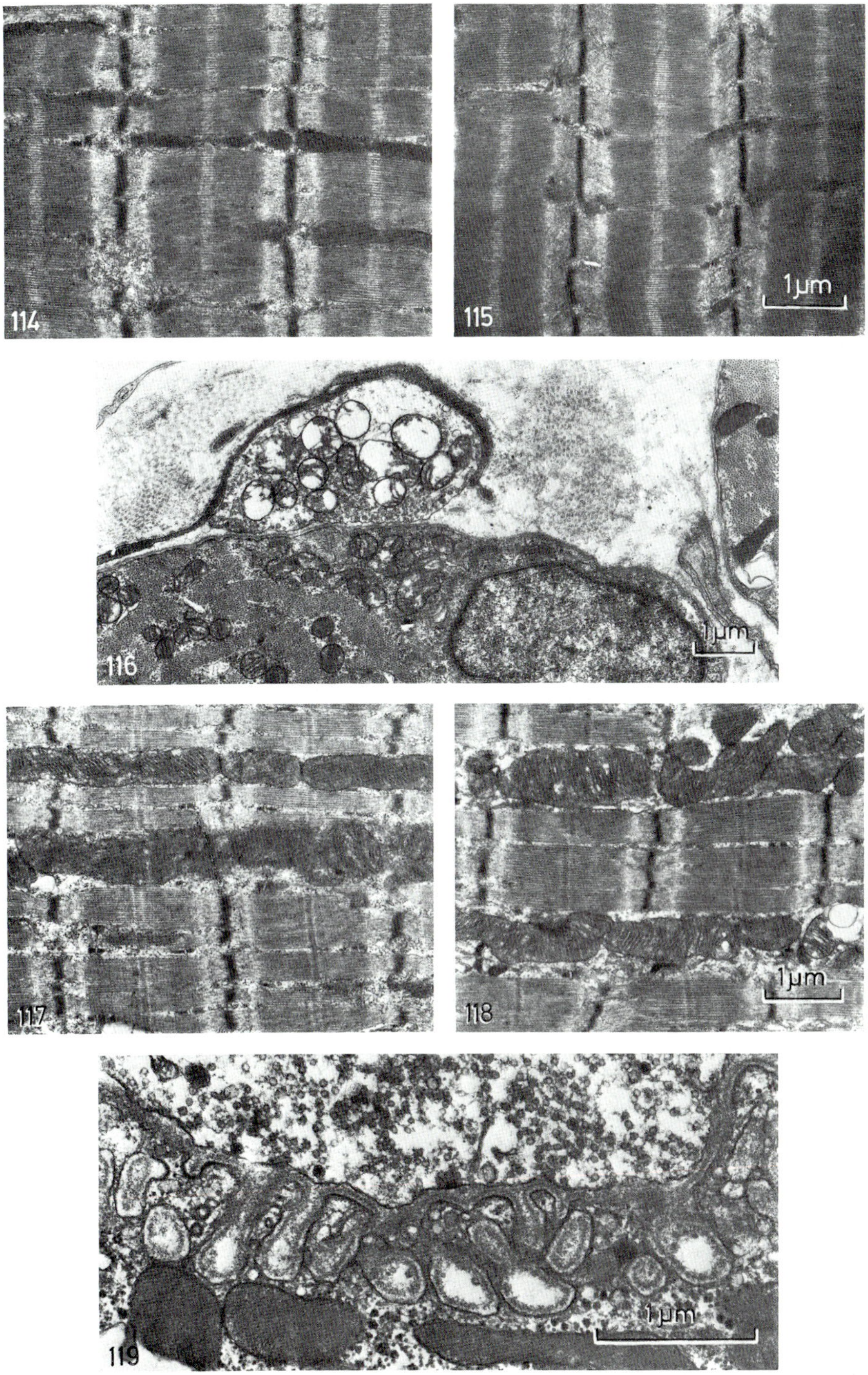

Plate 27

observed similar splitting and attachment in chicken iris muscles). Some of the daughter fibres pass through the spindles as bag fibres. The only difference between an extrafusal orbital rim G fibre and a nuclear-bag G fibre is that the latter is enclosed in a spindle capsule for approximately one-sixth of its length, and possesses a typical nuclear-bag innervated by a primary ending. All the other features are identical. They share the same histochemical profile and ultrastructure (Plate 27, Fig. 115), and both receive a grape motor innervation, which may often be seen to originate from common stem fibres. The ultrastructure of the grape terminals (Plate 27, Fig. 116) is also identical. The collateral grape innervation of the bag fibres is particularly striking, and is in fact illustrated by Cilimbaris (1910, see his Fig. 10, Plate 29), though he did not describe it in these terms. The central core G fibres do not subdivide and are not associated with spindles.

The chain fibres in the spindles are much shorter and thinner and taper off soon after leaving the capsule. Their histochemical profile and ultrastructure are the same as those of the extrafusal type C fibres (Plate 27, Figs. 117, 118), features of the latter being the presence of an M line, a well-developed sarcotubular system, and large, numerous mitochondria. They receive a sensory innervation from the primary ending and from numerous secondary endings. The motor innervation consists of plates provided by collateral branches of the motor fibres that innervate the type C muscle fibres. The postsynaptic membrane of both the extrafusal and the intrafusal plates (Plate 27, Fig. 119) is thrown into numerous junctional folds.

The motor innervation of these sheep superior rectus spindles is thus segregated and collateral, a grape innervation being distributed to the bag fibres and a plate innervation to the chains. It is evident from Whitteridge's (1959) experiments that the bag fibres may receive collaterals from up to six motoneurons supplying grape-innervated motor units. Each type of intrafusal muscle fibre closely resembles its extrafusal counterpart with which it shares a common type of motor innervation. Such spindles could, in fact, be regarded as small bundles of G and C fibres modified by receiving a sensory innervation comprising primary and secondary endings. A schematic diagram illustrating the arrangement is shown in Fig. 120B. Presumably a similar arrangement obtains in spindles located in the sheep tensor tympanic (Cilimbaris, 1910) and human striated oesophageal muscles (Slawik, 1942). A remarkable feature is the casual way in which the G fibres contribute the bag-fibre component of the intrafusal bundles. Some of them, instead of entering via the poles, penetrate the capsule to join the intrafusal bundle in the equatorial region (cf. Cilimbaris, 1910, Fig. 1). In such an organization the terms 'extrafusal' and 'intrafusal' have to be used to distinguish between different parts of the same muscle fibres. Tandem spindles occur frequently. There is no intermediate type of bag fibre.

The structure and innervation of the spindles in the sheep levator palpebrae muscles are very similar, except that an extrafusal grape-innervated component in the muscle is absent. In effect what happens is that the orbital rim type of G fibre that occurs in superior rectus is now restricted to the spindles where it is found as the nuclear-bag fibre. The grape innervation is exclusively fusimotor and supplies the bag fibres, while the chain fibres, as in superior rectus, are the counterparts of extrafusal type C fibres and share a collateral plate innervation with them (HARKER, 1972b) (see Fig. 120A).

A feature of spindles in the eye muscles of other mammals is the thinness of the capsule, as in those of man (COOPER and DANIEL, 1949), macaque monkey (GREENE and JAMPEL, 1966), and mouse (MAHRAN and SALKA, 1965), though this is not so in sheep. The spindles in human eye muscles are composed of from 1 to 15

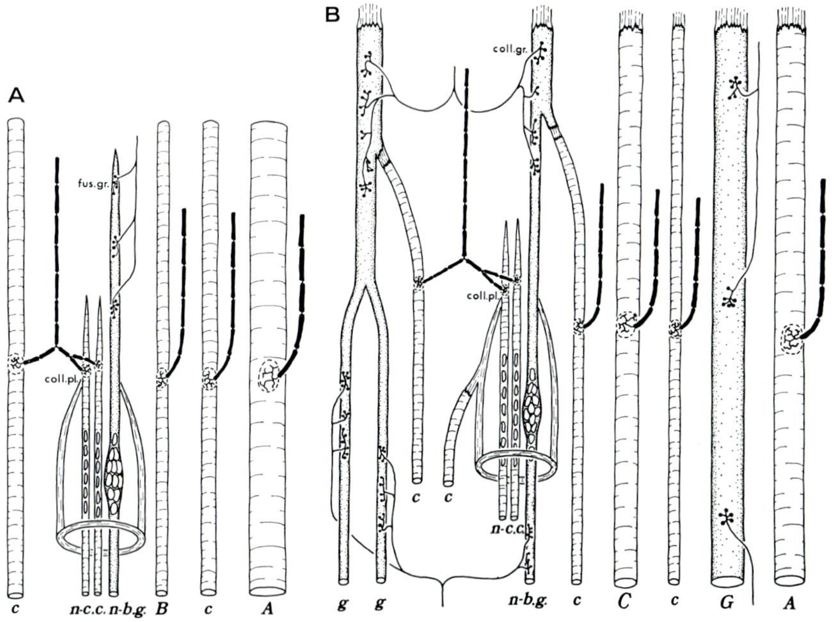

Fig. 120. Schematic diagrams of the pattern of motor innervation in A the levator palpebrae and B the superior rectus muscles of the sheep. *A* large A fibre; *B* small B fibre; *C* intermediate C fibre; *c* small C fibre; *coll.gr.* collateral grape endings; *coll.pl.* collateral plate endings; *fus.gr.* fusimotor grape endings; *G* large G fibre; *g* small G fibre; *n-b.g.* nuclear-bag G fibre; *n-c.c.* nuclear-chain C fibre. (From HARKER, 1972b)

(usually 2–4) muscle fibres, and are much shorter than those in other muscles, the majority being about 0.5 mm long with few longer than 1.0 mm (COOPER and DANIEL, 1949). The number of muscle fibres in mouse eye muscle spindles is 2–7 (MAHRAN and SALKA, 1965), in those of the macaque monkey 2–8 (GREENE and JAMPEL, 1966), and in those of sheep 3–15 (CILIMBARIS, 1910). Cilimbaris, incidentally, gave the first detailed description of the nuclear-bag region; he refers to the bag as a *bläschenförmige Gebilde* or *Bläschenspindel*.

There are afferents in some eye muscles that terminate in spindle-like endings; these are described in § IV E, 5b.

8. Number and Distribution

As a general rule it may be stated that every mammalian striated somatic muscle is provided with spindles. As with every rule there are exceptions, but many negative reports have later proved incorrect. For example, the human lateral pterygoid muscle was said to lack spindles by numerous investigators before their presence was established by HONEE (1966), GILL (1971) and RAKHAWY, SHEHATA and BADAWY (1971). Similarly after much controversy spindles were ultimately demonstrated in human intrinsic and extrinsic tongue muscles by COOPER (1953) and WALKER and RAJAGOPAL (1959). However, as with extraocular muscles, certain muscles may possess spindles in one species and lack them in another. They are present in the quadratus labii superioris facial muscle of the rabbit, but absent from this muscle in the cat and rat (BOWDEN and MAHRAN, 1956); the intrinsic tongue muscles of man and the rhesus monkey possess spindles,

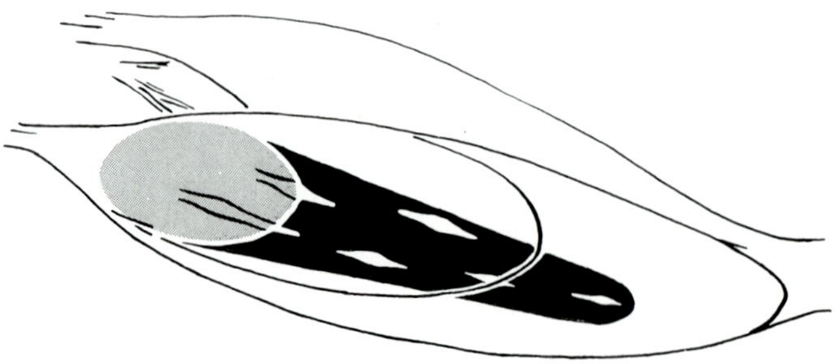

Fig. 121. Diagrammatic (medio-dorsal) view of the right gastrocnemius muscle of the rat, illustrating the general distribution of spindles with respect to extrafusal fibre types. The muscle origins are at the left, the insertion on the right. In the medial head of the muscle (foreground) the spindles are found within a core of deep, oxidative fascicles. The superficial and juxtaseptal glycolytic fascicles lack spindles. Normally the transition between regions is gradual. The precise number and disposition of the spindles are not depicted. (From YELLIN, 1969)

but they are lacking in the same muscles of the cat and sheep (COOPER, 1953); and so on. Such variation is not found in skeletal muscles, but among cranial muscles and other non-joint-moving muscles. In some muscles without spindles there is evidence that the role of the stretch afferent is taken over by sensory fibres that form simple spiral endings around non-encapsulated muscle fibres (see § IV E, 5b).

The distribution of spindles within a muscle follows that of the major intramuscular nerve trunks, and since the pattern of their distribution varies from muscle to muscle, concentrations of spindles do not necessarily occur in 'regions of nerve entry' or in the 'belly' of a muscle as has often been stated. 'Nerve-entry' concentrations occur only where there is a rapid subdivision of the intramuscular supply in this region, and 'belly' concentrations only where there is a supply of one or more intramuscular trunks to this region (BARKER and CHIN, 1960). This

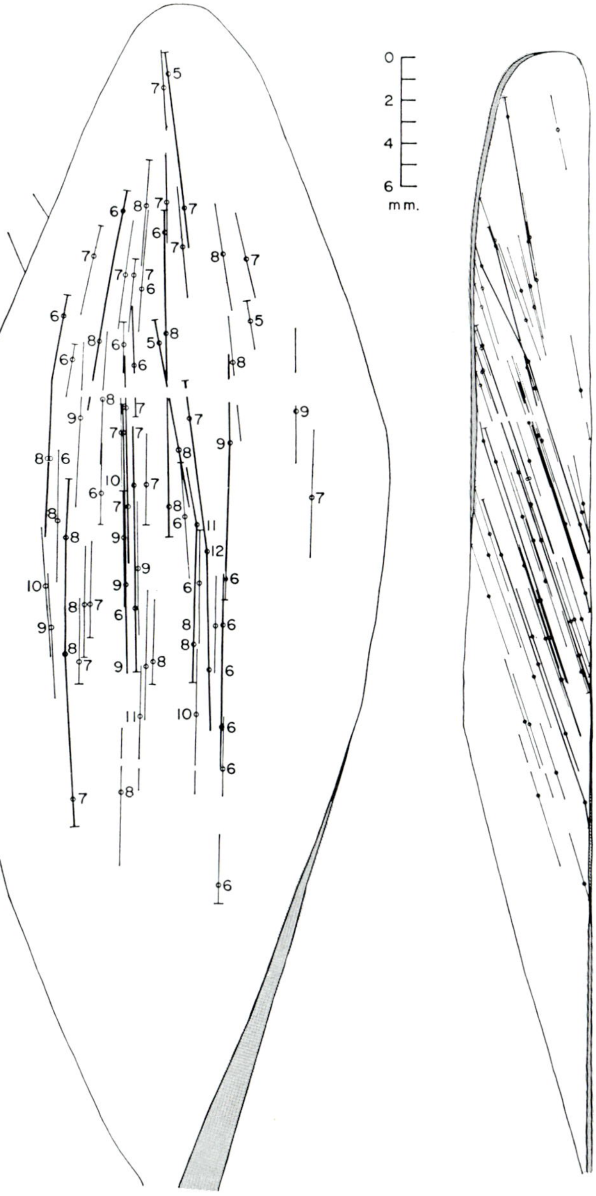

Fig. 122. Distribution of spindles in the left medial gastrocnemius muscle of the cat. In the *dorsal projection* each spindle is represented by a solid line bearing a circle indicating the position of the nuclear region. The number of intrafusal muscle fibres is given near each spindle. Spindles linked in tandem are represented by heavier lines. A short horizontal bar at the end of a spindle line indicates that the end is attached to an aponeurosis. Gaps seen in spindles in two regions of the muscle indicate transitions between paraffin blocks. In the *lateral projection* the spindles are projected onto an imaginary mid-sagittal plane. Superficial (dorsal) surface of muscle is to the left. In several sites three spindle units linked in tandem are seen to stretch from one aponeurosis to the other. (From Swett and Eldred, 1960a)

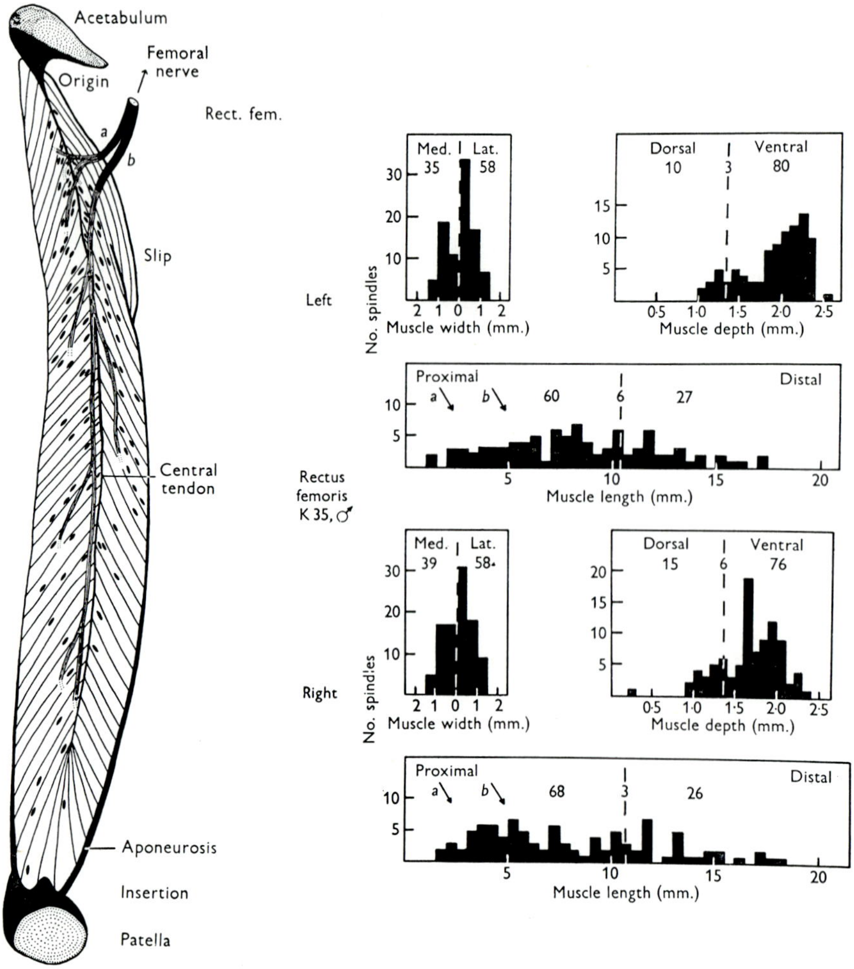

Fig. 123. Projection plan (left) of cat rectus femoris showing distribution of spindle capsules and nerve supply. Spindle capsules shown as black oval symbols; nerve branches *a*, *b* shown black up to point of entry into muscle. Histograms illustrate distribution of spindle capsules in cat rectus femoris by plotting their number against muscle width, depth, and length. The distribution is shown for one pair of recti from a kitten. *a*, *b* proximal and distal branches of supplying nerve. Arrows indicate nerve-entry points (tips of arrows) and direction of entry (angle of arrows). (From Barker and Chin, 1960)

feature follows naturally from the manner in which spindles develop, providing that the primary afferents are distributed generally throughout the intramuscular nerve supply (see § II A, 9). This is so in the majority of muscles, but there are exceptions. Clearly the distribution of spindles in extraocular muscles must necessarily be associated, as we have seen (§ II A, 7), with the distribution of the grape-innervated muscle fibres that contribute to their composition. A spindle distribution of comparable segregation has been demonstrated in certain fast hindlimb rat muscles by YELLIN (1969). Transverse sections of the medial gastrocnemius, plantaris, and tibialis anterior muscles, stained histochemically for enzyme profiles, show superficial fascicles of fast type A muscle fibres surrounding an axial core of deep fascicles composed mainly of slow type B and fast type C fibres. The spindles are restricted in their distribution to the fascicles of the axial core (Fig. 121). Yellin found no segregation of fibre types in the rat soleus (which lacks type A fibres) or hindlimb middle lumbrical muscles, and no specific spindle distribution with respect to extrafusal fibre types.

Yellin's work adds a new dimension to studies of the distribution of spindles within muscles, and more data of this kind are required. Previous studies have been made almost exclusively on serial transverse sections of muscles stained with routine dyes. This enables the length and position of the spindles to be recorded, and projection plans of their distributions to be prepared. These vary in their usefulness from the clear presentations of MERRILLEES, SUNDERLAND and HAYHOW (1950) and SWETT and ELDRED (1960a) to the misleading figures of HAGBARTH and WOHLFART (1952), which show the distribution of spindles not, as claimed, within the muscles containing them, but simply on the grid used by the authors to record them. Dorsal and lateral projection plans of spindle distribution prepared by SWETT and ELDRED (1960a) for the cat medial gastrocnemius muscle are reproduced in Fig. 122. While nerve entry and distribution could be mapped from a transverse section study, it would be a laborious undertaking and has not been attempted in the studies made. The alternative adopted by BARKER and CHIN (1960) and CHIN, COPE and PANG (1962) was to study longitudinal sections of muscles fixed shortly after birth. Their projection plans of various kitten muscles show muscle-fibre direction, distribution and orientation of spindles, and the entry and distribution of the nerve supply. The projection plan and histograms prepared by BARKER and CHIN (1960) for rectus femoris (Fig. 123) show that the nerve supply enters proximally on the medial side and is distributed mainly on the lateral side of the muscle. Most of the nerve branches are ventrally located, and the spindles are distributed in greater quantity in the ventral half of the muscle than the dorsal, more in the lateral half than the medial, and more in the proximal half than the distal. In other words spindle distribution closely follows the distribution of the nerve supply. This feature is particularly striking in the cat tenuissimus muscle where the spindles are arranged in chains alongside the intramuscular tibial and pelvic branches of the supplying nerve (BOYD, 1956).

There is some evidence that tonic muscles are equipped with longer spindles than phasic muscles (COOPER, 1960; SWETT and ELDRED, 1960b). In comparing the cat soleus and gastrocnemius muscles, SWETT and ELDRED (1960b) found the spindles in soleus to be on average 22% longer than those in gastrocnemius. Fascicle length, however, is twice as long in soleus, and its spindles, though longer, extend over only about one-fifth of this length, whereas gastrocnemius spindles extend over one-third. BARKER and IP (1961) showed that in rectus femoris the lengths of the spindles, and the relative lengths of their poles, are related to their position along the length of the fascicles, spindles located within the middle third of this length being longer than those located towards either end. Those spindles situated within the proximal third of the fascicles have longer distal poles than proximal, and those within the distal third have longer proximal poles than distal. Small muscles tend to be characterized by short spindles, as shown by VOSS (1937) in human lumbrical muscles, BOYD (1962c) in cat hindfoot interosseous muscles, and KUBOTA and MASEGI (1972) in the snout muscles of the Japanese shrew-mole. Voss found that about 70% of the spindles in human hand lumbricals measured 1–3 mm long (range 0.60–9.44 mm), and that the longest spindles are located in the belly of the muscle and the shortest at the ends.

Counts of spindles have been published for a variety of mammalian muscles, the most interesting being those in which the spindle density of a muscle is expressed in terms of the number of spindles per gram of the mean weight of the adult muscle. The most extensive information of this kind is that published by Voss and his co-workers for human muscles, which is summarized by COOPER (1960, 1966) and VOSS (1963). The paper by GREGOR (1904) is also relevant. This shows that spindle density is highest in the hand, foot and neck muscles, lowest in the shoulder and thigh muscles, and medium in the more distal muscles of the arm and leg, with relatively high values for the rotating muscles such as pronator quadratus and popliteus. Generally speaking, high spindle densities characterize muscles initiating fine movement (e.g. lumbrical and extra-ocular muscles) or maintaining posture (e.g. soleus), low densities those initiating gross movements (e.g. gastrocnemius). This is further illustrated by the spindle densities for various cat limb muscles published by OSHIMA (1938) and CHIN, COPE and PANG (1962) reproduced in Table 2. The figure for soleus (23) is higher than that for either the lateral (5) or medial (9) gastrocnemius, and the figures for the interosseous (88, 119) and lumbrical (173) muscles are much the highest in the series. The spindle content in Table 2B is expressed as the number of spindle capsules present, since, as pointed out by BARKER and IP (1961), an estimate of the number of spindles in any given muscle can only be made if the proportion of tandem spindles is known.

BARKER and CHIN (1960) and CHIN, COPE and PANG (1962) made their counts using pairs of kitten limb muscles, and showed that there is a close correlation between the number of spindle capsules in a given muscle and its partner in the opposite limb. They were able to derive formulae for predicting the best estimate of the spindle-capsule content of one member of a pair of muscles when the content of the other is known. A tendency towards asymmetry (right count higher than left) was evident among the muscles taken from female kittens.

Table 2. A. *Number and density of muscle spindles in various cat muscles* (OSHIMA, *1938, from* HOSOKAWA, *1961*)

Muscle	No. studied	Mean weight (g)	Mean no. spindles	No. muscle spindles/g
Rectus abd.	2	20.46	87	4
Sartorius	2	11.4	66	6
Biceps brachii	4	6.4	53	9
Flexor carpi radialis	4	1.3	18	14
Extensor carpi rad. long.	4	1.3	18	14
Flexor digit. superf.	2	2.2	35	17
Extensor digit. comm.	2	1.3	22	18
2nd external intercostal	3	0.7	18	27
3rd lumbrical (forelimb)	4	0.04	7	173

Table 2. B. *Number and density of spindle capsules in various cat limb muscles* (Chin, Cope *and* Pang, *1962*). Forty counts made of each kitten muscle (20 pairs); mean weight of each muscle derived from results of weighing 12 examples removed from six 2–3 kg cats

Muscle	Mean weight (g)	Spindle capsule-content range	mean and S.D.	No. spindle capsules/g
lat. gastroc.	7.61	21 (25 to 45)	35 ± 7	5
med. gastroc.	7.34	35 (46 to 80)	62 ± 9	9
[a]rect. fem.	8.36	56 (77 to 132)	104 ± 14	12
[a]tib. ant.	4.57	38 (52 to 89)	71 ± 9	15
semitendinosus	6.41	62 (80 to 141)	114 ± 14	18
soleus	2.49	31 (40 to 70)	56 ± 7	23
FHL	3.25	34 (58 to 91)	75 ± 8	23
tib. post.	0.78	19 (21 to 39)	31 ± 4	39
FDL	1.06	24 (36 to 59)	48 ± 6	45
[b]5th int. (hindlimb)	0.33	12 (22 to 33)	29	88
[a]5th int. (forelimb)	0.21	11 (21 to 31)	25 ± 2	119

[a] data from Barker and Chin (1960).
[b] four muscles counted (Ip, 1961).

9. Development

An essential requirement for studies of spindle development is a plentiful supply of accurately aged foetal material, and this has mainly restricted investigations to the embryos of man, pig and rat. The advantage of studying the rat is that, unlike the other species, the later stages of development are completed postnatally. The most detailed investigations are those of Tello (1917, 1922, man, cat, dog, rabbit, chick), Cuajunco (1927, pig; 1940, man), and Zelená (1957, 1959, rat) (see review by Werner, 1972). These studies were made with the light microscope; the use of the electron microscope has recently begun with studies of rat spindle development (Barker and Milburn, 1972; Landon, 1972a, b; Milburn, 1973).

The pattern of development that emerges from the earlier work is that the sensory fibres are the first to arrive in the muscle, spindle formation being initiated when a primary afferent makes contact with a bundle of developing myotubes (Sutton, 1915; Tello, 1917, 1922). The effect of this contact is to stimulate nuclear division in the myotubes and to initiate capsule formation in an area that will ultimately become the equatorial region of the spindle. As the primary ending develops, myofibril formation, though delayed intrafusally, proceeds in the polar regions and among the extrafusal myotubes, both of which assume the form of muscle fibres. When the motor and other sensory fibres arrive they produce no effect on development that is apparent in light microscope preparations. As Tello (1922) puts it: "The Weismann fascicle is thus nothing else but one of the primitive myotube bundles that form the muscle at the moment when the development of the spindle is initiated, and is maintained in a slightly modified embryonic form even in adult man ..." (translation, p. 407).

Nevertheless the number of intrafusal muscle fibres increases during development, and there are conflicting views as to how this is achieved. According to Cuajunco (1927, 1940) the adult pig and human spindles in the biceps muscles are composed of muscle fibres of large, medium, and small size, the polar diameters in pig being 18 μm, 12 μm, and 6 μm, respectively (cf. pp. 18–27). He found that the large fibres were the first to arise in development; that the medium fibres arose from somewhat younger myotubes grouped with the first; and that the small fibres were added later, either by the splitting of the original myotubes, or by the incorporation of extrafusal muscle fibres within the capsule. The problem has recently been investigated by MARCHAND and ELDRED (1969) in the rat. They observed, as reported by ZELENÁ (1957), that at birth rat spindles contain only two muscle fibres, the usual adult complement of four being achieved six days later. Of the two fibres present at birth, one is more profusely nucleated and has a larger diameter than the other. The former is recognized by Marchand and Eldred as a bag fibre, the latter as a chain fibre, and they maintain that each fibre splits longitudinally to produce the adult complement of two bag fibres and two chains. None of the intrafusal fibres contained labelled nuclei following the injection of radioactive thymidine into 0–12-day-old rats, which leads them to conclude that the nuclei present in the two muscle fibres at birth are simply distributed to the daughter fibres at the time of division.

The final events to occur in spindle development are further differentiation of the capsule and its extension to enclose parts of the polar regions; the appearance of the periaxial space; and the acquisition of the motor innervation. In pig embryos (length at birth 250–300 mm) the first sensory contact occurs at the 25–30 mm stage, and a distinct primary ending is formed by the 50 mm stage; the capsule begins to form at 80 mm and is obvious at 110 mm, by which stage it encloses 4–5 muscle fibres that receive both a sensory and a motor innervation. At 150 mm the periaxial space appears; at 180 mm the number of muscle fibres has increased to 5–12; and by the 190 mm stage the innervation is complete and the spindles are morphologically mature (CUAJUNCO, 1927). SUTTON (1915) recognized the presence of secondary 'flower-spray' endings at the 55 mm stage. In human development (gestation period approximately 40 weeks) the first neuromuscular contact occurs in the 7–11 week old foetus; all components are acquired by the 15th week; and the spindles are completely formed by the 24th week (FORSTER, 1894; TELLO, 1917, 1922; HEWER, 1935; CUAJUNCO, 1940; MAVRINSKAIA, 1960; BOWDEN, 1963). In the rat (gestation period 21 days) the first sensory contact, according to ZELENÁ (1957), occurs on the 19th day, and spindle development is complete 25 days after birth, the periaxial space appearing at 5 days postnatally and the motor innervation being acquired after 5–10 days during the period when the intrafusal muscle fibres increase in number. It should be noted that all these light-microscope investigations (except that of MARCHAND and ELDRED, 1969) were made before details of the fusimotor innervation were fully understood, and that most of them preceded recognition of the bag and chain types of intrafusal muscle fibre.

If neuromuscular contact by sensory fibres is prevented in developing rat muscle by intra-uterine severance of the sciatic nerve, no spindles are formed in the denervated muscles (ZELENÁ, 1957; ZELENÁ and HNIK, 1960a, b). Intra-uterine tenotomy, however, does not disturb the normal progress of spindle development, apart from retarding the longitudinal growth of the intrafusal muscle fibres (ZELENÁ, 1963). If rat muscle is denervated by crushing the sciatic nerve at birth, the half-formed spindles degenerate within 10 days, and on reinnervation few are found to have developed again 5 months later. However, if the same experiment is performed on 14-day-old rats (when spindle development is nearing completion), there is no spindle degeneration, and their subsequent reinnervation is successful as judged by the comparable numbers of spindles found in the experimental and control muscles 10 months later (ZELENÁ and HNIK, 1963a). These experiments demonstrate that the morphogenetic influence of the afferent innervation is essential for the development of spindles, indeed for a time their very existence depends upon it, but this dependence ceases when the final stage of development is reached. ZELENÁ (1964) also showed that de-efferentation in newborn rats does not have any effect on the number or type of intrafusal muscle fibres that develop, the normal complement of 2 bag and 2 chain fibres being achieved.

In an electron-microscopic investigation of the development of spindles in rat hindlimb muscles BARKER and MILBURN (1972) have shown that the adult complement of four intrafusal muscle fibres is achieved by a rapid sequential process completed within $5^1/_2$ days, starting in the 20-day-old foetus and finishing 4 days postnatally. Spindles were first detected in the $19^1/_2$-day-old foetus. They consist of a single bag fibre (fibre 1) bearing primary sensory terminals enclosed within a thin equatorial capsule. Myoblasts lie in close proximity to this fibre throughout its length, and in the 20-day-old foetus these are seen to have fused to form a smaller, less differentiated myotube with a nuclear bag containing fewer nuclei than fibre 1. The two fibres lie in close apposition, the deep surface of fibre 2 often extending into the sarcoplasm of fibre 1 in the form of pseudopodial inpushings, but never fusing with it. Half a day later myoblasts are seen lying in between these fibres along one side, and by birth (21 days) they have formed a third fibre, which lies in pseudopodial apposition with one or both of the others. Fibre 3 is smaller than the others and has a nuclear chain instead of a nuclear bag. Fibres 1 and 2 are now of equal maturity, but still myotubular at the poles. Myoblasts are still present in the capsule, which now extends to the poles. Motor terminals make their first appearance at this stage (MILBURN, 1973) with smooth myoneural junctions, as in developing extrafusal motor end-plates.

The 4-day postnatal spindle has the full complement of muscle fibres. The second chain fibre has appeared, though it is not always fully differentiated and may still lie in pseudopodial apposition with one of the other fibres. At this stage the nuclei of fibre 1 have become peripheral at the poles. Each fibre has a similar myofibrillar ultrastructure, with M line present. Adult ultrastructural variations do not become apparent until the 12-day postnatal stage, and these include the loss of the M line in fibre 1, the typical bag fibre (MILBURN, 1973). Several features of this development are illustrated in Plates 28 and 29, and the process is summarized schematically in Fig. 130 reproduced from MILBURN (1973).

LANDON (1972a, b) also reports that the development of rat intrafusal muscle fibres is a sequential process, and confirms their ultrastructural similarity in the early postnatal spindles. He observed simple contacts between large non-myelinated nerve terminals and apparently undifferentiated single myotubes, or pairs of myotubes, in the 18-day-old foetus. At this stage the perineural epithelial cells ensheathing the primary afferent extend to enclose the innervated region of the myotube, providing the first layer of the equatorial region of the future capsule.

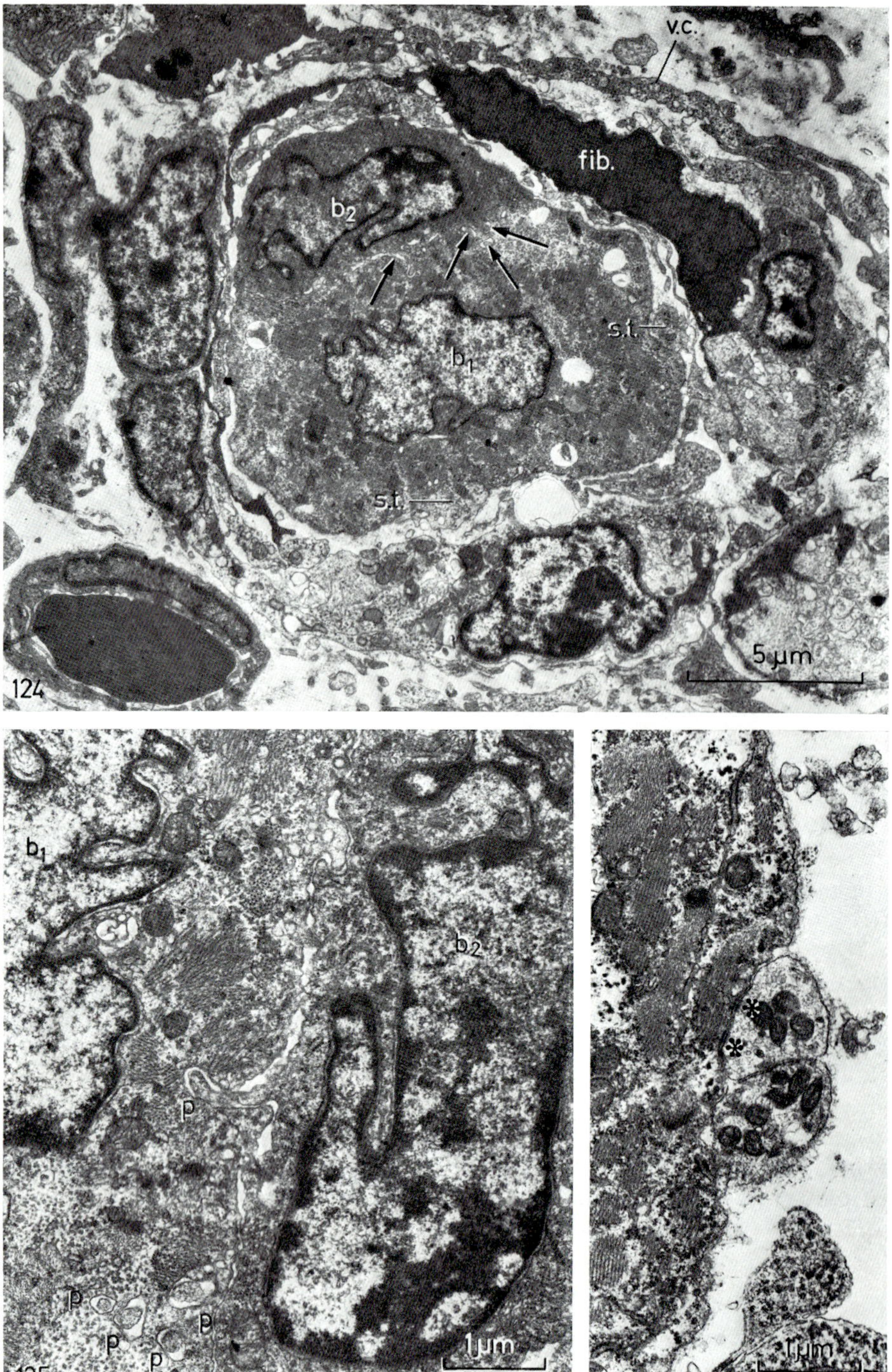

Plate 28

Landon describes how the developing intrafusal muscle fibres remain together as a tightly packed bundle until the latter part of the first postnatal week. Up to this stage the sensory axon terminals enwrap their common external surface, but thereafter they invade the bundle to form separate coils around individual fibres. This is accompanied by a progressive separation of the fibres one from another, starting at the poles and spreading to the equator. Fibrocytes appear during the second postnatal week to form the endomysial and axial sheaths, and in the 14-day postnatal spindle the beginnings of the periaxial space are evident. Landon observed the first definite intrafusal motor terminals 12h after birth, though developing extrafusal end-plates occur in the 18-day-old foetus.

Rat extrafusal myogenesis is sequential in nature, the development of secondary and tertiary generations of myotubes being characterized by pseudopodial attachment to existing myotubes (Kelly and Zacks, 1969; Landon, 1970, 1971). The manner in which rat intrafusal muscle fibres increase in number thus parallels extrafusal development. It is clear that the close apposition and separation of intrafusal myotubes that occurs during the sequential process led Marchand and Eldred (1969), using the light microscope, to believe that multiplication by splitting was taking place. Electron micrographs also establish that the motor innervation arrives considerably earlier than it proved possible to detect in light microscope preparations, i.e. at birth instead of at 5 days postnatally as observed by Zelená (1957). It is possible that collaterals from β fibres reach the spindle before birth, and that their terminals have not yet been detected. It seems probable, as Milburn (1973) suggests, that the motor innervation plays a key role in determining the ultrastructural characteristics of the intrafusal muscle

Plate 28, Figs. 124–126

Electron micrographs of transverse sections through developing rat spindles. *b1* typical bag fibre; *b2* intermediate bag fibre; *fib.* fibroblast; *p* pseudopodial extensions; *s.t.* sensory axon terminal; *v.c.* vesicular cell of the capsule (capsular sheet cell). (From Milburn, 1973)

Fig. 124. 20-day foetal rat. Transverse section through the juxta-equatorial region of a spindle. Note the appearance of a second intrafusal muscle fibre (*b2*), which is less differentiated and lies in pseudopodial apposition to the initial fibre (*b1*). Arrows indicate pseudopodial extensions

Fig. 125. Part of the region of apposition between *b1* and *b2* shown in Fig. 124, illustrating details of the pseudopodial extensions

Fig. 126. 19.5-day foetal rat. Typical early primary sensory terminals lying on the surface of *b1*. Asterisks indicate desmosomes at the junction of the upper terminal and the muscle fibre

Plate 29, Figs. 127–129

Electron micrographs of transverse sections through developing rat spindles. *b1* typical bag fibre; *b2* intermediate bag fibre; *c1* first chain fibre; *mb.* myoblast; *m.t.* motor axon terminal; *S.c.* Schwann cell. (From Milburn, 1973)

Fig. 127. 1-day postnatal rat. Transverse section through the equatorial region of a spindle showing the difference in size between the typical and intermediate bag fibres; the maturity of the first chain fibre; and the presence of myoblasts

Fig. 128. Newborn rat. Transverse section through the juxta-equatorial region of a spindle showing the first chain fibre differentiating in apposition to the two bag fibres. Myoblasts are still present within the capsule

Fig. 129. An early motor terminal in the polar region of a newborn rat spindle. Note smooth myoneural junction. The capsule has now extended into this region

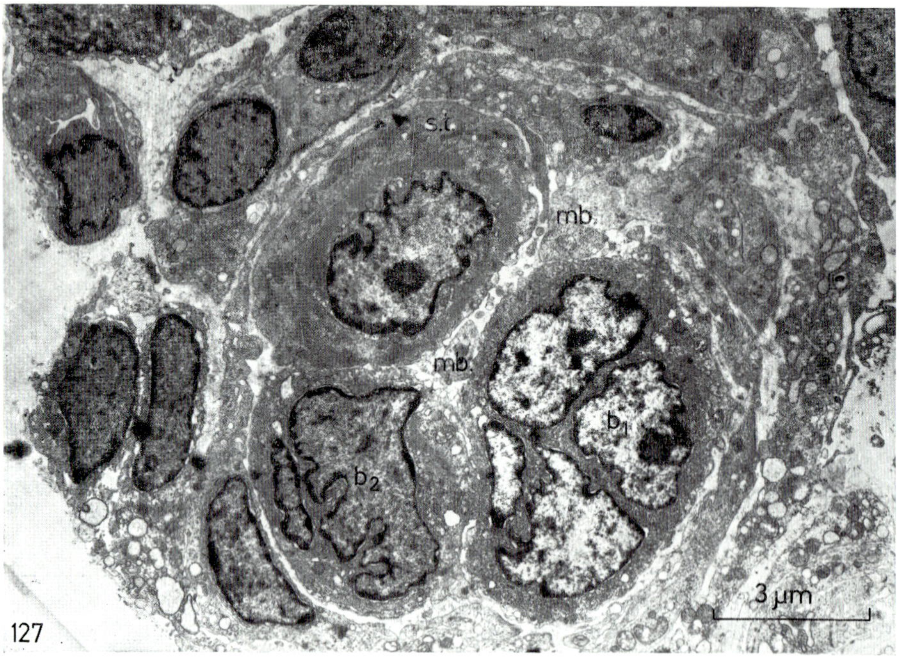

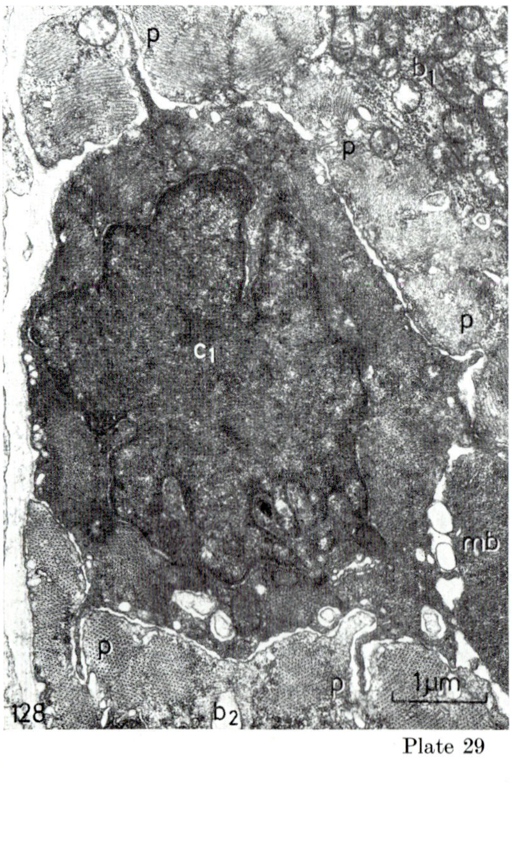

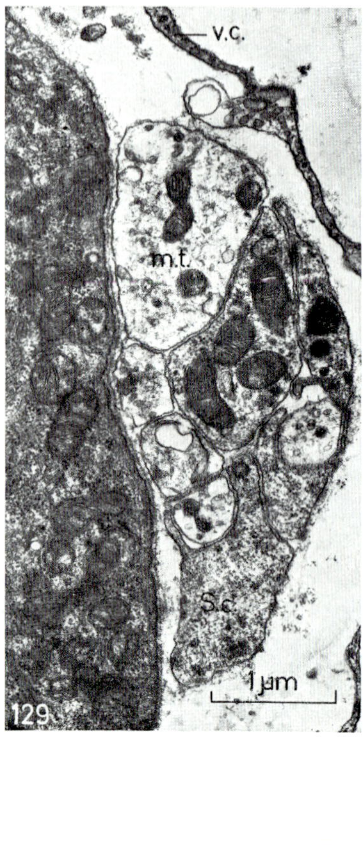

Plate 29

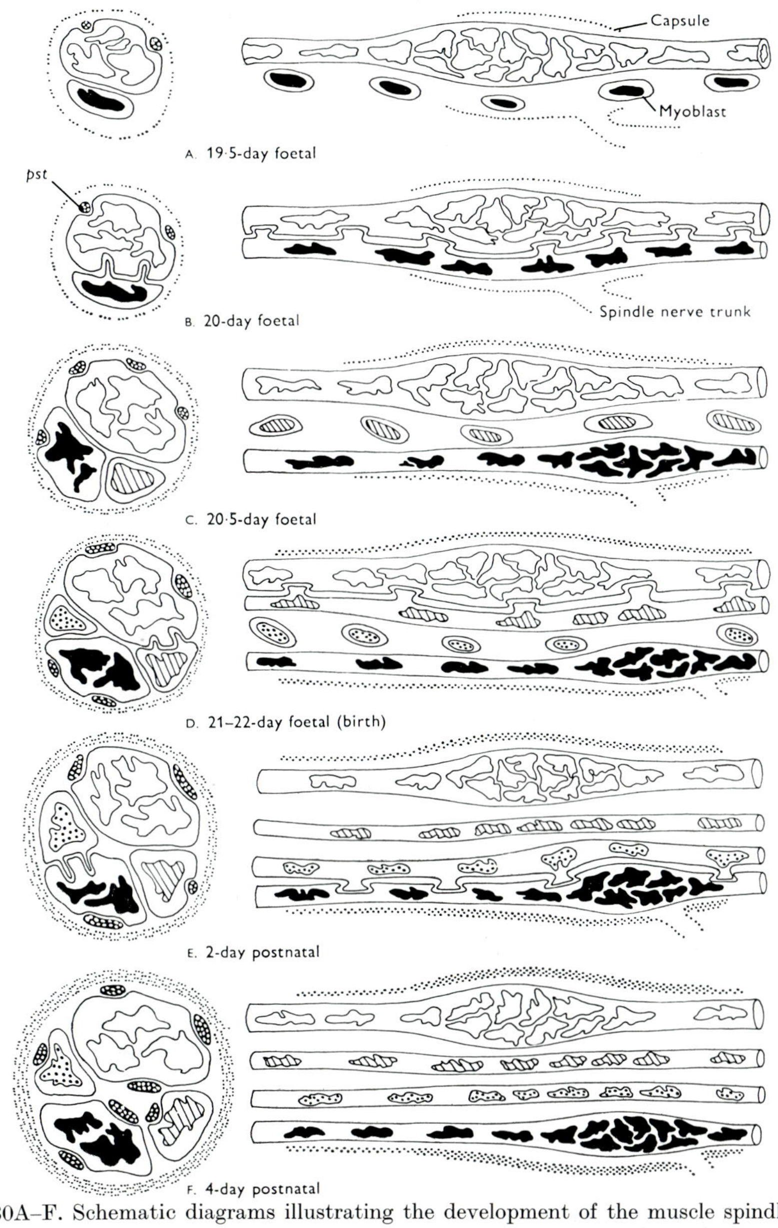

Fig. 130A–F. Schematic diagrams illustrating the development of the muscle spindle in the rat. Diagrams on the left represent transverse sections of spindles at each developmental stage, and those on the right, hypothetical longitudinal sections. Nuclei in the typical bag fibre are black; in the first chain fibre, cross-hatched; and in the second, stippled. Pseudopodial apposition between developing muscle fibres is represented by diagrammatic interlocking. The motor innervation is omitted. *pst.* primary sensory axon terminal. (From Milburn, 1973)

fibres, since these become evident during the same period (4–12 days postnatally) that the motor innervation is becoming fully established (cf. pp. 37–38).

B. Muscle Spindles in Other Vertebrates

Muscle spindles are absent in fishes, perhaps because these receptors evolved in the first instance to serve an antigravity function in limb musculature. Allen (1917) claims to have seen spindles in the caudal heart muscle of the cyclostome *Polistotrema*, but both Bone (1963) and I have failed to confirm this. The anuran Amphibia are the most primitive vertebrates to possess spindles. Because of the convenience of frogs and toads as laboratory animals, more is known about the spindles in these animals than in other non-mammalian vertebrates, though reptilian and avian spindles have recently begun to attract increasing attention.

1. The Amphibian Muscle Spindle

Spindles in the Amphibia are confined to the Anura. Very little work has been done on urodeles; spindles appear to be absent from newt limb muscles, though Mather and Hines (1934) describe non-encapsulated sensory endings associated with the mid-regions of small muscle fibres, which they regard as 'primitive' spindles. Nothing is known about muscle innervation in the worm-like Apoda.

In frogs and toads the axial musculature retains the basket-ending motor innervation characteristic of fishes, and spindles are absent (Ceccherelli, 1904). They are present in the abdominal, pectoral, and limb muscles, which are composed of two motor systems, namely, twitch muscle fibres focally innervated by large axons terminating in end-bush endings, and slow, non-twitch muscle fibres that receive a distributed grape innervation from small axons (Kuffler, 1953; Gray, 1957; Barker, 1968a). Only a small proportion of the muscle fibres in limb muscles are of the non-twitch type. In the frog's fourth toe extensor muscle Gray (1958) estimates that the proportion varies between 6 and 20%. Engel and Irwin (1967) give a figure of less than 1% for gastrocnemius. In sartorius there are none (Kuffler and Gerard, 1947). The twitch component consists of thick fast-contracting fibres and thinner more slowly-contracting ones, which in terms of diameter, contraction time, and histochemical profile fall into two groups (Lännergren and Smith, 1966). The fast-twitch fibres receive the typical end-bush type of motor ending, and the slow-twitch fibres receive a smaller more irregular type of plate (Type I and Type II plates of Pezard and May, 1937; see also Hess, 1960, 1967) (Plate 30, Fig. 135). The form and length of the plates on the slow-twitch fibres indicate that two categories may be recognized, which probably correspond to the two types of this fibre (Lännergren and Smith, 1966; Sterling, 1973).

The frog's fourth toe extensor muscle has been the subject of many histological and electrophysiological studies because of its easy accessability and small size. It is composed of about 50 muscle fibres that course longitudinally between the tendons of origin and insertion (Robertson, 1960). Studies of its spindles with the light microscope distinguish polar and equatorial regions, a thin capsule, and a periaxial space. Transverse sections through the equatorial region show the muscle fibres loosely enclosed in an axial sheath and suspended in the periaxial space by radial struts of connective tissue (Plate 30, Fig. 133). The axial bundle is composed of large and small intrafusal muscle fibres. Barker and Cope (1962a) give mean polar diameters of 11.0 μm for the large fibres (range 8–17 μm) and

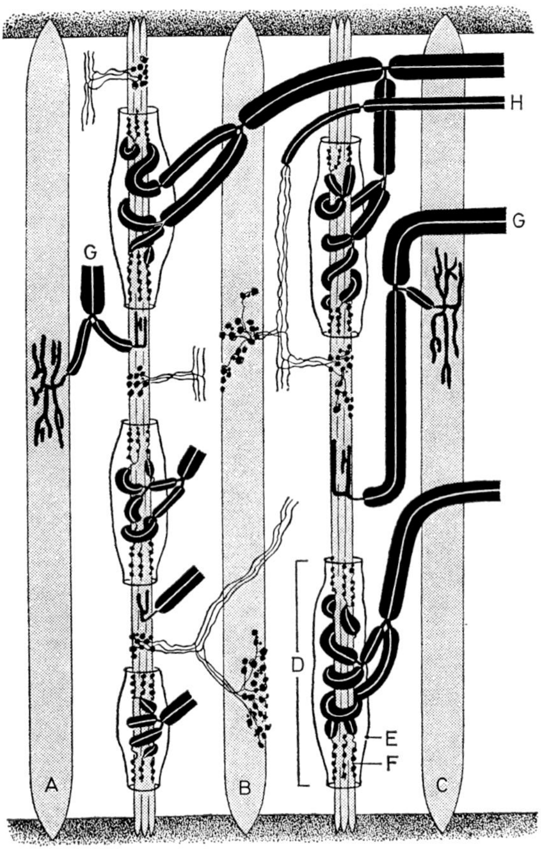

Fig. 131. Schematic presentation of the innervation of the frog's fourth toe extensor muscle according to GRAY (1957). *A* and *C* are twitch fibres, and *B* a non-twitch fibre. *D* spindle sensory region. *E* spindle capsule. *F* varicose terminals of spindle sensory ending. *G* plate-ending motor fibre. *H* grape-ending motor fibre

5.7 μm for the small (range 3–7 μm); SMITH's (1963) diameters for the two groups are 15 μm and 7 μm. The large fibres usually exceed the length of the small ones, but length and diameter are not so consistently related as in the mammalian spindle, and though equatorially there is a decrease in diameter accompanied by central nucleation, there is no differentiation into bags and chains (BARKER and COPE, 1962a).

Single and tandem types of spindle occur in the muscles as well as one or two 'spindle systems' (GRAY, 1957). The systems consist of two to four tandem spindles, and sometimes one or two single spindles, lying together in parallel to form a bundle of spindle units that usually extends throughout the length of the muscle as shown schematically in Fig. 140 (BARKER and COPE, 1962a). In such systems the several capsules (up to 13 in number) are so arranged as to constitute three or four areas of sensory innervation, each area being supplied by the branches of an apparently separate afferent fibre.

Only one sensory ending (the homologue of the mammalian primary) is distributed to each capsular region; it consists of long beaded chains formed by the terminal branches of the sensory fibre distributed longitudinally along the surface of the intrafusal muscle fibres (Fig. 139). The motor innervation is derived collaterally from the two extrafusal motor systems (Gray, 1957), the twitch motor system providing a collateral plate innervation for the large intrafusal

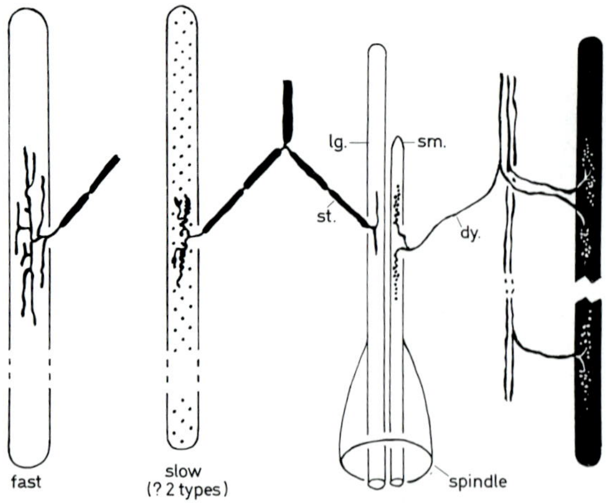

Fig. 132. Schematic diagram of the motor innervation of an anuran limb muscle such as iliofibularis. Twitch fibres on the left, non-twitch fibre in black on the right. *lg.*, *sm.* large and small intrafusal muscle fibres; *st.*, *dy.* static and dynamic fusimotor fibres. (From Barker, 1968a)

Plate 30, Figs. 133–138

Frog muscle spindles and motor end-plates in the fourth toe extensor muscle

Fig. 133. Thick (25 μm) frozen section through the mid-capsular region of a spindle. The intrafusal muscle fibres are loosely enclosed in an axial sheath (inner capsule) and suspended in the periaxial space. Aqueous haematoxylin stain. (From Barker and Cope, 1962a)

Fig. 134. Transverse paraffin section (10 μm) through two spindle units belonging to a spindle system. They are shown just prior to undergoing one of several separate encapsulations. Haematoxylin and eosin stain. *l.m.f.*, *s.m.f.* large and small intrafusal muscle fibres. (From Barker and Cope, 1962a)

Fig. 135. End-bush (*e.-b.pl.*) and irregular (*i.pl.*) types of extrafusal motor end-plates. Methylene blue preparation. Preparation by Mr. R.J. Sterling. (From Barker, 1968a)

Figs. 136 and 137. Teased, methylene blue preparation showing, in Fig. 136, one of three irregular motor end-plates supplied by a large motor axon to slow-twitch muscle fibres in the neighbourhood of a spindle. Fig. 137 shows two intrafusal plates supplied by a collateral branch of the same axon. A drawing of the complete preparation is shown in Fig. 141 A. (From Sterling, 1973)

Fig. 138. Teased, methylene blue preparation showing a large motor axon dividing into two branches (at asterisk), one of which supplies an irregular plate (*i.pl.*) to a slow-twitch muscle fibre, while the other supplies an intrafusal motor end-plate (*int.pl.*). A drawing of this preparation is shown in Fig. 141 B. (From Sterling, 1973)

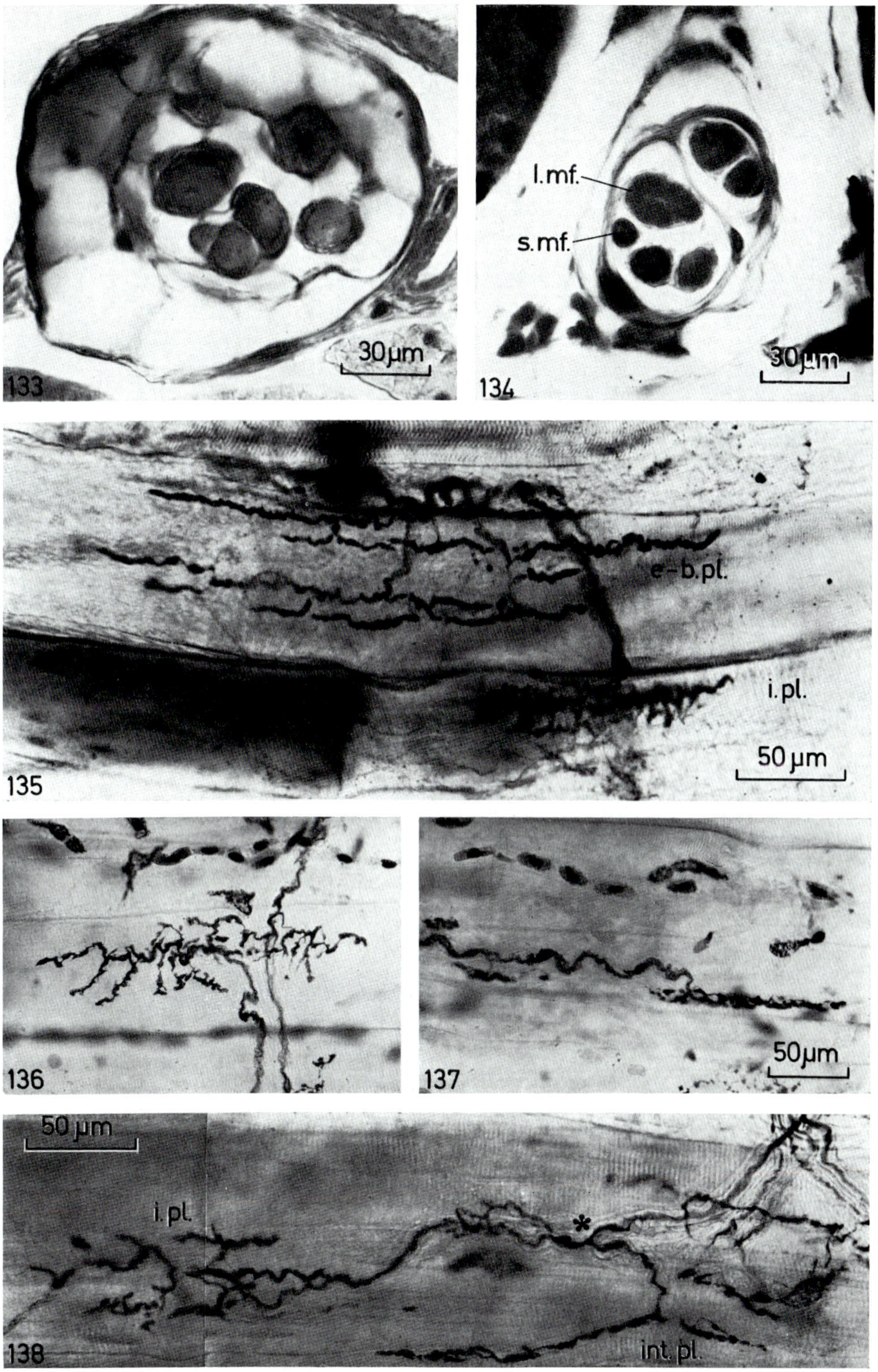
133
30μm
l.mf.
s.mf.
134
30μm
e-b.pl.
i.pl.
50μm
135
136
137
50μm
50μm
i.pl.
*
int.pl.
138

Plate 30

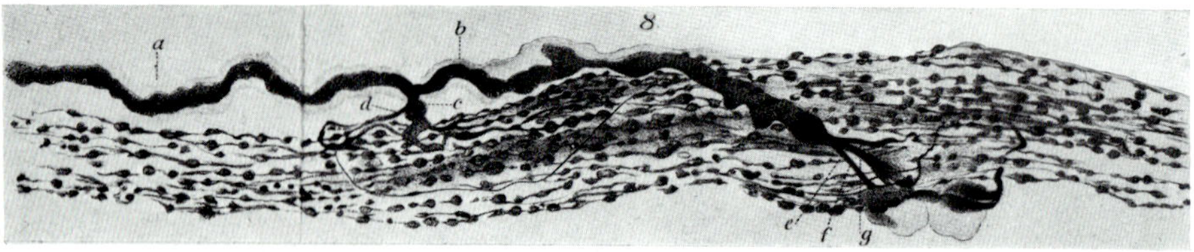

Fig. 139. A camera lucida drawing by DOGIEL (1890, Fig. 8) illustrating the sensory ending in a frog spindle. Pectoralis muscle. Methylene blue preparation

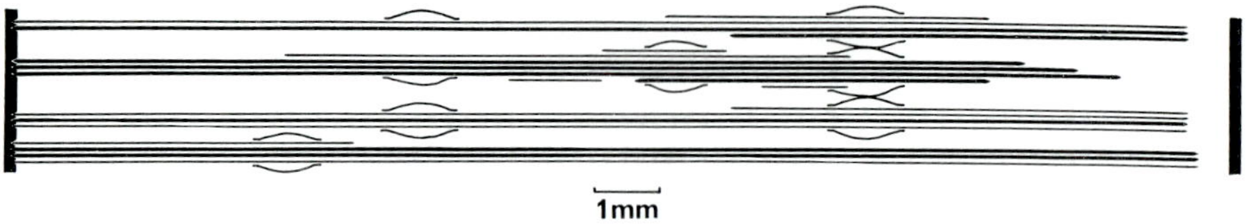

Fig. 140. Projection-plan of a spindle system from the frog's fourth toe extensor muscle whose musculo-tendinous junctions in the region of the system at each end are shown as horizontal bars (proximal end on the right). Length of muscle fibres and capsules accurately shown; fibre width standardized to 12 μm (large) and 6 μm (small), and shown on a scale $\times$ 5 that used for length. Capsules are indicated as diagrammatic fusiform swellings. (From BARKER and COPE, 1962a)

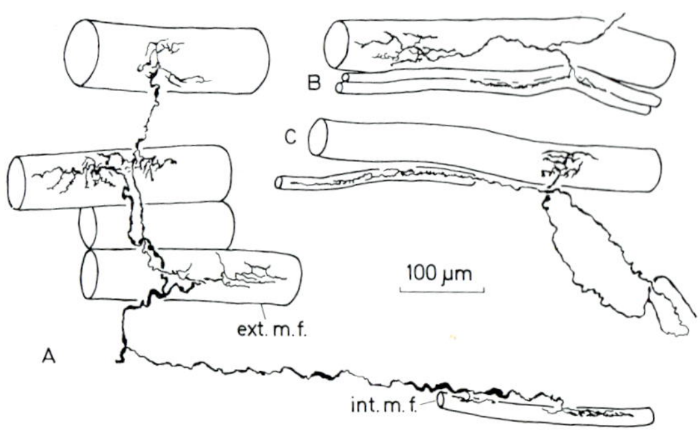

Fig. 141A–C. Drawings traced from photographs of teased, methylene blue preparations of the frog's fourth toe extensor muscle illustrating three examples of the collateral plate innervation of intrafusal muscle fibres (*int.m.f.*) by branches of large motor axons which supply irregular plates to slow-twitch extrafusal muscle fibres (*ext.m.f.*). Compare with Plate 30, Figs. 136 and 137 (photographs of A) and 138 (photograph of B). (From STERLING, 1973)

muscle fibres, and the non-twitch system providing a collateral grape innervation for the smaller fibres (Figs. 131 and 132). The collaterals supplying the plate innervation are always derived as branches of those motor fibres that terminate as irregular plates on extrafusal slow-twitch muscle fibres (STERLING, 1973; see Plate 30 and Fig. 141). MATTHEWS and WESTBURY (1965) have shown that the static and dynamic responses of the sensory ending are controlled by the large and the small motor fibres, respectively.

Electron microscopy has confirmed the presence of two types of fusimotor ending, the plates appearing as long axon terminals overlying a folded post-synaptic membrane, and the grape endings appearing as scattered small contacts in which junctional folds are absent (Page, 1966; Karlsson and Andersson-Cedergren, 1966). Page (1966) describes the intrafusal muscle fibres as being ultrastructurally of two types, one (plate-innervated) corresponding to the extrafusal twitch muscle fibres, though with a less regularly organized sarcoplasmic reticulum, the other (grape-innervated) being intermediate in structure between twitch and non-twitch extrafusal fibres. The latter resemble non-twitch extrafusal fibres in their lack of an M line and possession of ill-defined H zones, but their Z lines are the same as those of extrafusal twitch fibres, and their sarcoplasmic reticulum is similar to that of the plate-innervated intrafusal fibres.

Some of the toe extensor spindles sectioned by Page (1966) lacked the intermediate type of muscle fibre. That these grape-innervated fibres may quite often be absent from frog spindles is indicated by the recent work of Brown (1971) and Sterling (1973) on the iliofibularis and sartorius muscles. Iliofibularis is characterized by the 'tonus bundle' of Sommerkamp (1928), which is located as an inner zone of slow-twitch and non-twitch muscle fibres separated to some extent by intramuscular connective tissue from an outer zone consisting mainly of fast-twitch muscle fibres (Hess, 1960; Orkland, 1963; Lännergren and Smith, 1966). The proportion of twitch fibres in the tonus bundle is about 60% (Hess, 1967). Sterling (1973) found that about half of the ten spindles present in one muscle were located in the tonus bundle, mostly within the areas populated by non-twitch muscle fibres. The phosphorylase profile of the muscle fibres in these spindles showed that two types were present, one corresponding to the extrafusal slow-twitch fibres, the other to the extrafusal non-twitch fibres. Spindles in the outer zone generally lacked muscle fibres showing the non-twitch profile.

These results are in good agreement with those of Brown (1971), who studied the responses of spindles in iliofibularis to stretch during either small motor nerve fibre stimulation or the application of suxamethonium. Those spindles excited by stimulation of the small motor axons were also excited by suxamethonium, and Brown concludes that the drug stimulates prolonged contraction of the small intrafusal muscle fibres, which, as we have seen, are the ones supplied with grape endings by small motor axons. He found on average that a little under half of the maximum of 13 spindles isolated from the muscle in any one experiment responded to suxamethonium and therefore contained small grape-innervated muscle fibres. In 10 sartorius muscles none of the spindles responded to suxamethonium, and this agrees with Sterling's finding that in 4 sartorius muscles sectioned and stained for the phosphorylase reaction, all the intrafusal muscle fibres had a histochemical profile corresponding to that of extrafusal slow-twitch fibres. Sterling did nevertheless encounter four examples of sartorius spindles receiving a specifically fusimotor grape innervation in methylene blue preparations, which suggests that small motor fibres, though not engaged extrafusally, may occasionally participate in this way.

Ultrastructural studies of frog intrafusal muscle fibres have shown that in the region of equatorial nucleation the surface of each fibre is thrown into irregular, greatly contorted folds, so that in longitudinal section the fibre presents the

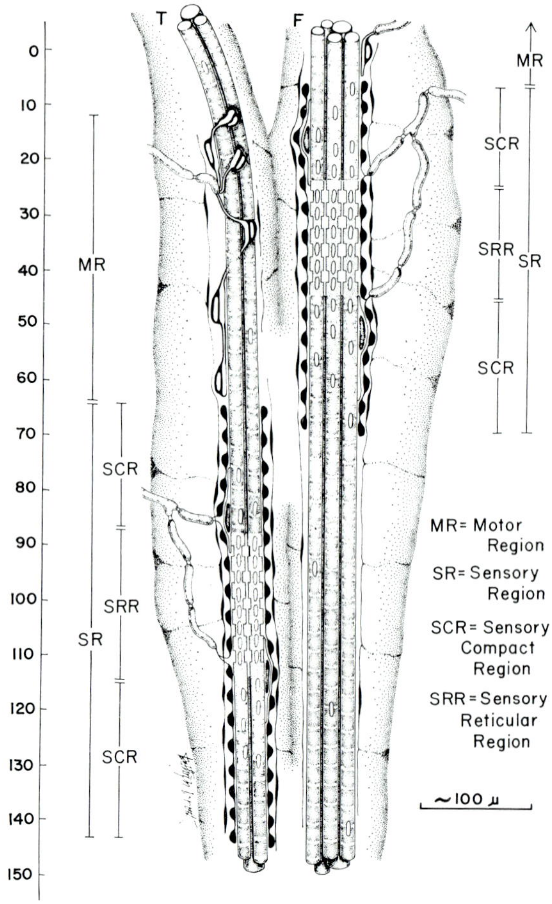

Fig. 142. Schematic drawing of a frog muscle spindle specimen that was serially cross-sectioned through its regions of sensory innervation and parts of its regions of motor innervation. The scale to the left displays the section groups analysed, each of which represents about 5 μm. The specimen consists of two bundles of intrafusal muscle fibres, *F* five fibres to the right and *T* three fibres to the left. The intrafusal muscle fibres display, when in phase, compact (*MR* plus *SCR*) and reticular regions (*SRR*) of the *F* and *T* muscle bundles. The reticular regions are represented by the heavily nucleated part of the muscle fibres, the plasma membranes of which are folded into regularly spaced processes. The rest of the intrafusal muscle-fibre bundles comprise the compact regions. The sensory regions are represented by the string of black beads that symbolize the sensory nerve endings. The myelinated axons, before giving rise to the sensory nerve endings, lose their myelin in the border regions between the sensory reticular and sensory compact regions, the transition region. In the sensory compact regions fusiform cells, the intrafusal satellite cells, are seen in intimate contact with the intrafusal muscle fibres. Parts of the non-sensory regions are occupied by motor nerve endings, a whole innervation region of which is illustrated in the bundle *T* and a part of which can be seen in the upper end of the bundle *F*. The intrafusal bundles are partly encapsulated in the same outer capsule (stippled). At the reticular regions they are, however, individually encapsulated. Processes from the outer capsule radiate in toward the intrafusal muscle-fibre bundles and approach the inner capsule. The inner capsule is predominant in the sensory regions, but ends before the outer capsule. (From KARLSSON, ANDERSSON-CEDERGREN and OTTOSON, 1966)

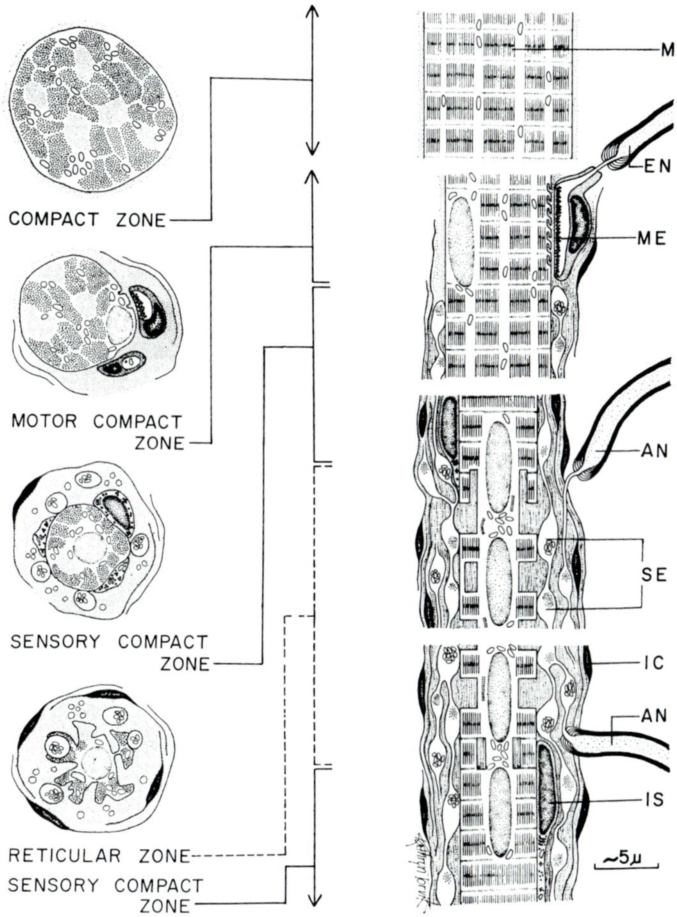

Fig. 143. Imaginary drawing of an intrafusal fibre from frog illustrating longitudinal (right) and transverse (left) aspects of its different zones (compact, motor compact, sensory compact and reticular zones). The muscle fibre is seen to change its diameter from compact to reticular zones. The ultrastructure of the compact zone does not differ significantly from that observed in extrafusal muscle fibres. The reticular zone displays regularly spaced muscular processes, each containing one sarcomere, many central nuclei, and relatively few myofilaments. The compact innervation zones are divided into two parts according to the mode of innervation. In the compact zone near the transition zone to the reticular zone, intrafusal satellite cells (*IS*) with nuclei darker than those of the muscle are seen in intimate contact with the intrafusal muscle fibre. Sensory nerve endings (*SE*) arising from the two lower nerves in the longitudinal aspect (*AN*) do not exhibit any Schwann investment in their peripheral course. The varicosities contact the intrafusal muscle fibre preferably on the muscular processes in the reticular zone. Their content is compartmental. The sensory nerve endings are surrounded by a longitudinally oriented, filamentous material (stippled) which is delineated by a heavy representation of inner capsule cells (*IC*). The outer capsule is not illustrated. The motor nerve endings (*ME*) are characterized by a Schwann investment and larger space between nerve ending and muscle at the contact. Accumulations of small vesicles occur in the nerve ending. (From Karlsson, Andersson-Cedergren and Ottoson, 1966)

appearance of a complicated framework with interstitial spaces filled with collagen. KATZ (1961) recognized this as the 'reticular zone', in contrast with the 'compact zones' on either side where the muscle fibres have the usual circular contour in transverse section and are filled with myofibrils. KARLSSON, ANDERSSON-CEDERGREN and OTTOSON (1966) have carried the terminology a stage further and distinguish the following regions as present in each intrafusal muscle fibre; a

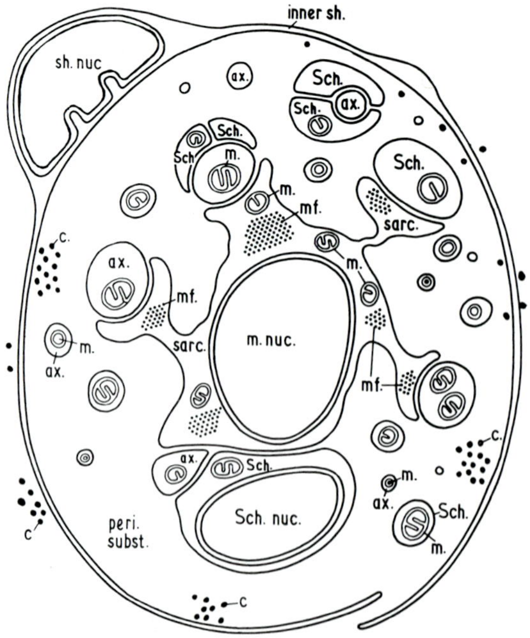

Fig. 144. Diagram of a frog intrafusal muscle fibre in the sensory region. A single inner sheath cell (*inner sh.*) is shown with its nucleus (*sh.nuc.*). There are moderate numbers of collagen fibrils (*c.*) associated with the inner sheath. The perimuscular substance (*peri.subst.*) lies between the inner sheath cell and the muscle surface. The surface of the muscle fibre is thrown into numerous folds and evaginations. A few myofilaments (*mf.*) are seen in groups in the sarcoplasm (*sarc.*). The section includes a muscle nucleus (*m.nuc.*). Numerous terminal axons (*ax.*) are seen in the perimuscular substance and in closer relationship with the muscle surface. These contain moderate numbers of mitochondria (*m.*), and are sometimes partially or completely surrounded by Schwann cells (*Sch.*). One Schwann cell nucleus (*Sch.nuc.*) is shown. (From ROBERTSON, 1960)

reticular zone flanked by sensory compact zones on either side, which collectively form the area of sensory innervation; and compact zones in the poles, designated as motor compact zones in the intra- and extracapsular regions where the motor innervation is received (Figs. 142 and 143).

The reticular zone is characterized by a chain of central nuclei and a reduction of myofilaments of up to 85% as compared with the compact zones (KATZ, 1961). ROBERTSON's (1960) schematic diagram of a transverse section through an intrafusal muscle fibre in the reticular zone is reproduced in Fig. 144. According to

Karlsson et al. (1966) the pseudopodial-like processes of the muscle fibre in this zone are arranged in a regular pattern, each containing one sarcomere of contractile material attached at each end to an electron-dense substance in close association with the plasma membrane. There is some indication that the sarcomere-containing processes (reticulomeres) are arranged in rows in a spiral fashion around the fibre (Karlsson, 1972). Karlsson et al. (1966) found the length of the reticular zone to vary from 70 μm to 140 μm. Microladders (leptomeres) often occur among the myofibrils in the region of sensory innervation and may span several sarcomeres (Katz, 1961; Karlsson and Andersson-Cedergren, 1968). Some fibres split into two in the reticular zone (Robertson, 1960; Katz, 1961), sometimes recombining with other fibres in the following compact zone (Katz, 1961).

According to Page (1966), the majority of the small intrafusal muscle fibres (i.e. those of intermediate ultrastructure receiving a grape innervation) do not possess a reticular zone. Katz (1961) also observed such fibres, noting that their myofilament content in this region is only slightly reduced. On the other hand, Karlsson (1972) has never observed a muscle fibre without a reticular zone, and is of the opinion that "there is still no conclusive evidence for ultrastructural differentiation into several types of fibre". The fact that many spindles are composed of large plate-innervated fibres only (see Page, 1966, Table 1) may explain the confusion on this point, which clearly requires further investigation.

Micrographs of the sensory ending show that it consists of chains of nerve bulbs 2–3 μm in diameter linked together by thin tubes of axoplasm of around 0.1 μm diameter. Such an arrangement of sensory terminals strung together like beads to form a varicose thread is common to a number of receptors in lower vertebrates (see § II B, 2; § IV). In the frog spindle ending the nerve bulbs synapse with the muscle fibres in shallow sockets on the surface (Katz, 1961). In the reticular zone they form contacts with the reticulomeres (Karlsson et al., 1966) (Figs. 143 and 144). Schwann cells are occasionally found in association with the nerve bulbs. Satellite cells are of frequent occurrence in the transition region between the reticular and sensory compact zones (Karlsson et al., 1966).

The organization of the capsule and axial sheath (the outer and inner capsules of Karlsson et al., 1966) is much the same as in the mammalian spindle, as is the disposition of the endomysial enclosures of the intrafusal muscle fibres. The capsule ends as a sheath one lamella thick in the compact zone, with the capsular sheet cells projecting irregular processes towards the surface of the muscle fibres in a region full of collagen bundles (Karlsson and Andersson-Cedergren, 1970).

2. *The Reptilian Muscle Spindle*

The muscles of reptiles, like those of amphibians, are composed of both plate-innervated and grape-innervated muscle fibres, the latter usually being relatively few in number and sometimes altogether absent (Hess, 1963, 1970; Proske and Vaughan, 1968; Ridge, 1971). Muscle spindles have been studied in snakes, lizards, tortoises, and alligators. Those of snakes and lizards are remarkable in that, wherever found, they consist of one muscle fibre only, rarely two. Histologists in the last century (see review by Regaud and Favre, 1904) recognized two types, distinguishing them by differences in the equatorial region. One type has a short, wide, and obvious capsule, which encloses a muscle fibre that over a short length loses its cross-striations, increases in diameter, and becomes filled with an accumulation of nuclei and granular sarcoplasm. The other type has a long, narrow, indistinct capsule, which encloses a muscle fibre that remains cross-striated throughout, the equatorial nucleation being in the form of a chain.

The spindles with short capsules receive a restricted type of sensory ending in which the axon ends abruptly in a few simple terminals on the non-striated region (Plate 31, Fig. 145). The spindles with long capsules receive an extensive sensory ending, either in the form of varicose threads, as in snakes; or filaments, as in the lizard *Tiliqua nigrolutea* (Proske, 1969) (Plate 32, Fig. 151); or annulospiral terminals, as in the lizards *Lacerta viridis* and *L. muralis* (Perroncito, 1901, 1902) (Fig. 152 and Plate 31, Figs. 146 and 148). Spindles of one type often lie in groups, and, as in frog spindles, a single afferent may branch to supply more than one capsule (Szepsenwol, 1960, Proske, 1969). Proske (1969) observed tandem arrangements to occur among spindles with long capsules, but not among those with short ones.

The distribution of the two types of spindle varies in different muscles. In *Tiliqua* Proske (1969) found the long type only in semimembranosus, which is exclusively plate-innervated; the short type only in scalenus, in which about one third of the muscle fibres are grape-innervated (Proske and Vaughan, 1968); both types in iliofibularis, in which about 10% of the muscle fibres are grape-innervated (Proske, 1967); and both types in semitendinosus, which is predominantly grape-innervated.

Plate 31, Figs. 145–148

Teased preparations of reptilian spindles. *cap.* capsule; *cp.* capillary; *ex.m.f.* extrafusal muscle fibre; *m.ax.* motor axons; *s.ax.* sensory axon

Fig. 145. Sensory ending in lizard (*Lacerta viridis*) spindle with short, wide capsule. Preparation by Miss B. Purdy. (From Barker, 1968a)

Fig. 146. Equatorial region of lizard (*Lacerta viridis*) spindle with long, narrow capsule. Note the width of the single intrafusal muscle fibre as compared with that of the capillary and extrafusal muscle fibres. Preparation by Miss B. Purdy. (From Barker, 1968a)

Fig. 147. Part of the equatorial region of a long-capsule garter snake (*Thamnophis sp.*) spindle. Living specimen photographed with Nomarski optics. Apparent oval indentations in the muscle-fibre surface are nuclei. Preparation by Prof. C.C. Hunt

Fig. 148. Part of the preparation illustrated in Fig. 146 photographed at higher magnification after further teasing to remove extrafusal muscle fibres

Plate 32, Figs. 149–151

Photographs from Proske (1969) illustrating the innervation of muscle spindles in the lizard *Tiliqua nigrolutea*. Teased, silver preparations. *cap.* capsule; *ex.gr.e.* extrafusal grape ending; *ex.m.e-p.* extrafusal motor end-plate; *int.gr.e.* intrafusal grape ending; *int.m.e–p.* intrafusal motor end-plate; *s.ax.* sensory axon

Fig. 149. A motor axon branches to supply an extrafusal and an intrafusal muscle fibre with grape endings

Fig. 150. A motor axon branches to supply an extrafusal and an intrafusal muscle fibre with motor end-plates

Fig. 151. Sensory ending in a long-capsule spindle. The axon terminals consist of fine filaments

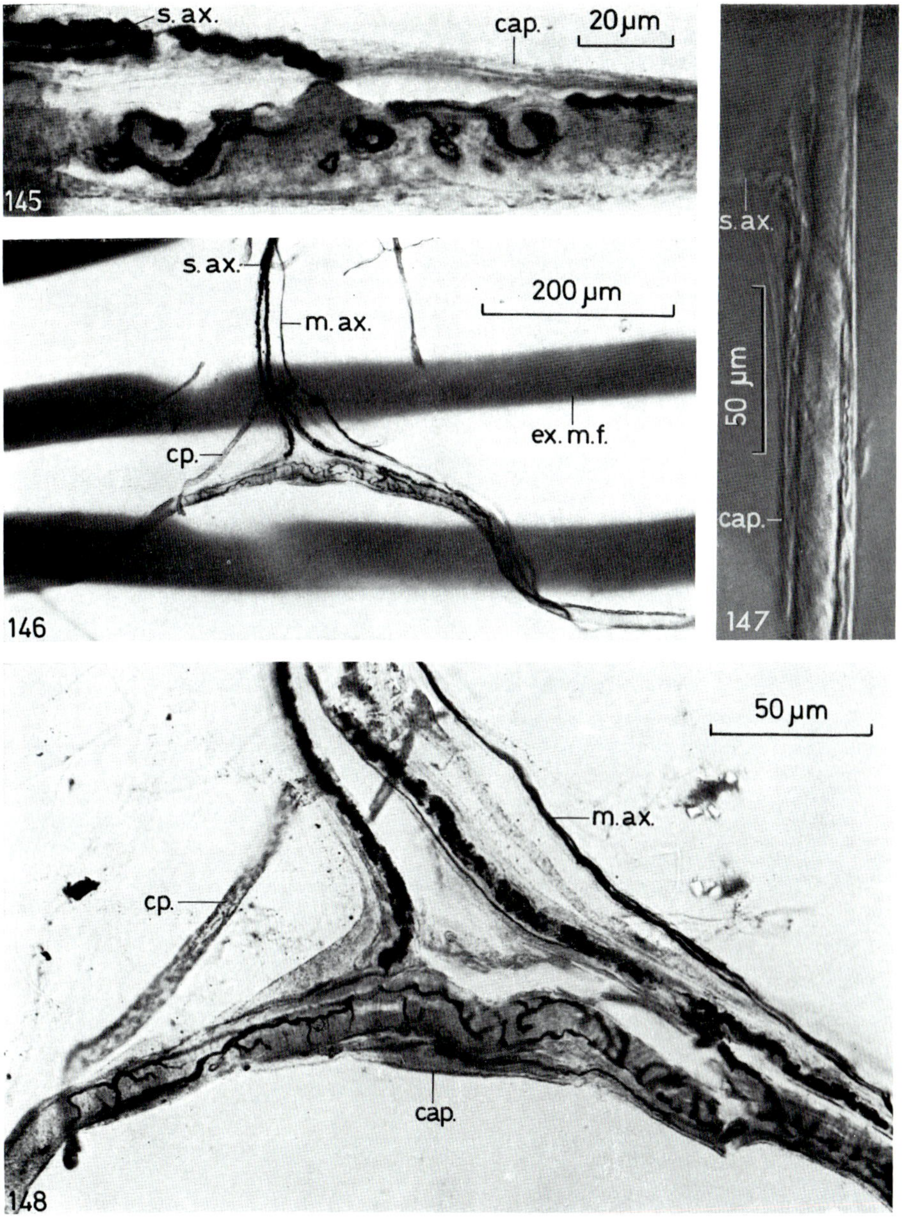

Plate 31

Fukami (1970) has shown that the sensory ending in snake spindles with short capsules is very sensitive to rate of stretch (phasic-type spindles), in contrast to that in the spindles with long capsules, which responds principally to amplitude of stretch (tonic-type spindles). Intrafusal contraction in both phasic and tonic spindles is of the twitch type, and according to Hunt and Wylie (1970) does not preferentially enhance either the dynamic or the static response to stretch.

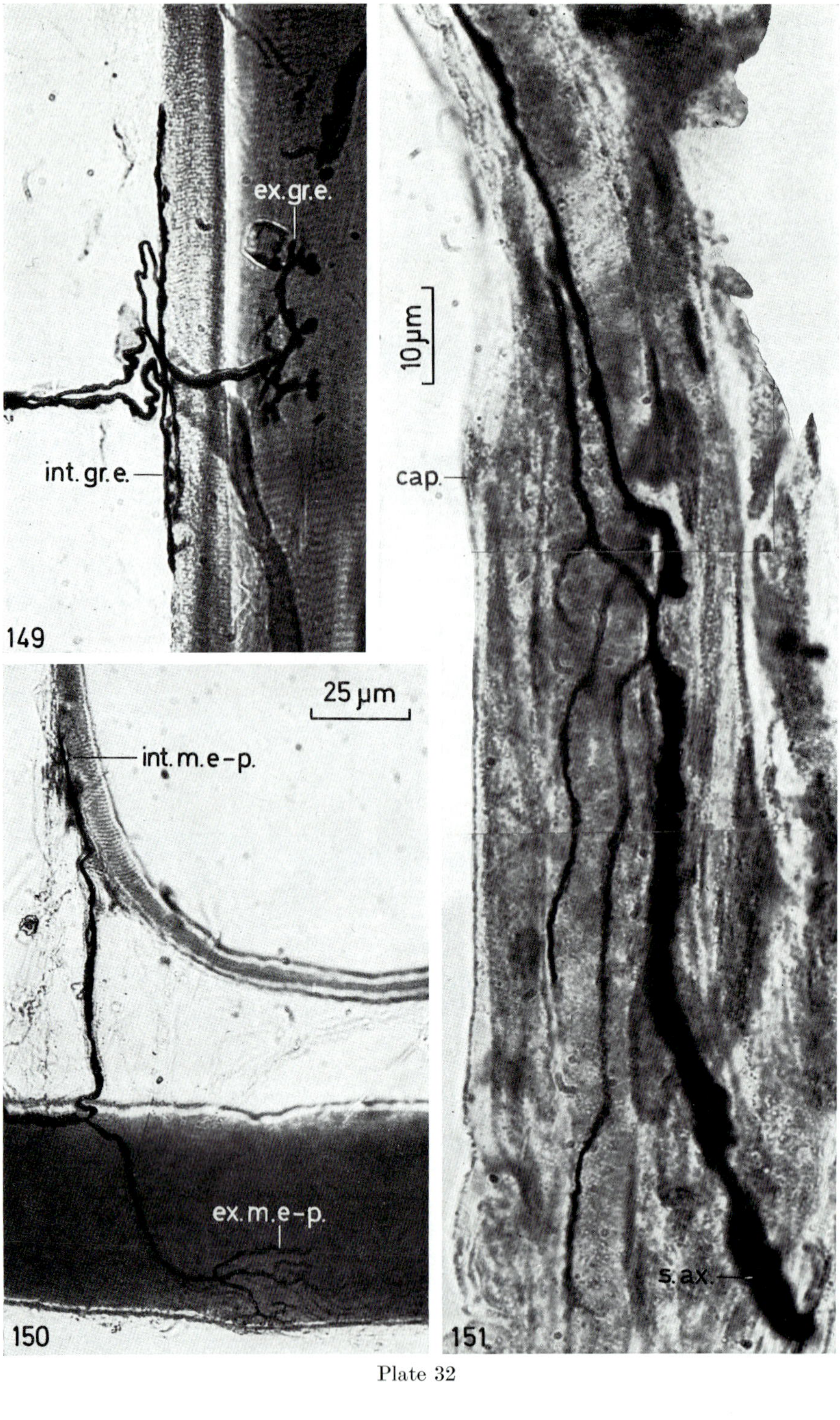

Plate 32

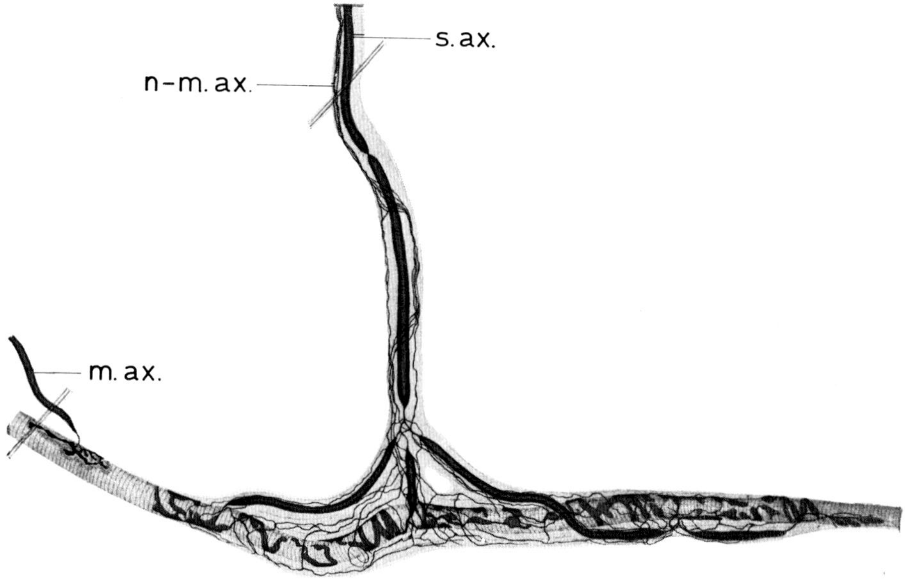

Fig. 152. A drawing by PERRONCITO (1902, Fig. 7) illustrating the sensory ending in a long-capsule lizard spindle (*Lacerta viridis* or *muralis*). A bundle of thin non-myelinated axons (*n-m.ax.*) is shown entering the spindle together with the sensory axon (*s.ax.*) to ramify within the capsule wall. A motor axon (*m.ax.*) enters on the left to supply a motor end-plate. Gold chloride preparation. (Compare with Plate 31, Figs. 146 and 148)

However, PROSKE (1973) has shown that repetitive motor stimulation during stretch produces different effects on the afferent discharge of the two types of spindle in the lizard *Tiliqua*, the firing of phasic spindles being increased much more than that of tonic spindles.

The motor innervation of these monofibral spindles is collaterally derived from the two extrafusal motor systems. PROSKE (1969), extending the earlier study by SZEPSENWOL (1960), has shown that the phasic (short capsule) spindles receive a collateral grape motor innervation, and that the tonic (long capsule) spindles may be innervated either by grape collaterals or plate collaterals. Some tonic spindles evidently receive both types of collateral innervation since CLIFF and RIDGE (1973) have observed a single tonic intrafusal muscle fibre to twitch in critically curarized muscle on stimulating single motor units that are either of the twitch (plate-innervated) or tonic (grape-innervated) type. PERRONCITO (1901) illustrates the collateral plate innervation of long-capsule spindles in the lizard by non-myelinated preterminal, ultraterminal, or nodal branches, as well as by a myelinated branch. He also describes (1902) the innervation of these spindles by two or three very thin fibres that approach the capsule together with the sensory fibres, apparently running within its perineural sheath. On reaching the equatorial region they ramify and spread out to form an anastomosing network within the walls of the capsule (Fig. 152).

The ultrastructure of snake spindles has been studied by FUKAMI and HUNT (1970), PALLOT and RIDGE (1971, 1972, 1973) and PALLOT (1973). According to FUKAMI and HUNT (1970) the muscle fibre in each type of spindle has a poorly developed sarcotubular system, thick Z lines, and lacks an M line, i.e. an ultrastructure resembling that of vertebrate 'tonic' (non-twitch) extrafusal muscle fibres. PALLOT (1973), however, maintains that such an ultrastructure is characteristic of phasic spindles, but that in tonic spindles the muscle fibre has a well-defined H zone with M line present and a variable sarcoplasmic reticulum (dyads and triads plentiful in some fibres, scarce in others).

In the phasic spindles the sensory region is 100–150 μm long and the diameter of the muscle fibre increases from 7–10 μm to about 40 μm. As this region is approached the myofilaments decrease in number and lose their regular orientation, while nuclei become centrally located, rounder in shape, and more numerous. In the sensory region itself myofilaments are rarely seen, and the sarcoplasm becomes filled with nuclei, mitochondria, microtubules, lipoid droplets, and aggregations of electron-dense granular material. An unusual feature described by FUKAMI and HUNT (1970) is the occurrence of tubular canals that open on the surface of the fibre into the extracellular space enclosed by the endomysial sheath (inner capsule) (Plate 33, Fig. 154). The sensory region of the tonic spindle is longer (250–300 μm),

---

Plate 33, Figs. 153 and 154

Ultrastructure of snake spindles

Fig. 153. Transverse section through the capsular region of a short-capsule spindle in the costocutaneous muscle of a grass snake (*Natrix* sp.). The single muscle fibre is enclosed by fibrocytes forming the endomysial sheath and inner capsule (*i.cap.*), and is suspended in a periaxial space bounded by the outer capsule (*o.cap.*). Nerve bulbs (*n.b.*) of the sensory ending and satellite (probably Schwann) cells (*sat.c.*) are located on the surface of the muscle fibre and bound on by the collagen crescent (*coll.cr.*). *coll.* collagen. (Electron micrograph by Dr. D.J. PALLOT)

Fig. 154. Longitudinal sections through the sensory region of a short-capsule spindle in the costocutaneous muscle of a garter snake (*Thamnophis* sp.) showing the opening of large tubules (*T*) into the extracellular space. MN, muscle-fibre nucleus. (From FUKAMI and HUNT, 1970)

Plate 34, Figs. 155 and 156

Ultrastructure of the sensory region of long-capsule snake (*Natrix* sp.) spindles (electron micrographs published by PALLOT and RIDGE, 1972). *coll.cr.* collagen crescent; *i.cap.* inner capsule; *n.b.* nerve bulb of sensory ending; *n.e.* nerve link of sensory ending; *sat.c.* satellite cell; *sat.c.n.* satellite cell nucleus; *s.ax.* sensory axon

Fig. 155. A nerve link of the varicose sensory ending is sandwiched between the intrafusal muscle fibre and satellite cell, and both are bound on to the surface of the muscle fibre by the collagen crescent

Fig. 156. A nerve bulb of the sensory ending formed from the link shown in Fig. 155

Plate 35, Figs. 157 and 158

Ultrastructure of snake fusimotor endings

Fig. 157. Transverse section through a motor terminal in a short-capsule snake (*Natrix* sp.) spindle. Mitochondria are numerous in the muscle fibre in the region of the myoneural junction, and the postsynaptic membrane shows some irregular shallow folding. (Electron micrograph by Dr. D.J. PALLOT)

Fig. 158. Transverse section through a motor terminal in a snake (*Thamnophis* sp.) spindle. The myoneural junction is smooth. (From FUKAMI and HUNT, 1970)

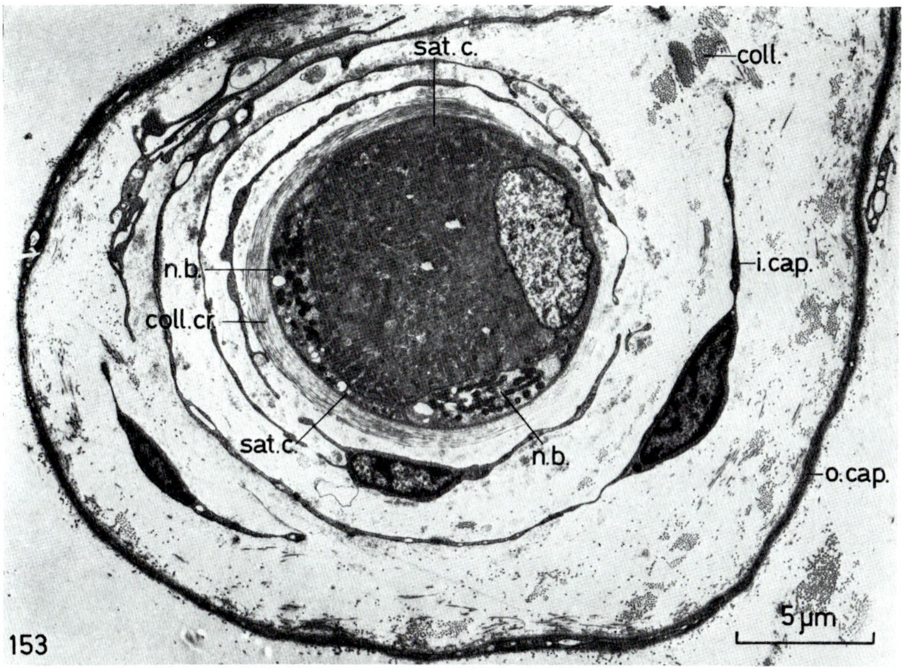

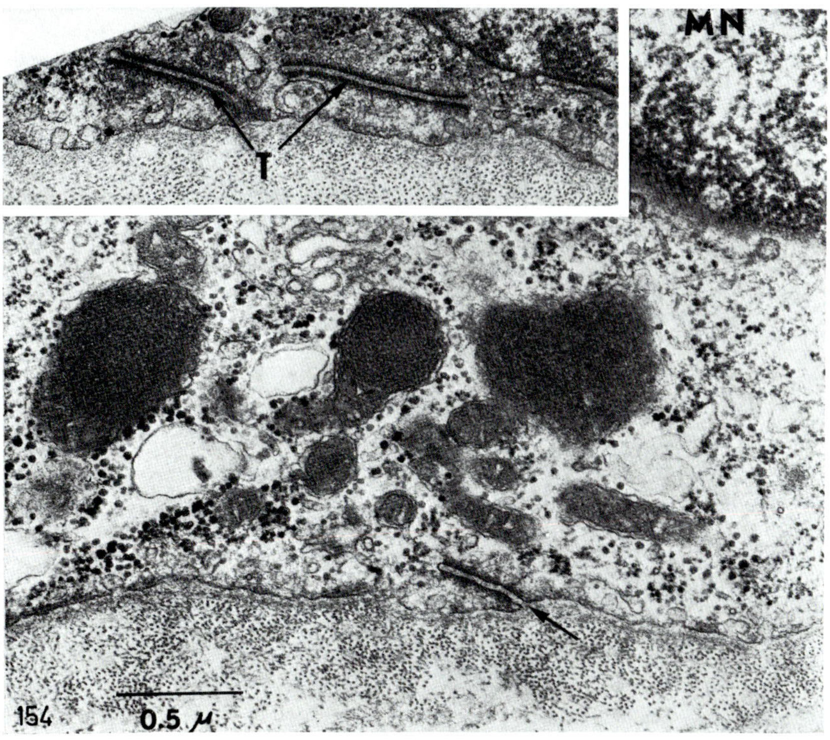

Plate 33

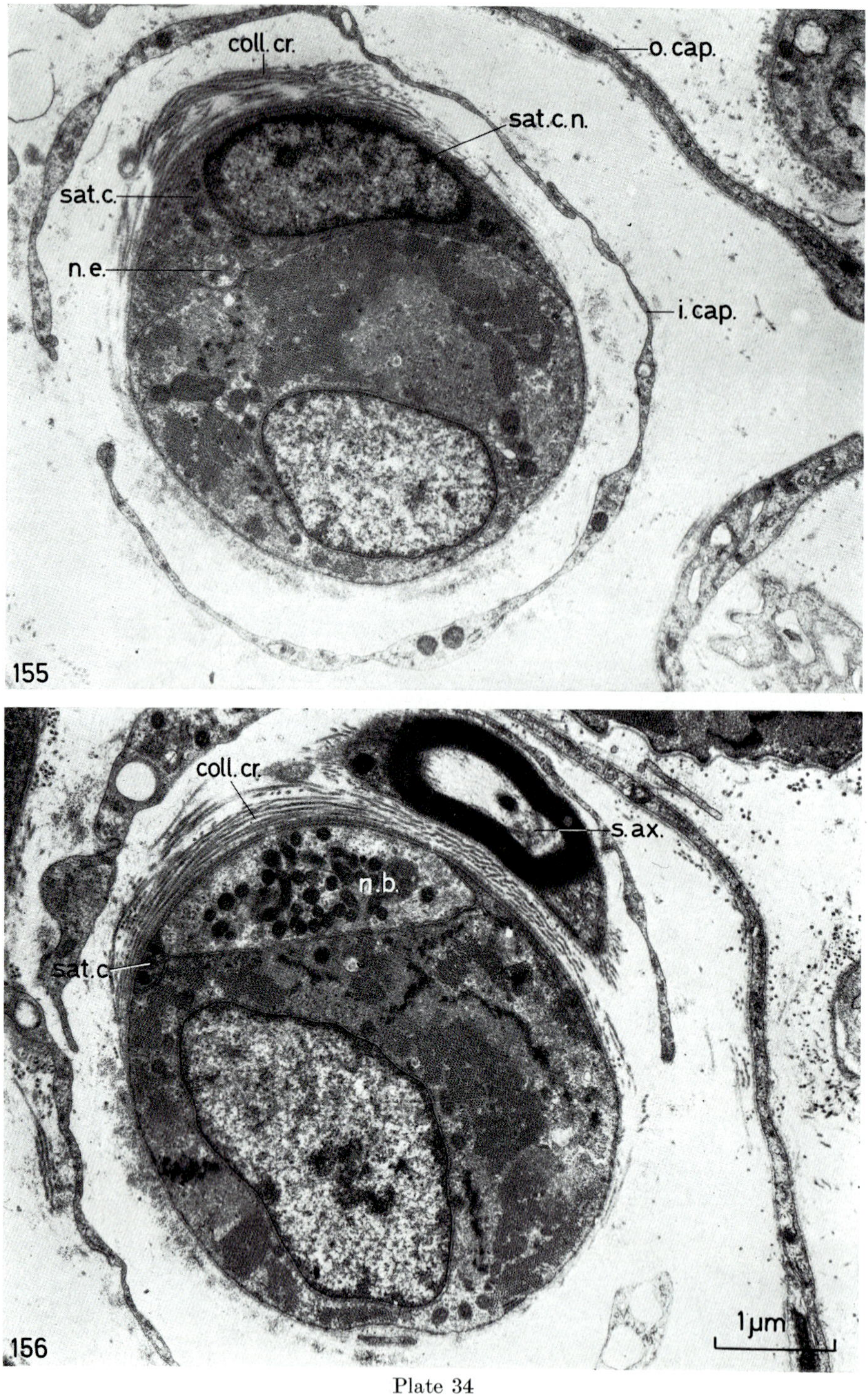

Plate 34

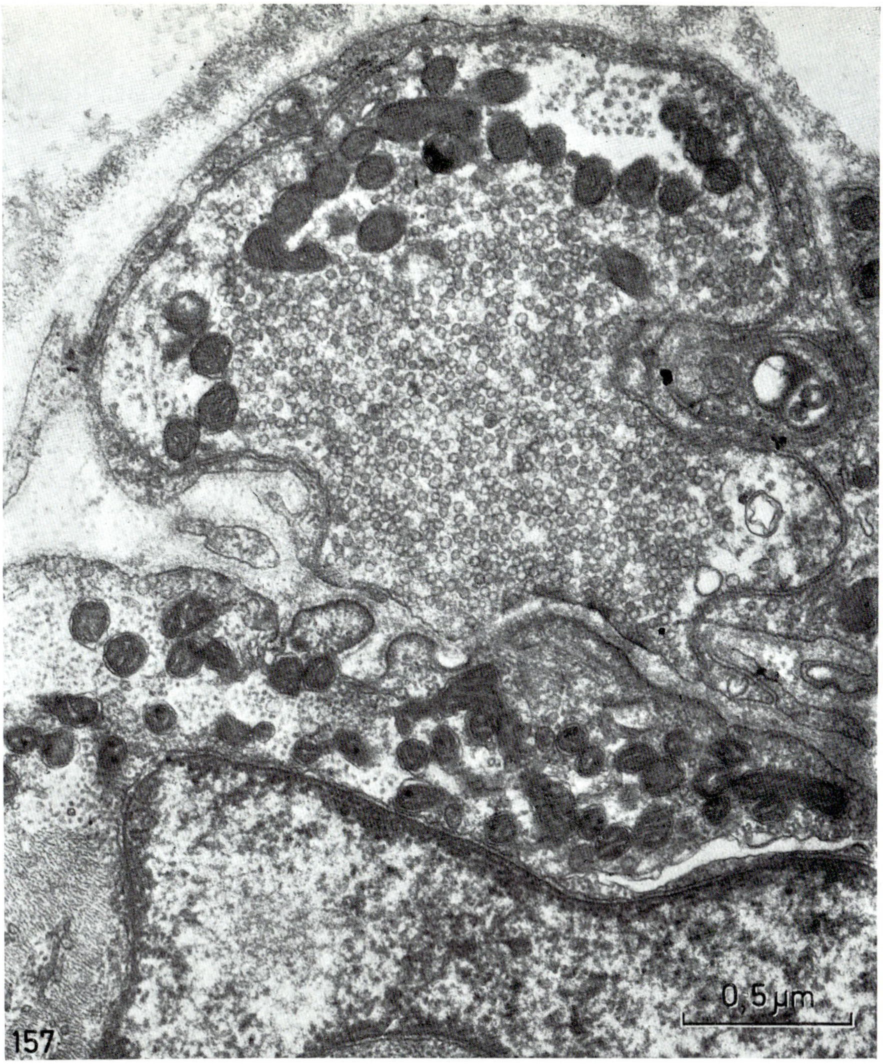

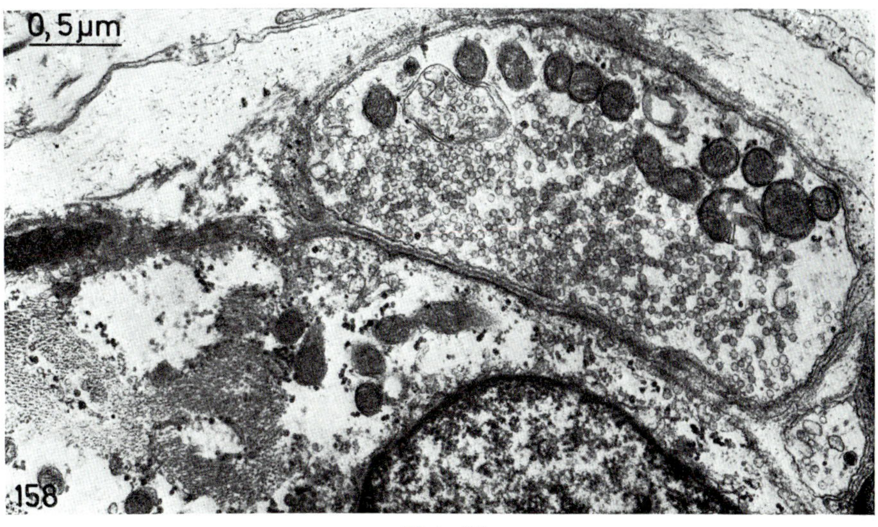

Plate 35

and the increase in diameter of the muscle fibre to 15–20 μm is much less. Myofibrils extend throughout the region enclosing a chain of nuclei in a core of sarcoplasm rich in mitochondria, much as in a mammalian nuclear-chain fibre. Satellite cells occur in the sensory region of both types of spindle. Microladders have not been observed. Transverse sections through the sensory regions of phasic and tonic spindles are shown in Plates 33 and 34.

FUKAMI and HUNT (1970) made their observations on spindles in the costocutaneous muscle of the garter snake (*Thamnophis* sp.); PALLOT and RIDGE (1971, 1972, 1973) and PALLOT (1973) worked on the same muscle in the grass snake (*Natrix natrix* or *N. tessellata*). Differences occur in the descriptions of the sensory innervation, which may be due to the different species studied. FUKAMI and HUNT (1970) observed that the terminals in *Thamnophis* run longitudinally as 1 μm thick processes for distances of 20 μm or more making contact with up to 7 μm of the muscle-fibre perimeter. PALLOT and RIDGE (1971, 1972, 1973), on the other hand, find that the terminals in *Natrix* tonic spindles consist of a series of bulbs (lengths 14–40 μm) connected by tubular links 0.5–1.0 μm in diameter and 10–30 μm long. In phasic spindles they found a similar arrangement, but with more nerve branches and shorter chains of bulbs connected together by shorter (2–3 μm long) links.

Transverse sections in both studies show crescent-shaped sensory terminals bound tightly onto the surface of the muscle fibre by a thick layer of transversely orientated collagen fibres. The outline of the collagenous layer is also crescentic roughly duplicating that of the terminals. It is thicker in phasic than in tonic spindles, and the basement membrane material beneath the layer is thicker than elsewhere on the muscle-fibre surface. A similar collagenous layer is found in association with the sensory terminals in avian spindles (see § II B, 3). It seems probable that such binding is a device whereby sensory endings of the varicose or filamentous types are held in close myoneural contact. The annulospiral type of ending evidently has no need for such binding for it is not associated with the sensory terminals of mammalian spindles. It is also presumably lacking from the sensory regions of lizard tonic spindles such as those illustrated in Plate 31, Fig. 148 and in Fig. 152.

The ultrastructural features of the motor innervation await full investigation. FUKAMI and HUNT (1970) report the presence of myoneural synapses with junctional folds, but their published micrographs show only smooth synapses, and the type of spindle, phasic or tonic, is not specified (Plate 35, Fig. 158). PALLOT (1973) has observed some motor endings with smooth synapses and others in which there are some irregular shallow junctional folds (Plate 35, Fig. 157), but finds no clear correlation between type of synapse and type of spindle. Motor endings may occur in the sensory region of tonic spindles and be located among the sensory terminals (PALLOT and RIDGE, 1972). Pallot's micrographs have not yet revealed any small axons ramifying within the capsule wall, as described by PERRONCITO (1902).

Detailed studies of tortoise (*Testudo graeca*) spindles have recently been published by CROWE and RAGAB (1967, 1969, 1970a, b, 1972). According to the brief reports of HUBER and DE WITT (1897) and GIACOMINI (1898), the spindles in these reptiles occur in both mono- and multifibrillar form, and are innervated by a

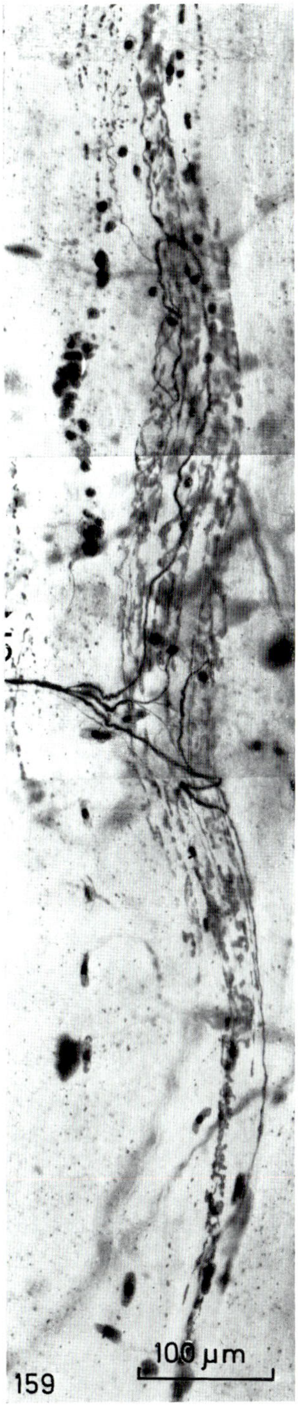

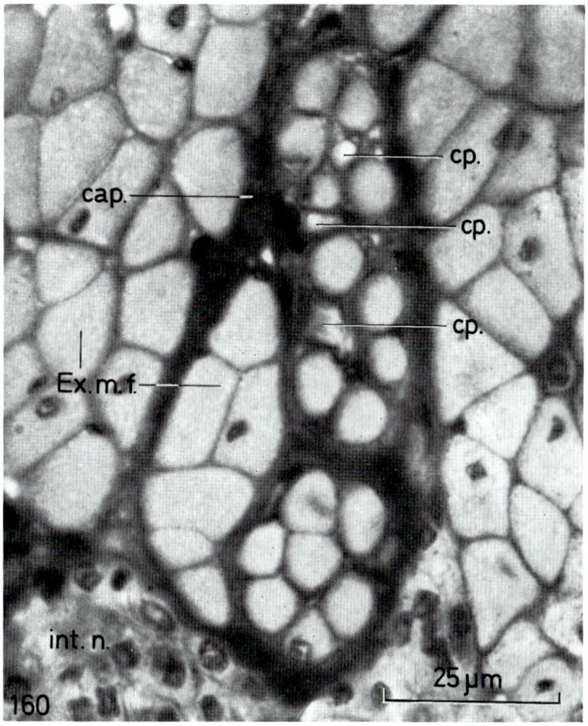

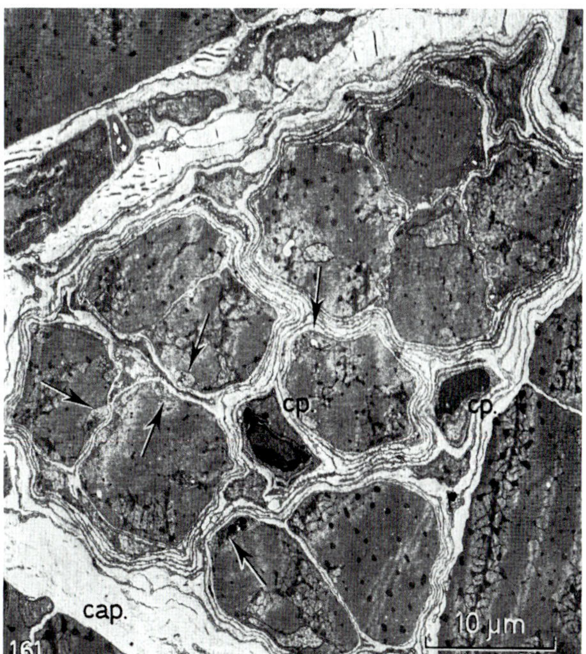

Plate 36

varicose type of sensory ending and either plate or grape motor endings. Crowe and Ragab studied the first extensor digitorum muscle and found it to be composed of both plate and grape-innervated muscle fibres. They describe the spindles as consisting of 2–17 muscle fibres (mean 10.4) of similar length (mean 9.1 mm) and diameter (mid-polar mean 10.6 μm). The equatorial region lacks a periaxial space (Plate 36, Figs. 160, 161), and the muscle fibres in this region show only a slight increase in nuclear density. All the fibres share the same histochemical profile and ultrastructure, the latter being identical with that described by Page (1968) for the extrafusal twitch fibres in the tortoise iliofibularis muscle, i.e. M line and N bands present, well-developed sarcotubular system, and numerous mitochondria. Regions of close apposition may occur between pairs of muscle fibres as in mammalian nuclear-chain fibres (p. 28). Elastic fibres are absent.

The sensory ending consists of a number of varicose threads among which forked, triangular and leaflike terminals occur (Plate 36, Fig. 159). A sensory fibre may branch to supply two or three capsular regions of adjacent spindles. The terminals are generally ellipsoidal in shape and lie in troughs on the muscle-fibre surface or may be completely enclosed within the myofibrils (Plate 37). Satellite

---

Plate 36, Figs. 159–161

Tortoise muscle spindles. First extensor brevis muscle

Fig. 159. Methyleneblue preparation of the equatorial region showing the sensory ending. (From Crowe and Ragab, 1970a)

Fig. 160. Transverse section (10 μm) through the equatorial region. The intrafusal muscle fibres are arranged to form three groups, and a small fascicle of extrafusal muscle fibres (*ex.m.f.*) is included within the capsule (*cap.*). Note absence of periaxial space. *cp.* capillary; *int.n.* intramuscular nerve. Masson's triple chrome stain. (From Crowe and Ragab, 1970a)

Fig. 161. Low-power electron micrograph of a transverse section through the equatorial region. Note disposition of capsule (*cap.*), lack of periaxial space, and sensory axon terminals (arrows). *cp.* capillary. (From Crowe and Ragab, 1970b)

Plate 37, Figs. 162 and 163

Ultrastructure of tortoise spindles. (From Crowe and Ragab, 1970b)

Fig. 162. Transverse section of a single intrafusal muscle fibre, which is associated with a number of sensory axon terminals. Two of these are completely enclosed underneath the sarcolemma. The fibre is surrounded by several layers of fibrocytes. Although the section was not stained with uranyl acetate, collagen fibres (*coll.*) can be seen between the fibrocytes. Two of the axon terminals contain vacuolar structures (arrows) that have a dense border of glycogen granules (? leached fat droplets)

Fig. 163. Longitudinal section of an intrafusal muscle fibre. Note sensory axon terminals, M line (*M*), satellite cell processes (*sat.c.*)

Plate 38, Figs. 164–167

Photographs from Crowe and Ragab (1970a) illustrating the motor innervation of tortoise muscle and muscle spindles. Teased, silver preparations of first extensor brevis muscle

Fig. 164. Extrafusal plate ending. A linear array of synaptic contacts forms a T-junction with the terminal axon

Fig. 165. Plate endings on two intrafusal muscle fibres. The endings are shorter than extrafusal plate endings

Fig. 166. Grape endings on extrafusal muscle fibres. The synaptic contacts tend to lie across a muscle fibre rather than along its axis

Fig. 167. Grape ending on an intrafusal muscle fibre

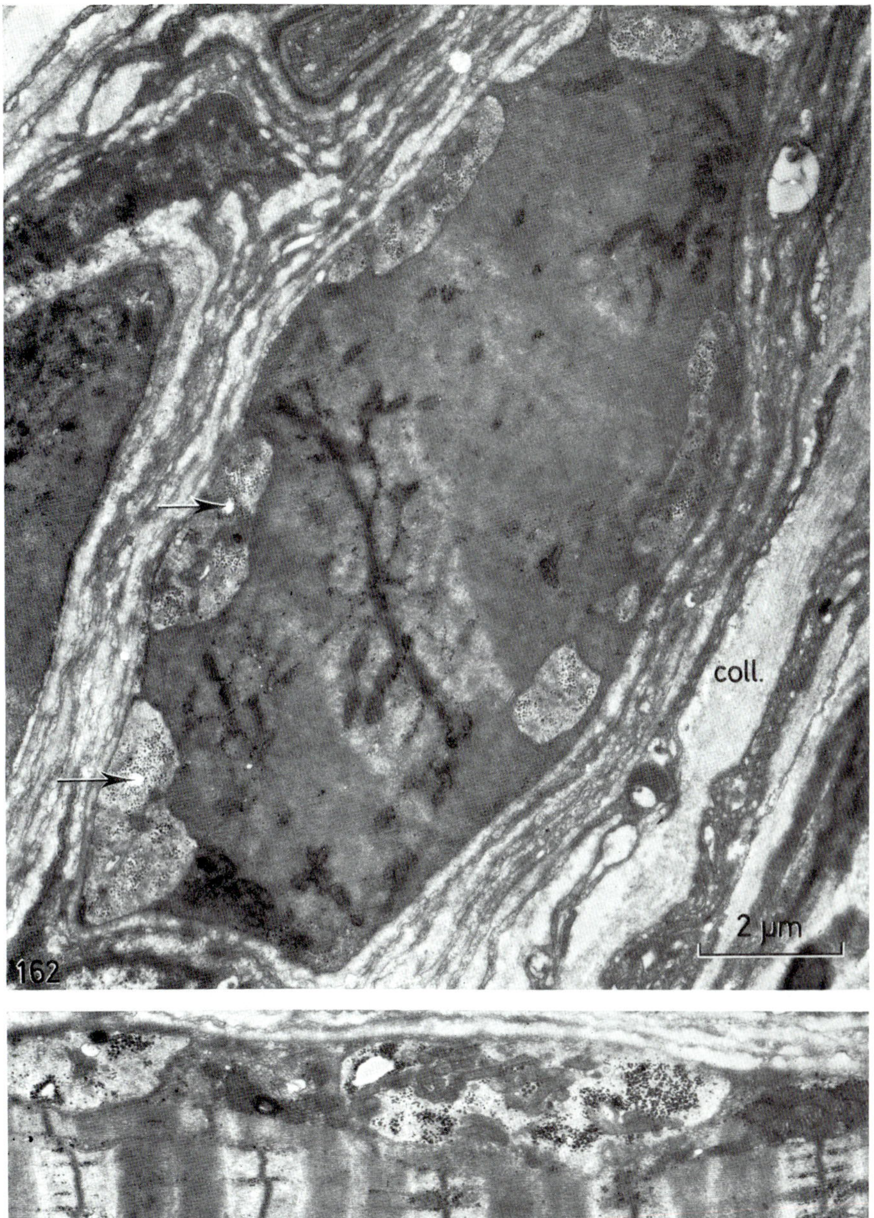

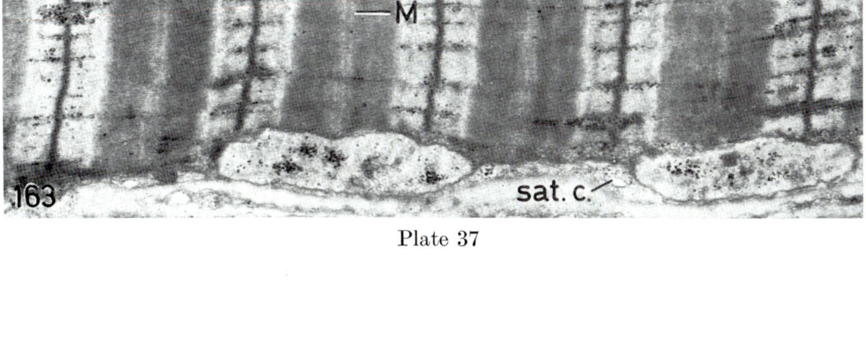

Plate 37

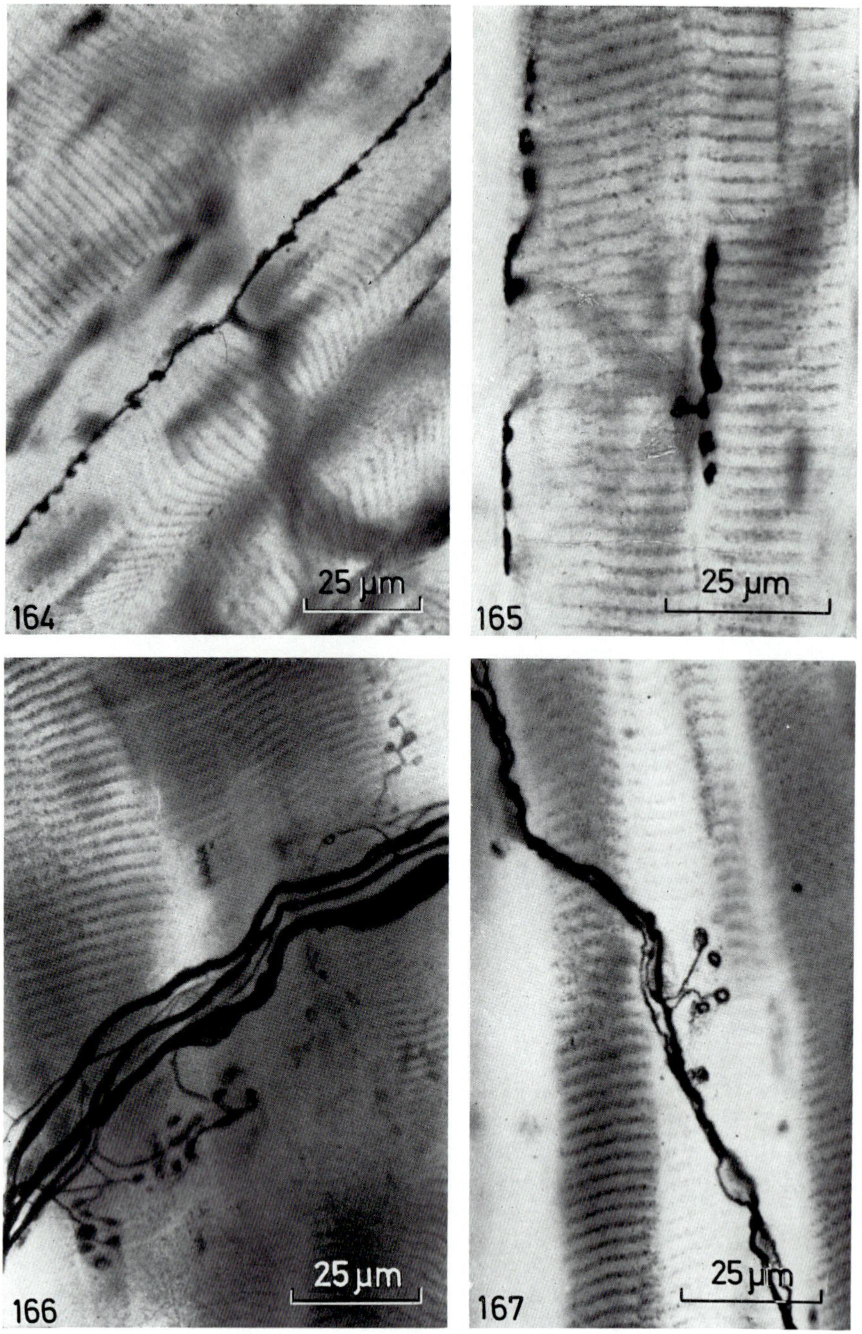

Plate 38

cells[5] frequently occur in the sensory regions, and their processes may partially envelope or wholly enclose a sensory terminal. Sensory cross-terminals do not occur.

The motor innervation consists of collateral plate and grape endings located in the extracapsular regions of the muscle fibres (Plate 38). Of 20 spindles, 13 received both types of ending, and 7 received plate endings only (RAGAB, 1970). Several plates may occur on one intrafusal muscle fibre, but Ragab was unable to ascertain whether both types of ending, plate and grape, occur on the same muscle fibre. The ultrastructure of the motor endings remains for future investigation.

HINES (1930) reports that the spindles of the alligator are composed of 2–5 muscle fibres, and that they receive a collateral grape innervation. COLE (1955) adds that the sensory ending is occasionally varicose but 'generally compact', and that there is little difference in diameter between the intrafusal and extrafusal muscle fibres.

3. *The Avian Muscle Spindle*

The avian spindle has so far received little attention. A few observations made in the last century, summarized by REGAUD and FAVRE (1904), established that both mono- and multifibrillar forms occur, and that they receive a single sensory

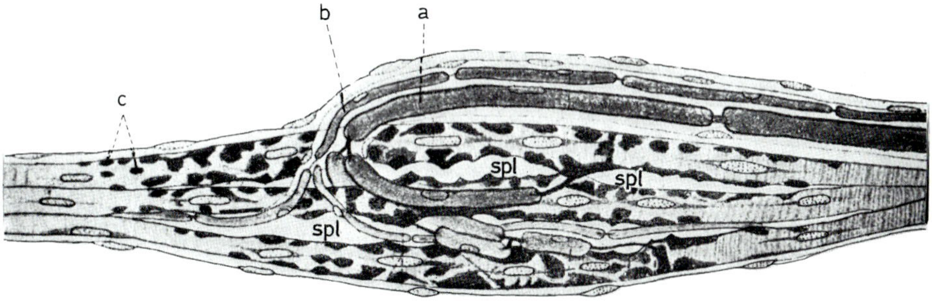

Fig. 168. A drawing by CIPOLLONE (1897, Fig. 6) illustrating the sensory ending in a pigeon spindle. Teased, gold chloride preparation, gastrocnemius muscle. *a* sensory axon; *b* motor axon that enters the capsule together with the sensory axon and branches to supply motor endings beyond the limits of the figure (possibly the juxta-equatorial multiterminal endings described by CHIN, 1970, see text); *c* sensory axon terminals; *spl* periaxial space. (Reproduced from REGAUD and FAVRE, 1904)

ending (Fig. 168) and an extracapsular motor innervation of plate or grape endings. TELLO (1917, 1922) includes the chick in his study of spindle development (see § II A, 9), and his Fig. 30 illustrates a spindle from a 14-day-old chick embryo at the time when the motor fibres arrive to form plates in the polar regions.

5 The satellite cells in reptilian and avian spindles are not comparable with those in mammalian spindles. They are more likely to be Schwann cells than stem myoblasts.

More recently reports of investigations with the light microscope have been published by Germino and D'Albora (1965), de Anda and Rebollo (1967), Rebollo and de Anda (1967), Sağlam (1968), and Maier and Eldred (1971); there is also a thesis by Chin (1970). The only ultrastructural report so far is an abstract by James and Meek (1972). The following account is based on these investigations, and also upon unpublished observations made in my own laboratory[6].

The most detailed study made with the light microscope is that by Maier and Eldred (1971), who cut serial transverse sections of 264 spindles in the sartorius and adductor profundus muscles of the chicken, quail, pigeon, and canary. The number of intrafusal muscle fibres present ranged from 1 to 8; 17% of the spindles were monofibral. The fibres appeared to be all of one type, there being no significant differences in length, diameter or equatorial nucleation. Equatorially they observed that a reduction in myofibrils over a length of 100–300 μm was accompanied by a moderate increase in number of nuclei, though there were no specific aggregations in the form of nuclear bags. Similarly Sağlam (1968), who studied woodpecker spindles in the dorsal and lateral rectus capitis and biventer cervicis muscles, found that the intrafusal muscle fibres lacked nuclear bags, though some of the fibres were thicker than others. However, de Anda and Rebollo (1967), studying chicken gastrocnemius spindles, distinguished muscle fibres of thick, thin, and intermediate diameter that were characterized by an equatorial nucleation comparable to that in mammalian bag and chain fibres. Their oxidative and glycolytic enzyme profiles as given by Rebollo and de Anda (1967) correspond with mammalian extrafusal fibre types as follows: thick (C), thin (B), and intermediate (G) (cf. p. 19).

It is possible that the nature of the fibre types in avian spindles may vary in different muscles and be influenced by the presence or absence of the extrafusal grape-innervated component, as in frog spindles (see p. 101). As to whether nuclear bags are present would seem to be a matter of interpretation; Rebollo and de Anda (1967) regard three nuclei within a fibre in a single transverse section as constituting a bag, but Maier and Eldred (1971) do not. However, the data presented by Maier and Eldred (see especially their Figs. 3–6) certainly make it clear that the density of the equatorial nuclear aggregation in the avian intrafusal muscle fibre is considerably less than in the typical bag fibres of cat or rabbit spindles.

A striking feature of the avian spindle is the manner in which each muscle fibre becomes individually encapsulated in the equatorial region. In addition to the axial and endomysial sheaths, each fibre is enclosed within 'tight rings' of collagen fibres (de Anda and Rebollo, 1967). These are eccentrically arranged so that in transverse section they assume the form of a crescent (Plate 39). The collagen crescent is always found in association with the sensory terminals, and rapidly thins down and disappears in their absence. The deepest part of the crescent (2.5–3.0 μm) corresponds with the deepest part of the sensory terminal that it

6 Miss A.M. Saunders and Miss C.M. West investigated the ultrastructure of chicken muscle spindles under the supervision of Dr. M.J. Stacey in a small research project that formed part of their second-year Zoology Honours Course programme at Durham in the summer of 1971.

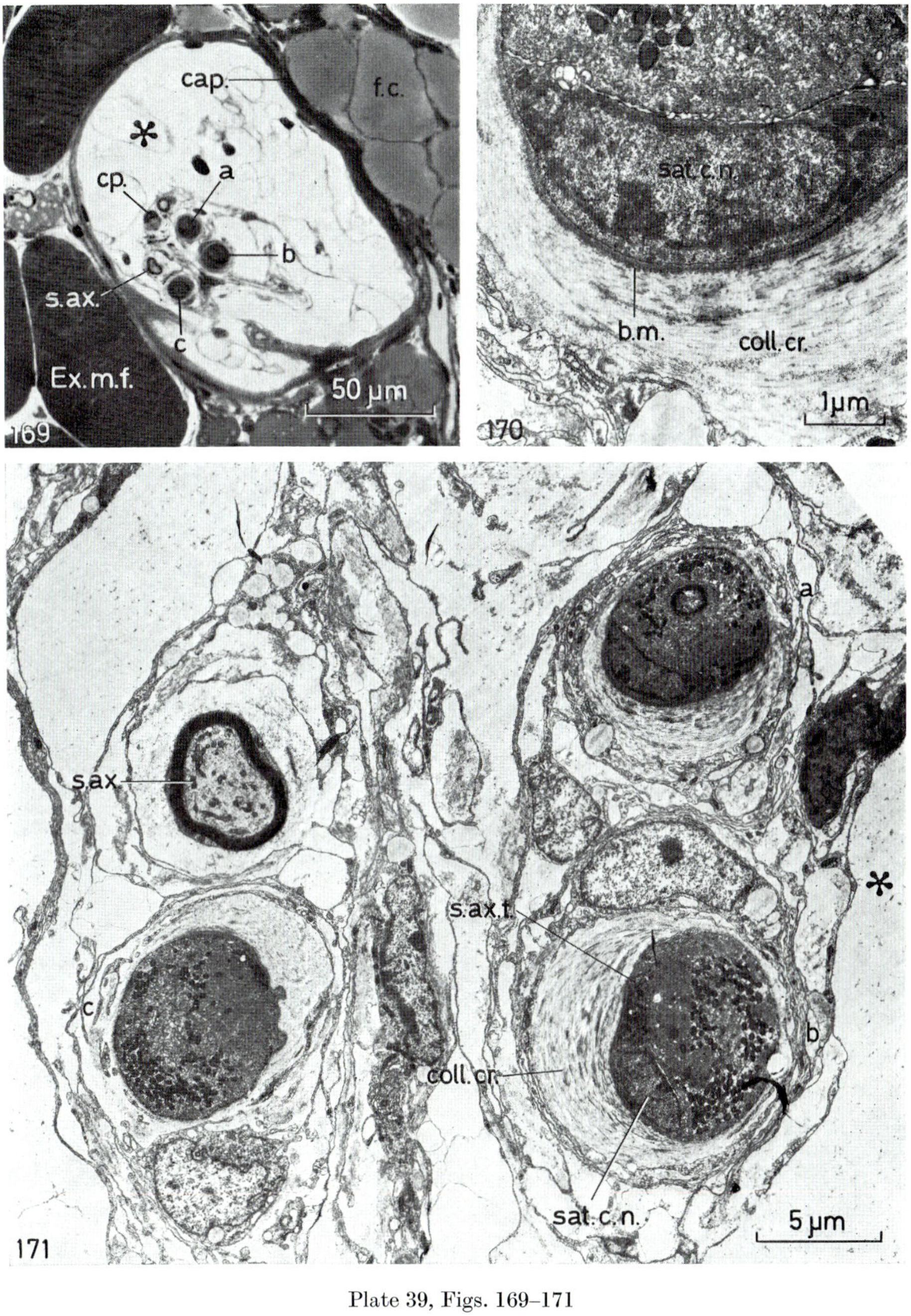

Plate 39, Figs. 169–171

Structure of chicken (*Gallus domesticus*) spindles in posterior latissimus dorsi muscle. Preparations by Miss C.M. West. *a*, *b*, *c* intrafusal muscle fibres; *b.m.* basement membrane; *cap.* capsule; *coll.cr.* collagen crescent; *cp.* capillary; *ex.m.f.* extrafusal muscle fibre; *f.c.* fat cell; *sat.c.n.* satellite cell nucleus; *s.ax.* sensory axon; *s.ax.t.* sensory axon terminal. Asterisk indicates periaxial space

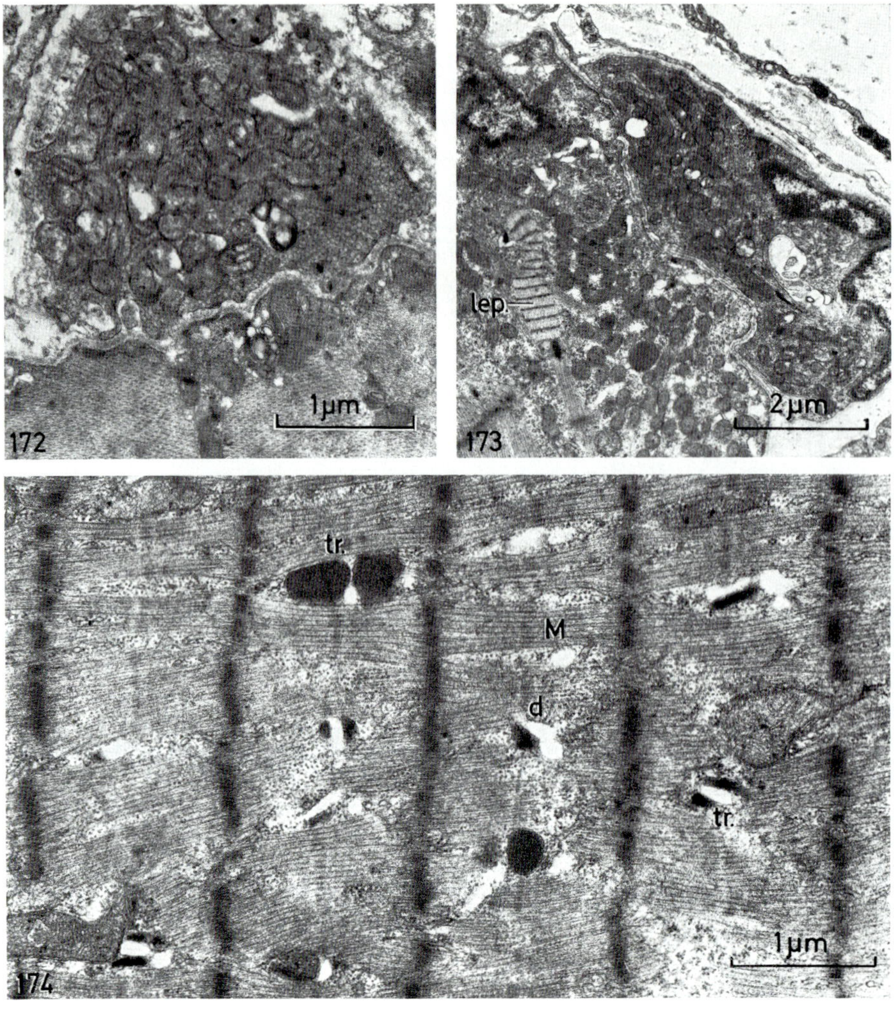

Plate 40, Figs. 172–174

Ultrastructure of chicken (*Gallus domesticus*) spindles in posterior latissimus dorsi muscle. Preparations by Miss A.M. Saunders (Figs. 72, 173) and Miss C.M. West (Fig. 174)

Figs. 172 and 173. Transverse (Fig. 172) and longitudinal (Fig. 173) sections through juxta-equatorial motor terminals. Note smooth myoneural junctions and leptofibril (*lep.*)

Fig. 174. Longitudinal section of intrafusal muscle fibre in juxta-equatorial region. Note M line (*M*), triads (*tr.*) and dyads (*d*)

Fig. 169. Transverse section (1 µm) through the equatorial region of a spindle containing three muscle fibres (*a*, *b*, *c*). Toluidene blue stain

Fig. 170. Electron micrograph of part of the intrafusal muscle fibre *a* showing collagen crescent with underlying thickened basement membrane and satellite (probably Schwann) cell

Fig. 171. Low-power electron micrograph of the three intrafusal muscle fibres *a*, *b*, *c* enclosed within endomysial sheaths. Note well-developed collagen crescents

encircles. Satellite cells occur frequently among the terminals and are similarly enclosed. The arcs of collagen fibres that form the crescent become thicker and less tightly packed together towards its outer border. The basement membrane of the muscle fibre extends over the surface of the sensory terminals and satellite cells, and thus lies in contact with the inner border of the crescent. In doing so it increases in thickness from 0.037 μm to 0.2 μm. It will be recalled that a similar, though less well-developed collagen crescent is found in association with the sensory terminals of snake spindles (see p. 114).

Chin (1970) has studied the motor innervation of spindles in the anterior and posterior latissimus dorsi (ALD and PLD) muscles of the chicken in teased, silver and cholinesterase preparations. ALD is a grape-innervated muscle, whereas PLD usually receives plates only, but may occasionally possess a few fibres with grape endings (Ginsborg and Mackay, 1961; Hess, 1961c). Chin found that the spindles in both muscles are supplied with multiterminal endings derived from an axon that enters the capsule together with the sensory axon and branches to terminate in each juxta-equatorial region (see Fig. 168). In a preliminary stage of the work it was thought that these endings were sensory and corresponded to mammalian secondary endings (see Barker, 1968a, and cf. Cooper, 1960). Subsequent work suggested that they are motor, since in a preparation of an ALD spindle Chin observed a branch to leave one of the juxta-equatorial sites of termination to supply an adjacent extrafusal muscle fibre with a grape ending.

In addition to this presumed motor innervation, the spindles in ALD receive a collateral grape innervation in both poles, but no plates; and the spindles in PLD receive a fusimotor grape innervation together with a collateral plate innervation supplied to the longer of the two poles. Monofibral spindles in PLD were observed to receive both a plate and a grape innervation (including juxta-equatorial multiterminal endings), a condition resembling that found in snake tonic spindles (see p. 109).

In their ultrastructural study of chicken PLD spindles Miss Saunders and Miss West found only one type of intrafusal muscle fibre present. Apart from its smaller diameter, this is very similar in structure to the extrafusal muscle fibres, there being an M line present and a well-developed sarcotubular system (Plate 40, Fig. 174). Equatorially the myofibrils are reduced in number and the nuclei lie within a myotube of sarcoplasm rich in mitochondria. The nuclei and myotube are eccentrically placed and located nearest to the perimeter of the fibre that lies opposite to the sensory terminals and collagen crescent (Plate 39, Fig. 171). Microladders frequently occur, usually located peripherally, often beneath sensory and motor terminals. Motor terminals with smooth myoneural junctions were observed juxta-equatorially (Plate 40, Fig. 173). These probably correspond to the multiterminal motor endings described by Chin (1970) in this location.

Some ultrastructural features of spindles in the budgerigar sartorius and the biventer cervicis muscle of a 2-day-old chick are briefly reported by James and Meek (1972). The intrafusal muscle fibres were all found to be of one type with respect to equatorial nucleation, there being no differentiation into bags and chains as in mammals.

The density of spindles in avian muscles is generally very high. The figures obtained by Maier and Eldred (1971) for the adductor profundus and sartorius muscles of the chicken, quail, pigeon, and canary range from 0.025 to 5.3 spindles per cubic millimetre, the greatest density being in the canary sartorius. By way of comparison, the muscle that has the highest spindle density in man is longissimus capitis with 63.3 spindles per gram (Voss, 1963), approximately equivalent to 0.06 spindles per mm^3. Spindles are absent from the extraocular muscles (Maier, de Santis and Eldred, 1971).

III. The Tendon Organ

The afferent nerve fibres that innervate tendons and aponeuroses are derived from both muscle and cutaneous nerves, and terminate in a number of different receptors (see § IV E). Among these is the encapsulated receptor found in birds and mammals that was first described by Golgi (1878, 1880) and is generally known as the tendon organ. Golgi himself called it a 'musculo-tendinous end-organ'. Other terms that have been used include 'neuro-tendinous end-organ', 'Golgi organ', and 'tendon spindle'.

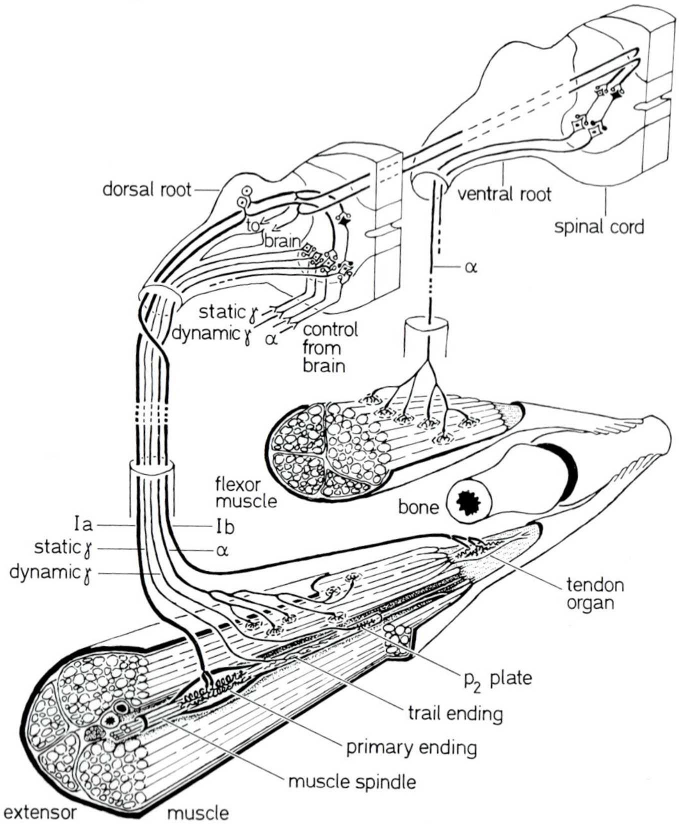

Fig. 175. Schematic diagram showing the location and innervation of a muscle spindle and a tendon organ in a cat extensor muscle, and the synaptic connexions made in the spinal cord between their Ia and Ib afferent fibres and the alpha motoneurons innervating the synergic (extensor) and antagonistic (flexor) muscles. Excitatory nerve cells and synaptic knobs drawn as open structures, inhibitory cells and knobs filled in. Ia and Ib pathways as postulated by Eccles (1957, Fig. 62)

Tendon organs engaged the attention of a number of Italian histologists before the turn of the century, the most detailed reports being those of CATTANEO (1888) and CIACCIO (1890). Cattaneo established the sensory nature of the terminals by cutting the ventral roots of nerves supplying the dog's hindlimb, and showing that the tendon organ and its nerve supply remained intact four months later. He also cut the sciatic nerve in rabbits and guinea-pigs and followed the degeneration of the tendon organ over a period of 63 days. The work during the latter part of the century is reviewed by HUBER and DE WITT (1900), who add some observations of their own, and there is also a comprehensive review by REGAUD (1907).

A tendon organ consists of a thinly encapsulated bundle of small tendon fascicles that is generally fusiform in shape and located at the musculo-tendinous junction, though occasionally some occur wholly within tendon. The receptor thus lies 'in series' with the muscle fibres, and is considered to be specifically sensitive to their tension. The 'in series' arrangement of the tendon organ, and the 'in parallel' arrangement of the muscle spindle, are illustrated schematically in Fig. 175. This also shows the synaptic connexions that, according to ECCLES (1957), are made in the spinal cord between the Ia (primary) and Ib (tendon organ) afferents and the alpha motoneurons innervating synergic and antagonistic muscles.

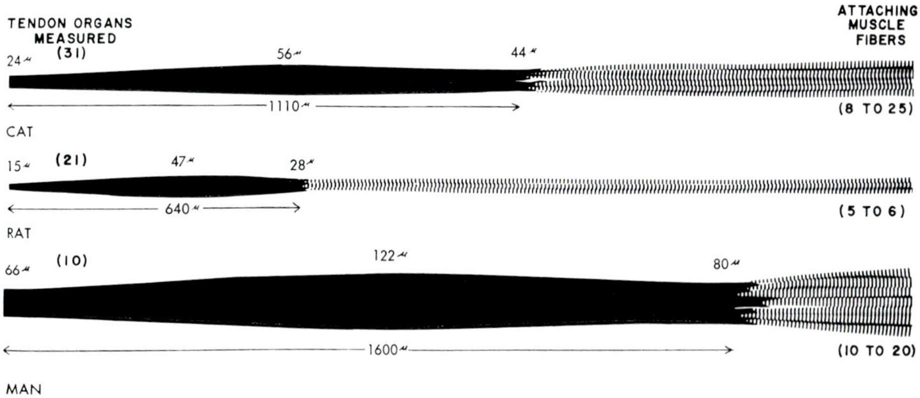

Fig. 176. Comparisons of size and proportions of tendon organs from the rat, cat, and man. Diameters and lengths are means of 31 tendon organs from the cat's extensor digitorum brevis muscle; 21 from the rat's gastrocnemius; and 10 from the human extensor digitorum brevis. (From BRIDGMAN, 1970)

The dimensions of tendon organs vary according to species and the conditions of histological preparation. Cat tendon organs in teased, gold chloride preparations on average measure 500 μm long and 100 μm across at the widest point. Using formalin fixation with the muscle held at its maximum physiological length, BRIDGMAN (1970) obtains a mean length of 1110 μm and a maximum mean width of 56 μm. His comparable measurements for rat tendon organs are 640 μm and 47 μm, and for human ones 1600 μm $\times$ 122 μm (Fig. 176). The shape of most tendon organs is fusiform; in others either the muscular or the tendinous end, or both ends, are branched to form bifid or trifid shapes.

The perineural sheath of the nerve supply is continuous with the capsule, which internally is usually prolonged into septa that divide the tendon fascicles into longitudinal compartments. The capsule is not as well developed as in muscle spindles, and though there are internal spaces there is nothing corresponding to the periaxial space. The fascicles have more nuclei between them and appear less differentiated than ordinary tendon. As they course through the receptor they subdivide into smaller bundles which then join together again, so that the fascicles in the mid-region are smaller and more numerous than those at either end (BRIDGMAN, 1968). Capillaries run between the layers of the capsule, often being derived from a blood vessel that approaches the receptor together with the nerve supply.

The afferent fibre innervating the receptor is derived solely from intramuscular nerve trunks, and its diameter falls within the range of 8 to 15 μm (WEDDELL and HARPMAN, 1940; BOYD, 1954; WOHLFART and HENRIKSSON, 1960; BRIDGMAN, 1968). Usually each afferent supplies one tendon organ, but some may branch so as to innervate several (GOLGI, 1880; CATTANEO, 1888; DOGIEL, 1906a), and occasionally the sensory endings in one tendon organ may be derived from two apparently independent afferents (HUBER and DE WITT, 1900). After dividing into two or three branches prior to entering the receptor, the afferent fibre gives rise to further branches within it that ultimately terminate in a number of arborizations. These ramify among the small, subdivided fascicles clasping and encircling them in an irregular fashion (Plate 41, Fig. 179). A few spiral turns occur here and there, but they are not typical of the general mode of termination, which bears some resemblance to the flower-spray type of secondary ending in mammalian muscle spindles.

According to REGAUD (1907), the muscular end of the tendon organ is generally attached to 3–8 muscle fibres, though in birds instances occur of attachments to one muscle fibre only. BRIDGMAN (1970) found cat tendon organs attached to 8–25 muscle fibres, those of rat to 5–6, and those of man to 10–20. In human tendon organs some of the muscle fibres attach to the outer surface of the capsule (BRIDGMAN, 1970). Though it is known that two motor units may excite the same tendon organ (HOUK and HENNEMAN, 1967), no histochemical investigation has yet been undertaken relating muscle-fibre types to individual receptors.

During the course of a quantitative study of mammalian proprioceptors carried out by my research group in Hong Kong in 1960–1961, Miss M. Pang gathered the

---

Plate 41, Figs. 177–179

Teased, silver preparations by Dr. M.J. STACEY illustrating tendon organs and their association with muscle spindles and paciniform corpuscles in cat muscles. *cp.* capillary; *P* primary ending; *pf.a.* paciniform afferent; *pf.c.* paciniform corpuscle; S_1 secondary ending; Ia primary afferent; Ib tendon-organ afferent; II group II secondary afferent

Fig. 177. Tendon organ from de-efferented extensor digitorum longus muscle with five paciniform corpuscles on its surface (four visible) derived from an afferent with a stem diameter of 10.5 μm. (From STACEY, 1969)

Fig. 178. Spindle-tendon-organ dyad from de-efferented and sympathectomized peroneal muscle. Two paciniform corpuscles are shown, one of which is associated with the tendon organ. (From BARKER, 1967)

Fig. 179. Tendon organ from de-efferented peroneus longus muscle. The proximal end, attached to extrafusal muscle fibres, is at the top of the photograph

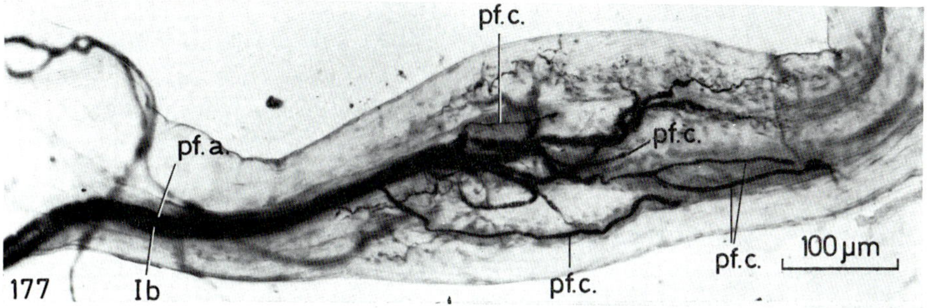
pf.c.
pf.a.
pf.c.
pf.c.
100 µm
177
Ib
pf.c.

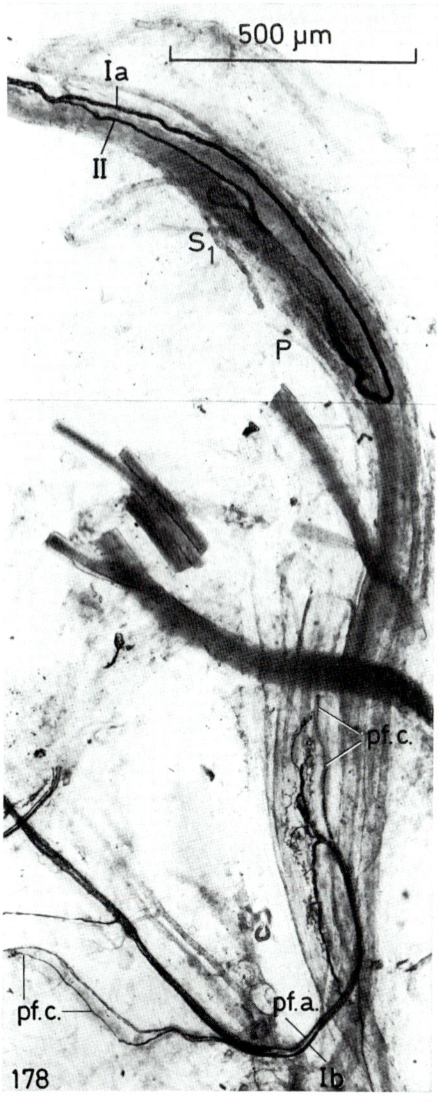
500 µm
Ia
II
S_1
P
pf.c.
pf.c.
pf.a.
178
Ib

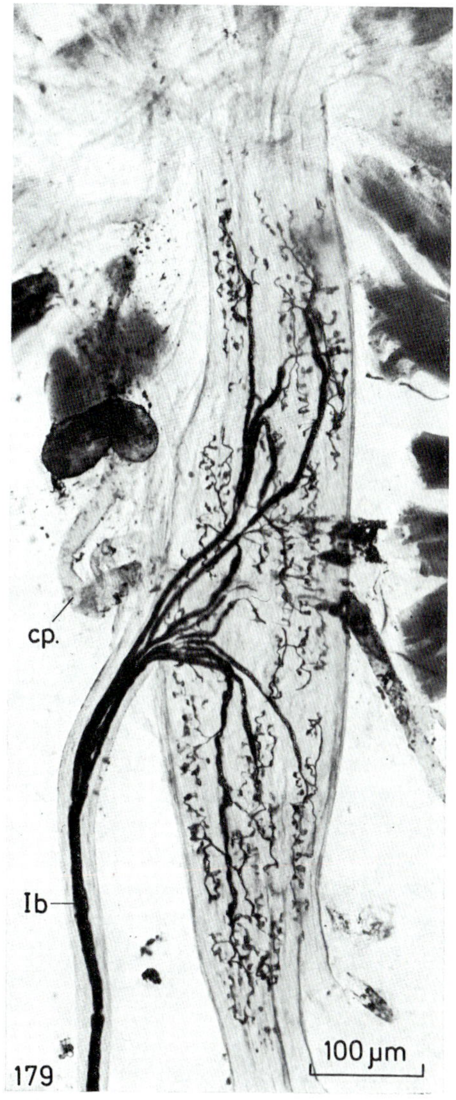
cp.
Ib
100 µm
179

Plate 41

following information from gold chloride preparations of 1,557 cat tendon organs teased from the quadratus femoris, rectus femoris, semitendinosus, FDL, FHL, soleus, extensor digitorum brevis, pes interossei, and external and internal intercostal muscles. In a sample of 1,337, 92.4% were located at musculo-tendinous junctions, the rest wholly within tendon. The mean number of muscle fibres attached to the receptor was 10.5 in a range of 3–25; about 5% of the receptors were attached to a number of muscle fibres in excess of 25 that could not be counted, but was estimated to be below 50. Of 1,278 tendon organs, 63% were fusiform, the rest bifid or trifid at one or both ends. The mean length of 100 tendon organs was 521.4 μm (range 242–1045 μm), and their mean width at maximum was 125.5 μm (range 66–220 μm). The mean total diameter of 100 tendon-organ (Ib) afferent fibres was 10.4 μm in a range of 7–14 μm. The majority of the afferents entered the receptors at their widest point in the middle of their length. Only 1.8% of 1,062 tendon organs were innervated by two apparently independent Ib afferents. In order to observe the frequency with which Ib afferents branch to innervate more than one tendon organ, preparations were made using a compressorium to give a wide field of 50–100 sq. mm. Of a total of 103 afferents, 88 each supplied one tendon organ, 13 each supplied two, and 2 each supplied three, giving an ending-to-fibre ratio of 1.2:1.

Tendon organs are part of the standard proprioceptive equipment of striated muscles in birds and mammals, and generally speaking a muscle that is provided with spindles also has tendon organs. However, tendon organs may be absent in small muscles that have few spindles, e.g. rat tail muscles, and cat tenuissimus (Palmer and Stilwell, 1958) and deep lumbrical muscles. Also muscles that lack spindles may nevertheless possess tendon organs, as in cat and rabbit eye muscles (Tozer and Sherrington, 1910).

A muscle is generally provided with more spindle primary endings (Ia fibres) than tendon organs (Ib fibres). The total number of primary endings found in teased, gold chloride preparations of nine cat rectus femoris muscles was 920 as compared with 705 tendon organs, the ratio of primaries to tendon organs varying over a range of 0.7–2.4:1, mean 1.3:1 (Barker, 1962b). The mean ratio for three soleus muscles was 1.8:1 (range 1.7–1.9:1) and for three semitendinosus muscles 1.6:1 (range 1.2–1.85:1). Similar findings are reported by Wohlfart and Henriksson (1960) and Eldred, Bridgman, Swett and Eldred (1962).

The tendon organs of a muscle are found wherever the muscle and tendon fibres composing it are united. Hence they occur not only at the musculo-tendinous junctions at either end of a muscle, but also in the places where these junctions are intramuscular, e.g. under aponeuroses, and in association with the intramuscular tendons of pennate muscles. Thus Swett and Eldred (1960) show that the tendon organs in cat soleus and medial gastrocnemius are located underneath their aponeuroses of origin and insertion. In cat rectus femoris, which is a pennate muscle, Barker and Ip (1961) found that the greatest concentration of tendon organs occurs in the proximal half of the muscle, as in the case of spindle capsules (see p. 87).

Tendon organs are occasionally found in intimate association with other muscle receptors, namely, paciniform corpuscles, free endings, or muscle spindles. In Pang's analysis, 7.6% of 1332 cat tendon organs were associated with paciniform corpuscles, 1.2% received free endings, and 1.2% were associated with spindles. In teased, silver preparations of cat hindlimb muscles, STACEY (1967) found that 5 (6.3%) of 126 tendon organs received a free-ending innervation, and that 24 (15%) of 158 were associated with paciniform corpuscles.

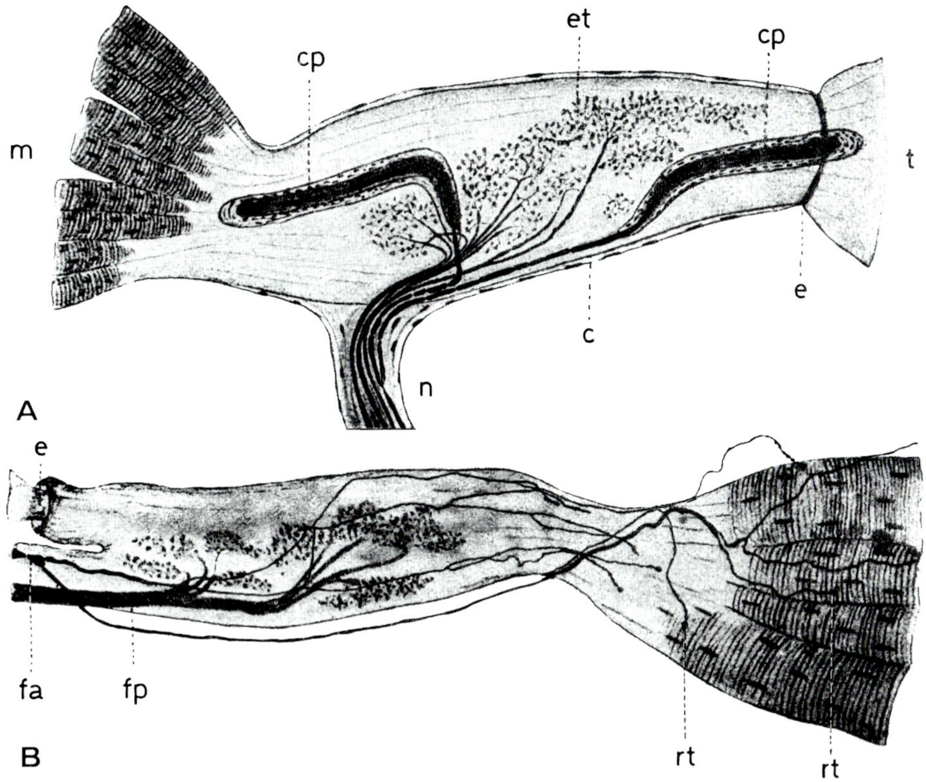

Fig. 180A and B. A Cat tendon organ with two associated paciniform corpuscles. *c* capsule of tendon organ; *cP* paciniform corpuscle; *e* capsule collar; *et* axon terminals of Ib afferent; *m* muscular, and *t* tendinous extremity of tendon organ; *n* nerve trunk carrying Ib and paciniform afferents. Gold chloride preparation. Compare with Plate 41, Fig. 177. (Reproduced from REGAUD, 1907, after RUFFINI, 1893a). B Cat tendon organ associated with afferent fibre terminating in free endings. *e* capsule collar; *fa* free-ending afferent; *fp* Ib afferent; *rt* free-ending axon terminals. Gold chloride preparation. Compare with Plate 43, Fig. 195. (Reproduced from REGAUD, 1907, after RUFFINI, 1897)

The relationship between tendon organs and paciniform corpuscles was first noted by CATTANEO (1888) and subsequently described in detail by RUFFINI (1893a, 1897) (see Fig. 180A). The corpuscles may lie either outside the capsule, or be enclosed within it and lie in troughs on the surface of the fascicles. They are longitudinally orientated, pointing towards either the muscular or the tendinous end. From 1 to 5 corpuscles may be associated with a single tendon organ; they are supplied by the branches of a single myelinated afferent fibre that enters the receptor together with the Ib fibre (STACEY, 1967) (Plate 41, Figs. 177, 178). Pang found that the corpuscles were located underneath the capsule with about twice the frequency of being located externally. Further details about paciniform corpuscles are given in § IV E, 2.

Tendon organs, like muscle spindles, occasionally receive free endings from small (2–3 μm) afferents called 'concomitant' fibres by RUFFINI (1893a, 1897), who first described the innervation; 'accessory' fibres by TELLO (1922) and WEDDELL and HARPMAN (1940); and 'satellite' fibres by WOHLFART and HENRIKSSON (1960). According to WEDDELL and HARPMAN (1940, man, rabbit) and WOHLFART and HENRIKSSON (1960, mouse) such fibres are a constant feature, and STILWELL (1957b, rabbit, cat, monkey, man) also reports them as present in all the spindles and tendon organs he examined. However, observations on de-efferented and sympathectomized cat spindles and tendon organs emphasize the casual rather than the specific nature of the innervation (STACEY, 1967, 1969). Such afferents may or may not enter the tendon organ together with the Ib afferent. Their branches and terminals may lie within the tendon organ; outside it on the external surface of the capsule; in and among the connective tissue surrounding the muscle fibres that are attached to the organ; in the surrounding tendon or blood vessel adventitia; or in any combination of these situations. The usual condition in cat tendon organs is for the Ib afferent to approach and enter the receptor as the only nerve fibre concerned with its innervation, but at maximum it may also be accompanied by a paciniform afferent, a free-ending afferent, and vasomotor fibres.

Intrafusal muscle fibres that arise or insert at the level of musculo-tendinous junctions do so by means of a spindle tendon. Occasionally such a tendon attaches to the muscular or tendinous end of a tendon organ, so that in teased preparations the two receptors remain linked together. More frequently, as MARCHAND, BRIDGMAN, SHUMPERT and ELDRED (1971) have shown in a serial transverse-section study, the spindle tendon makes its attachment to the main tendon or aponeurosis immediately alongside a tendon organ. MARCHAND et al. refer to such a linkage or juxtaposition of the two receptors as a 'spindle-tendon organ dyad', or simply 'dyad' (see Plate 41, Fig. 178). They found that the proportion of tendon organs in the cat's extensor digitorum brevis muscle forming dyads with spindles was 50%; in medial gastrocnemius the proportion was 20%, in soleus 27%. The tendon organs in such dyads were larger than solitary ones and had more muscle fibres attached to them (mean of 15 as compared with a mean of 11). The equatorial cross-sectional area of dyad spindles was also larger than that of solitary ones, though their mean number of intrafusal muscle fibres was about the same (7.2 as compared with 7.3).

The ultrastructure of tendon organs has been studied by MERRILLEES (1962, rat, forelimb lumbrical muscle), SHIMIZU (1971, mouse, gastrocnemius), and SCHOULTZ and SWETT (1972, cat, extensor carpi ulnaris). Micrographs show that the perineural epithelial cells that ensheath the Ib afferent fibre are identical and continuous with the cells forming the tendon-organ capsule. Similarly the subperineural space and its fluid are confluent with the fluid-filled capsule lumen. The capsule encloses the bundles of collagen, which form the main body of the organ, in several concentric tubular layers, and forms tight-fitting collars around them at the proximal and distal ends. The collagen bundles are smaller and not so tightly packed as those forming ordinary tendon. Septal cells (fibrocytes) incompletely divide the capsule lumen into longitudinal compartments through which the collagen bundles travel separated from each other by fluid-filled spaces. Reconstructions by SCHOULTZ and SWETT (1972) show how the bundles tend to twist around each other like the strands in a rope. They also divide and reunite over short distances. The non-myelinated terminal axon branches of the Ib afferent thread a sinuous course between and around the bundles, sometimes completely encircling them (Fig. 181 and Plate 42). Schwann cells and basement membrane cover all but the most distal parts of the terminals. Basement membrane material may be found forming concentric lamellae around preterminal axons as they thread their way through the longitudinal compartments (SHIMIZU, 1971). SCHOULTZ and SWETT (1972) suggest that the increased tensile forces on the colla-

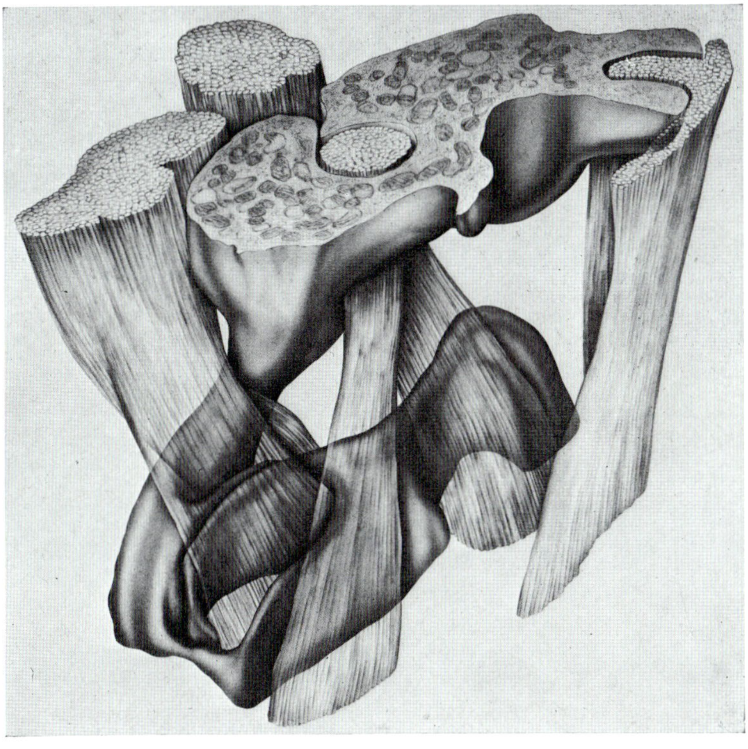

Fig. 181. A three-dimensional reconstruction by SCHOULTZ and SWETT (1972) demonstrating the relationship between an axonal branch of a Ib afferent and longitudinally orientated collagen bundles in a septal cell compartment of a tendon organ. SCHOULTZ and SWETT describe the reconstruction as follows. "Note the spiraling and clasping form of the axon as it threads its way down through the collagen bundles. Most collagen bundles, like these, do not run a straight and parallel course through the tendon-organ lumen. They twist like the strands of a braided rope and often split into two bundles, as in the case of the bundle at the far left, to surround a portion of an axon. Increased tensile forces on the collagen strands may well cause them to straighten and physically twist and pinch large surface areas of the axonal branches trapped between them"

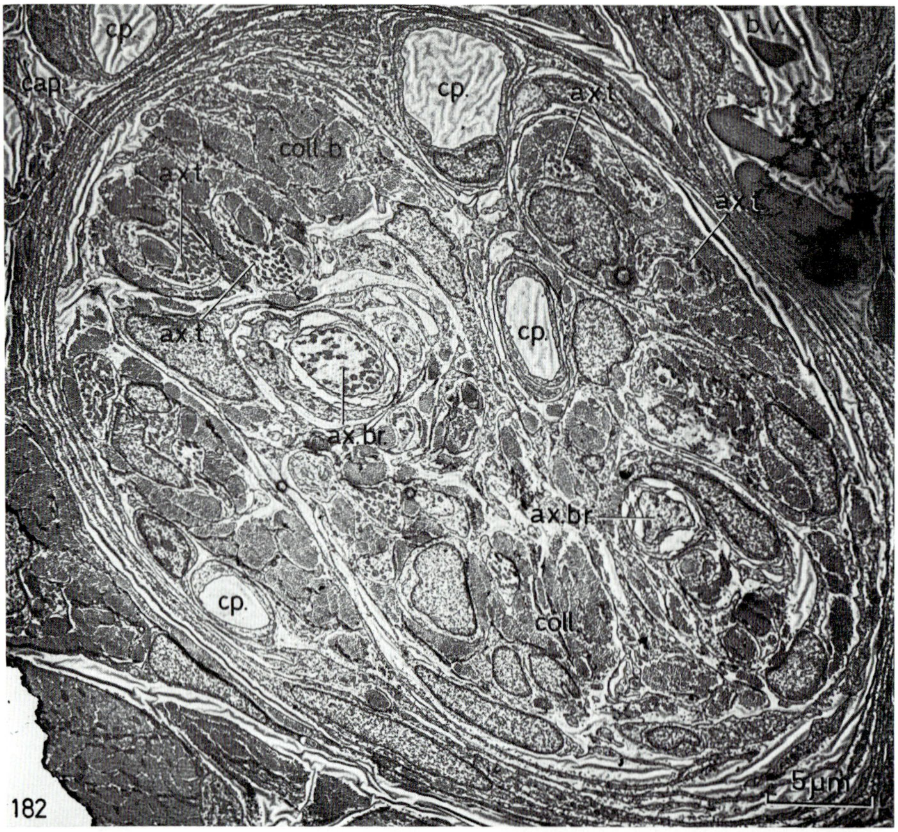
cp.
b.v.
cap.
cp.
ax.t.
coll.b
ax.t.
ax.t.
ax.t.
cp.
ax.br.
ax.br.
cp.
coll.
5µm
182

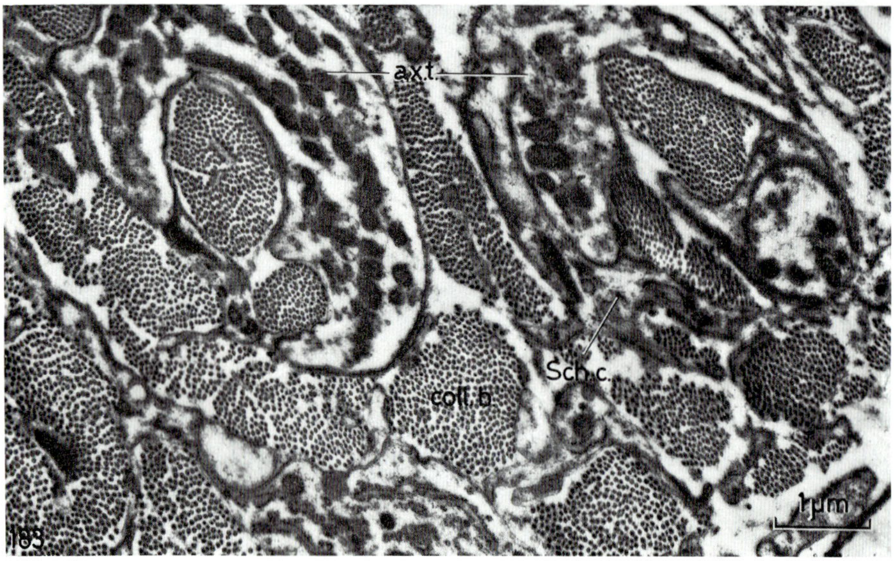
ax.t.
Sch.c.
coll.b
1µm
183

Plate 42

gen bundles caused by muscle contraction or passive muscle stretch tighten the entwined collagen bundles and squeeze the axon terminals.

The only study of the development of tendon organs is that by TELLO (1917, 1922), who followed the process in chick embryos, and also made some observations on cat and human embryos. The afferent fibres destined to initiate tendon-organ formation arrive in the musculo-tendinous regions of chick hindlimb muscles on the 9th day of incubation. They leave the small intramuscular nerve trunks and spread out over the region, being mostly directed towards the ends of the myotube bundles, though some are directed towards the tendon. Branching occurs, and the general pattern of the innervation at this stage resembles the branches of a weeping willow. The ends of the myotubes, according to Tello, appear to exercise a neurotropic influence, since the branches of individual Ib afferents establish contact with them terminating in small bulbs or rings. From the 12th day onwards a 'condensation' of connective tissue occurs in the future tendon-organ area, and the afferent fibre undergoes further branching, spreading terminals among the connective tissue mass. From the 15th day the contacts between the terminals and the ends of the myotubes are lost. The ramification and elaboration of the terminals among the bundles of collagen proceeds, and the receptor assumes its adult form. The developmental sequence is beautifully illustrated by Tello in his Figs. 19–23.

IV. Other Vertebrate Muscle Receptors

This section describes receptors, other than muscle spindles and tendon organs, that occur in the muscles and associated tissues of vertebrates. Much of the relevant histology was done before and around the turn of the century and is reviewed by REGAUD and FAVRE (1904) and REGAUD (1907). BONE (1964) has more recently reviewed work on the sensory innervation of muscles in the Agnatha and fishes.

A. Agnatha

BONE (1963) describes thin axons in the hagfish *Myxine* that branch to end freely under the connective tissue sheet covering the visceral face of the myotomes at the level of the myosepta. In some cases he was able to trace the axons back to their cell bodies in the dorsal root ganglion. Their final branches terminate as fine varicose threads. The form and disposition of the innervation is reminiscent of the mammalian free ending. A similar innervation appears to be present in lampreys (BONE, 1964). STEFANELLI (1932) describes endings in the lamprey that he supposes to be the equivalent of the Golgi tendon organ, but according to BONE (1964) these are motor endings disposed around the ends of muscle fibres.

Plate 42, Figs. 182 and 183

Ultrastructure of tendon organs. Electron micrographs from the study by MERRILLEES (1962) of a tendon organ in rat lumbrical muscle. Material fixed in 1956 and embedded in methacrylate. *ax.br.* axonal branch of Ib afferent; *ax.t.* axon terminal; *b.v.* blood vessel; *cap.* capsule; *coll.b.* collagen bundle; *cp.* capillary; *Sch.c.* Schwann cell

Fig. 182. Transverse section through the entire organ

Fig. 183. Transverse section at higher magnification illustrating axon terminals in association with collagen bundles

B. Fishes

No sensory endings are located within the myotomal musculature of fishes. There is, however, both physiological (FESSARD and SAND, 1937) and histological POLOUMORDWINOFF, 1898; BARETS, 1956) evidence for the presence of stretch receptors in the fin musculature of rays. These consist of a mass of varicose terminals about 500 μm long supplied by a thick myelinated fibre (diameter 11–14 μm). They lie in the connective tissue between the muscle fibres. The terminals consist of a series of bulbs connected together by thin tubular links, very similar in appearance to the sensory endings in frog spindles, frog tendons, and some snake spindles (see Fig. 184). These 'terminaisons en pinceaux' (POLOUMORDWINOFF, 1898) are abundant in the fins of *Torpedo*, common in those of *Raja*, and scarce in those of the dogfish *Scyliorhinus* (BONE, 1964).

Another receptor occurring in elasmobranchs consists of large coiled endings ('corpuscles of Wunderer') lying in the connective tissue at the bases of the fins of rays and dogfish. These are pressure receptors, discharging impulses when the fin is bent and subjects the connective tissue surrounding the endings to stretch and compression (LOWENSTEIN, 1956).

CIACCIO (1890) has described sensory endings in the fin tendons of the carp, tench, and ray. His figures of gold chloride preparations show axons branching to form plaques of terminals among the tendon fascicles. The plaques are disposed singly in the carp, but occur in compound form in the tench and ray. The terminals are similar in appearance to those found in Golgi tendon organs (Fig. 185). BONE (1964) has not seen such receptors in his methylene blue preparations of ray tendons. He describes "simple branching sensory endings" (presumably similar to mammalian free endings) located in connective tissue at the bases of the walking fin rays of gurnards, and has seen similar endings in the pectoral fins of *Torpedo*. This type of ending is also present in the pectoral and pelvic fins of the lungfish *Protopterus* (W. HOLMES, personal communication). Each fin is a slender tapering process containing a segmented rod enclosed within a connective tissue sheath (perichondrium). Holmes' preparations of the fins, stained with his silver method (1967), show that the perichondrium is innervated by thick axons (diameter 5.0–10.5 μm) that branch extensively and ultimately terminate as fine threads. These are often varicose with a small bulb at the tip and similar bulbs strung along behind like beads (Fig. 186).

C. Amphibia

The sensory fibres innervating the tendons of frogs and toads terminate in large bush-like endings that consist of numerous parallel varicose threads (Fig. 187). Each thread consists of a number of bulbs connected together by thin tubular links, as occurs in the spindle sensory endings. These 'terminaisons en buissons' have been described by CIACCIO (1890), SMIRNOFF (1890, 1893), and HUBER and DE WITT (1900). They may be very extensive and consist of as many as 12 to 15 individual tufts of threads derived from the branching of a single axon. In the picturesque language of Huber and de Witt each tuft "looks not unlike a small tree-like bush with long, slender nearly parallel branches, tapering toward the top, on which the buds are somewhat swollen, but which are unadorned with flowers, fruit or foliage" (p. 178). The endings are located wholly within tendon and lie among and between tendon fascicles that are smaller and more densely nucleated than ordinary tendon.

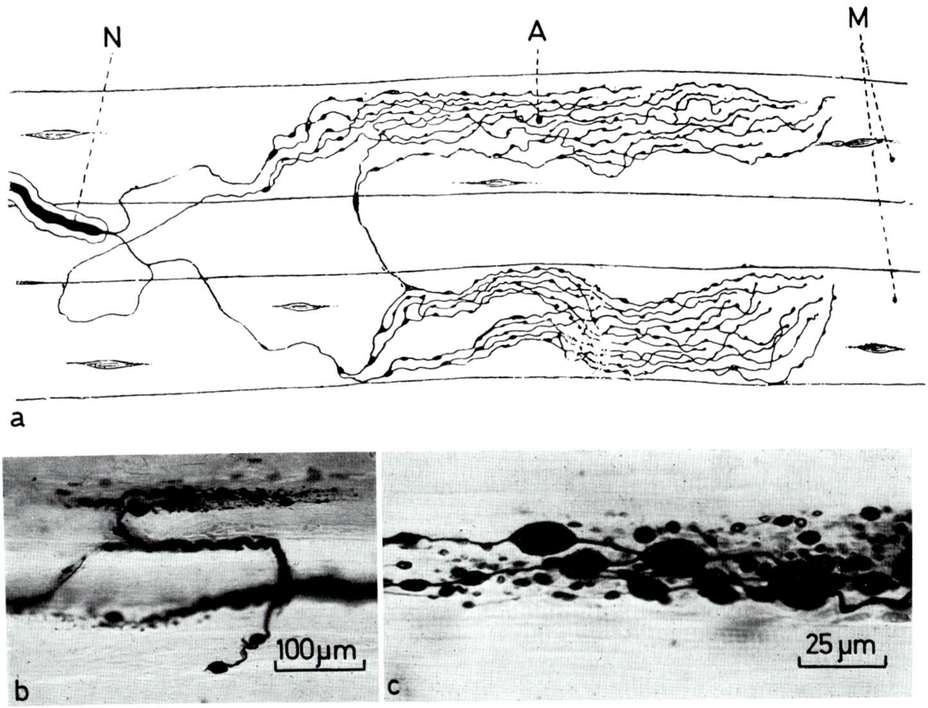

Fig. 184a–c. Intramuscular stretch receptors (*terminaisons en pinceaux*) in rays. a A drawing of the receptor by POLOUMORDWINOFF (1898) in the fin musculature of *Torpedo*. Methylene blue preparation. *A* axon terminals; *M* muscle fibres; *N* afferent fibre. (Reproduced from REGAUD and FAVRE, 1904). *b* and *c* Low- and high-power photographs of the receptor as seen in a Bielschowsky-Gros silver preparation of the pelvic fin muscles of *Raja clavata*. (From BARETS, 1956)

KULCHITSKY (1924b) has described leaf-like endings in the frog that lie upon the surface of the muscle fibres or in the interstitial connective tissue. He illustrates them as being formed by non-myelinated preterminal axons derived from the nodes of myelinated fibres, and also by axons that remained non-myelinated as far as he could trace them (Fig. 188). Some of the preterminal axons were quite short, others were of considerable length, but all terminated in "a bush of ramifying threads with leaf-like expansions at their ends" (p. 11). Kulchitsky believed that these endings were sensory and that they were formed by both myelinated and non-myelinated stem fibres.

ITO (1968a, b, 1971; ITO, TOYAMA and ITO, 1964) has studied extrafusal stretch receptors in the frog's sartorius muscle supplied by afferent fibres having a

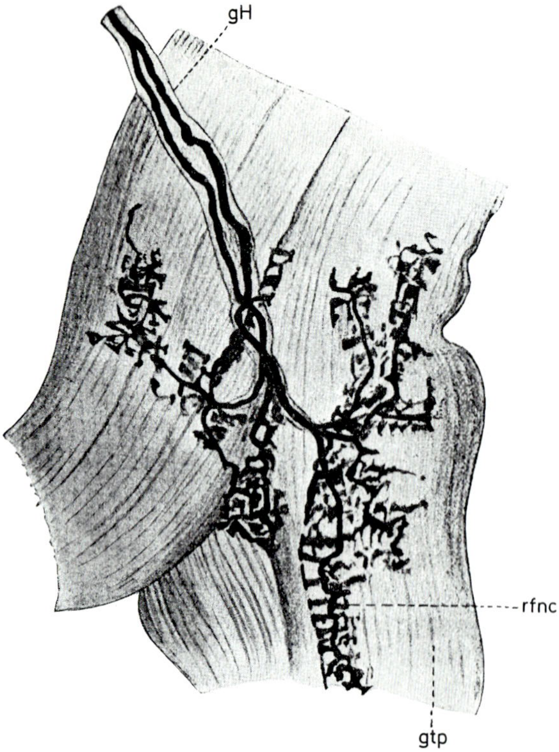

Fig. 185. A drawing by Ciaccio (1890) of plaques of sensory terminals innervating the tendons of tail muscles in *Raja marginata*. Gold chloride preparation. (Reproduced from Regaud, 1907)

conduction velocity of 10.2–20.0 m/sec. In osmic preparations he finds that they have a diameter of 3–11 μm and innervate a receptive field covering 2–4 muscle fibres over a length of 2–4 mm. After some initial uncertainty (see Discussion in Ito, Toyama and Ito, 1964), Ito (1971) concludes that the afferents terminate in Kulchitsky's leaf-like endings. This is most probably correct providing the correlation is made with the endings that Kulchitsky observed to originate from *myelinated* fibres. Judging from his figures it appears that most of the endings he supposed to originate from non-myelinated sensory fibres (his Figs. 11–15) are in fact grape motor endings. Though these are derived from myelinated stem fibres, their non-myelinated preterminal axons are often very long. Kulchitsky's Fig. 10 (reproduced in Fig. 188B) suggests that this may also be true of some of the afferents that supply the leaf-like receptors.

Ito, Toyama and Ito (1964) found that "ten or more" afferent fibres supply the leaf-like receptors in the frog sartorius. The number of spindle afferents is evidently of the same order since Pezard and May (1937) counted 19 spindle sensory regions in this muscle, and frog spindle afferents may branch to supply several capsules. The histology of the leaf-like receptors clearly merits further investigation, and the frog sartorius would be an excellent muscle to study, not only because of Ito's work, but also because it is free of any confusion that might be caused by the presence of grape motor endings.

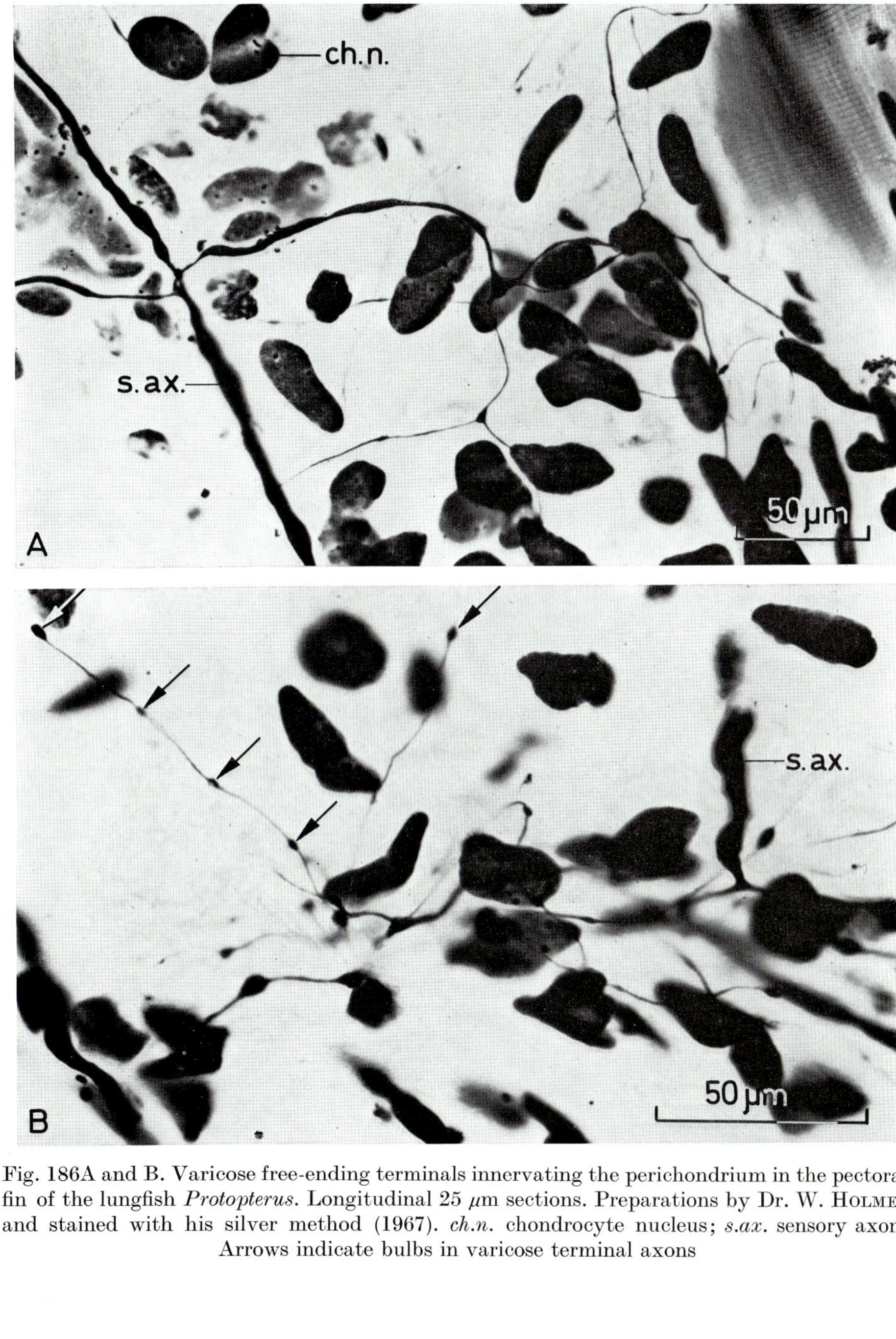

Fig. 186A and B. Varicose free-ending terminals innervating the perichondrium in the pectoral fin of the lungfish *Protopterus*. Longitudinal 25 µm sections. Preparations by Dr. W. Holmes and stained with his silver method (1967). *ch.n.* chondrocyte nucleus; *s.ax.* sensory axon. Arrows indicate bulbs in varicose terminal axons

Fig. 187A and B. Bush-like sensory endings (*terminaisons en buissons*) innervating frog tendons. Drawings of methylene blue preparations. A From the tendon of tibialis posterior. (From REGAUD, 1907, after HUBER and DE WITT, 1900). B From the tendon of supracoracoideus. (From REGAUD, 1907, after SMIRNOFF, 1893)

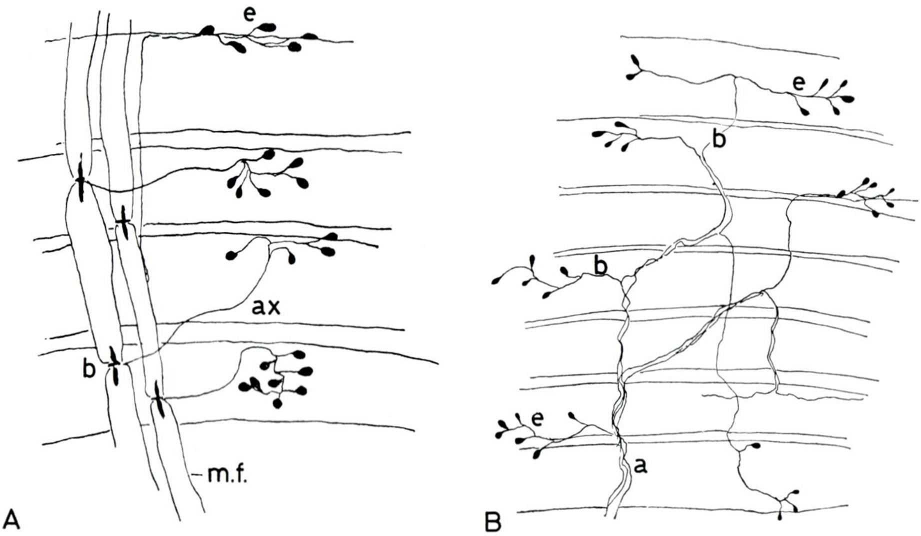

Fig. 188A and B. Extrafusal stretch receptors in frog muscle. Drawings by KULCHITSKY (1924b, Figs. 9 and 10) of methylene blue preparations of leaf-like endings *e*. In A they are formed by short, non-myelinated preterminal axons *ax* derived from the nodes *b* of myelinated afferent fibres *m.f.* In B Kulchitsky shows the endings being formed by a plexus of non-myelinated axons *a*, probably long preterminal axons of myelinated stem afferents

D. Reptiles and Birds

The tendons of reptiles are innervated by large afferent fibres that branch to terminate in a number of plaques. In lizards these measure 60–100 μm long and 40–50 μm wide (REGAUD, 1907). They consist of irregular leaf-like blobs of axoplasm distributed among the tendon fascicles, which in the area of termination are smaller and more densely nucleated than the surrounding tendon (Fig. 189). HUBER and DE WITT (1900) found that in the turtle *Emys* some of the receptors are invested with a thin capsule so as to form rudimentary tendon organs. These generally have more elaborate endings and are confined to the musculo-tendinous junction, while the non-encapsulated simpler plaques occur wholly within tendon.

KULCHITSKY (1924a) describes thin non-myelinated fibres in python muscle that branch extensively to terminate as free endings in the intramuscular connective tissue.

In addition to tendon organs, the tendons of birds are provided with non-encapsulated plaques such as occur in the lizard (HUBER and DE WITT, 1900), and

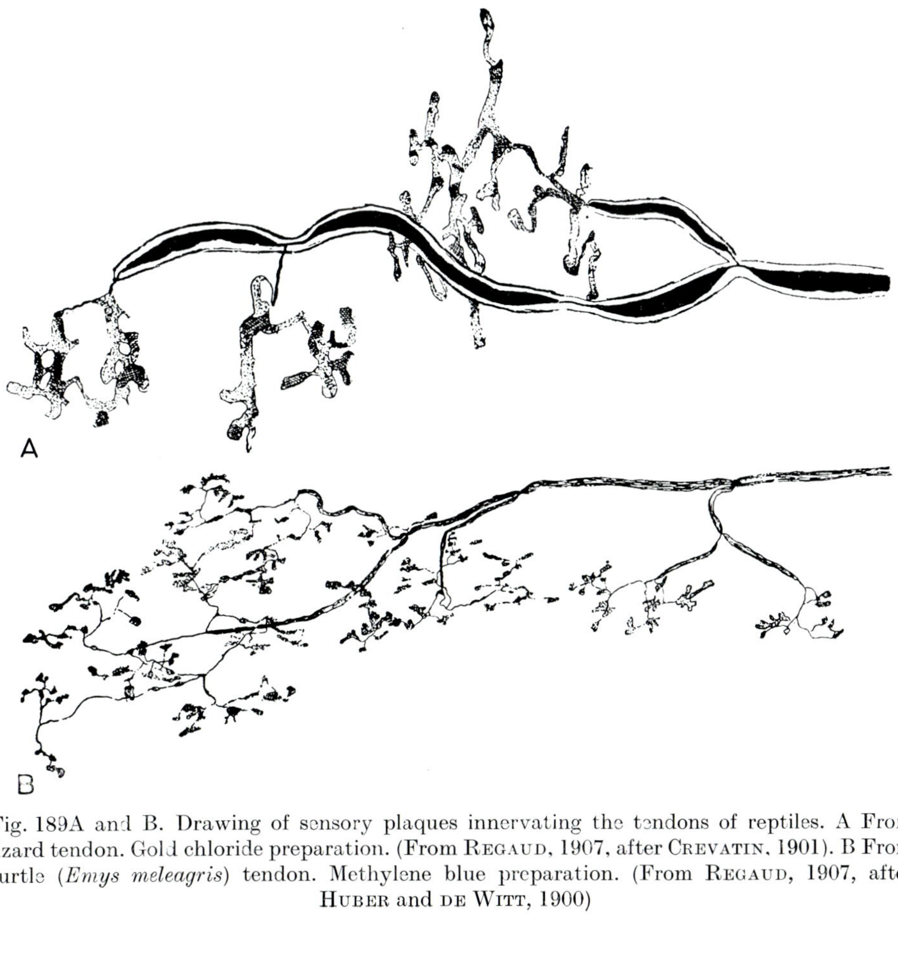

Fig. 189A and B. Drawing of sensory plaques innervating the tendons of reptiles. A From lizard tendon. Gold chloride preparation. (From REGAUD, 1907, after CREVATIN, 1901). B From turtle (*Emys meleagris*) tendon. Methylene blue preparation. (From REGAUD, 1907, after HUBER and DE WITT, 1900)

we may note here, for convenience, that similar plaques are also found in the limb tendons of bats (CIACCIO, 1890; CREVATIN, 1901). Paciniform corpuscles occur in birds and are frequently associated with tendon organs, as in mammals.

E. Mammals

1. Free Endings

As already noted (p. 50), free endings innervate all types of tissue within mammalian muscle except the capillary network (STACEY, 1969). They also innervate tendons, aponeuroses, fasciae, ligaments, and retinacula, the nerve fibres being supplied by muscular, cutaneous, and articular nerves (WEDDELL and HARPMAN, 1940; FEINDEL, WEDDELL and SINCLAIR, 1948; STILWELL, 1957a, b, c, d). Their distribution in muscle has recently been studied by STACEY (1967, 1969), who has shown that the afferent fibres involved comprise the whole of the non-myelinated group IV, and some of the myelinated fibres in groups III, II, and I. The diameter range of 357 group IV afferents in two cat tibialis posterior nerves was 0.15–1.25 μm, with a peak at 0.35 μm. Five myelinated fibres that

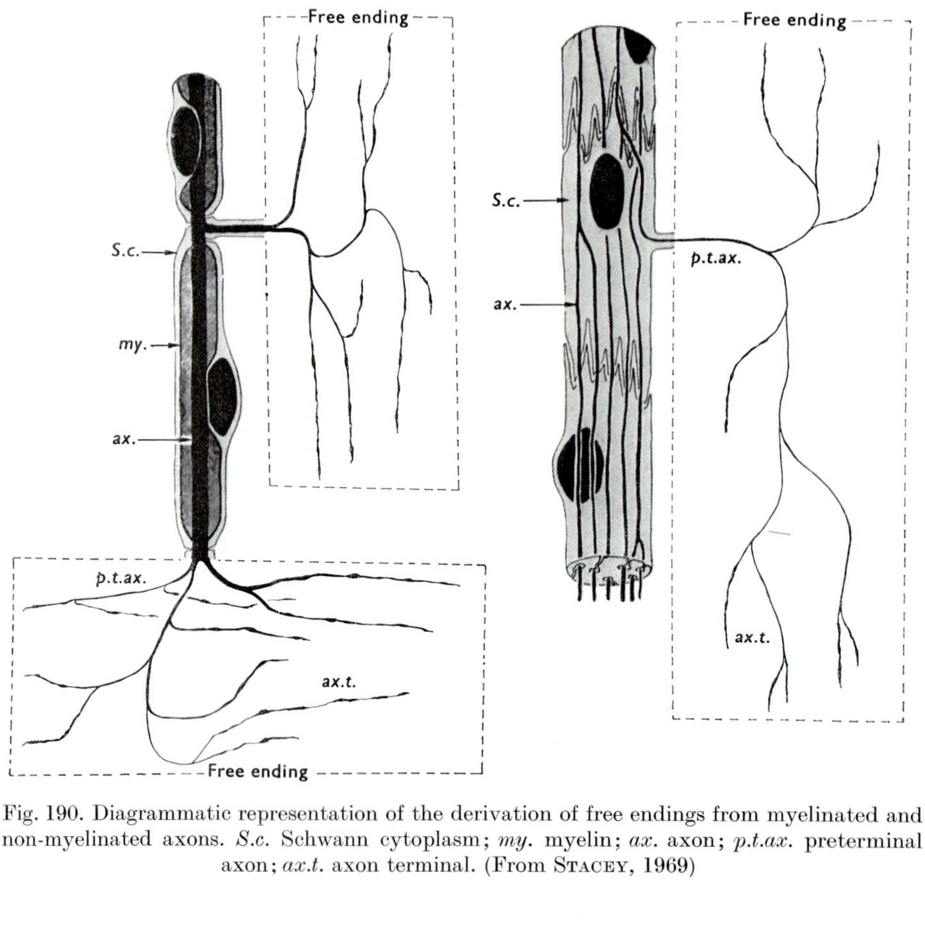

Fig. 190. Diagrammatic representation of the derivation of free endings from myelinated and non-myelinated axons. *S.c.* Schwann cytoplasm; *my.* myelin; *ax.* axon; *p.t.ax.* preterminal axon; *ax.t.* axon terminal. (From STACEY, 1969)

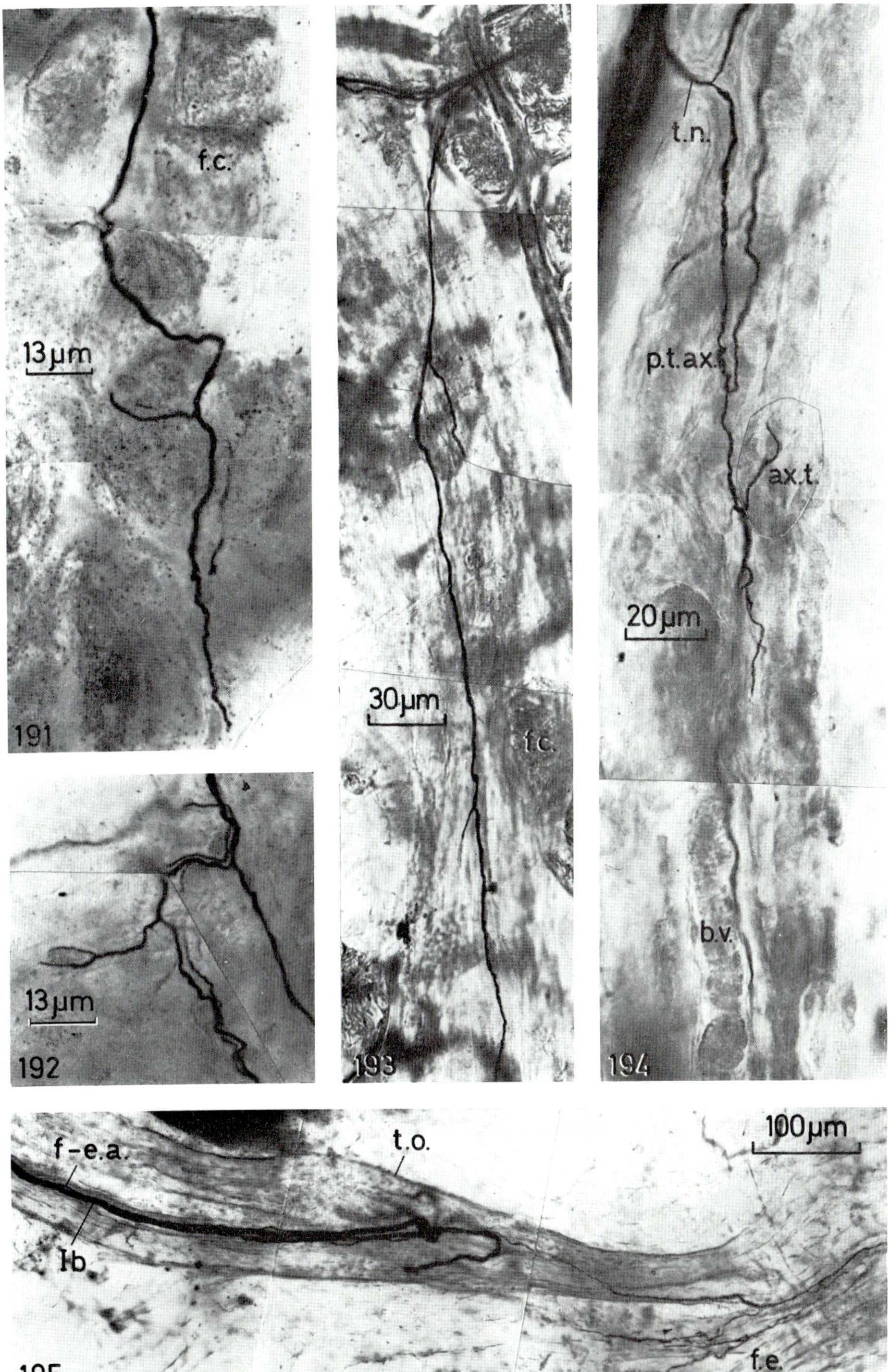
f.c.
13µm
191
13µm
192
30µm
f.c.
193
t.n.
p.t.ax.
ax.t.
20µm
b.v.
194
f-e.a.
t.o.
100µm
Ib
f.e.
195

Plate 43

were traced in teased, silver preparations to their free-ending terminations, or to the level of their 0.5–1.0 μm diameter branches, had nerve-entry stem diameters of 3.0 μm and 4.0 μm (goup III), 7.0 μm and 7.0 μm (group II), and 14.0 μm (group I) (STACEY, 1969).

The participation of large afferents in free-ending innervation is not typical, but has also been observed by STILWELL (1957b) in the Achilles tendon. The fibres were supplied by cutaneous nerves and were "equal in size to any muscle spindle afferent" (i. e. axon diameters of 5–7 μm in his fixed and dehydrated methylene blue preparations, as compared with 7–8 μm for spindle primary afferents, and 4–6 μm for tendon-organ afferents). They branched extensively to cover an area of up to a square millimetre, 15–25 groups of terminals being derived from a single fibre.

STACEY's (1969) diagrammatic representations of free-endings are reproduced in Fig. 190, and some of his photographs are illustrated in Plate 43. He found that the endings derived from non-myelinated fibres cover larger areas (e.g. 300×1000 μm) than those derived from myelinated ones (e.g. 10×110 μm, 150×850 μm). The areas normally overlap, but are widely separate in fasciae, which have few free endings (STILWELL, 1957a). The axon terminals consist of thin, 0.5–1.0 μm diameter varicose threads that usually taper abruptly to a fine point. The terminals are randomly dispersed among e.g. fat cells and intramuscular connective tissue, but are disposed in a clasping fashion among collagenous bundles, as in e.g. tendon and aponeuroses (STILWELL, 1957b). CAUNA (1966) has shown that the free-ending terminals in human skin are enfolded by Schwann cells and contain an accumulation of mitochondria and small vesicles.

Free-ending afferents are not specific as to the type of tissue they innervate; a single fibre may distribute terminals among a variety of different tissues, e.g. fat, blood vessel adventitia, intramuscular connective tissue, tendon, and so on. Free endings located within muscle are not necessarily all supplied by fibres from the muscle nerve. Fasciae are innervated by a plexus of fibres derived from cutaneous nerves and branches of nerves from the underlying muscles. In the interchanges that occur, some of the free-ending afferents derived from cutaneous nerves may enter and terminate within the muscle (STILWELL, 1957a; see his Fig. 1).

Plate 43, Figs. 191–195

Examples of free endings in de-efferented and sympathectomized cat hindlimb muscles. Figs. 191–194 from STACEY (1969); Fig. 195 from STACEY (1967). *ax.t.*axon terminal; *b.v.* blood vessel; *f.c.* fat cell; *f.e.* free ending; *f-e.a.* free-ending afferent; *p.t.ax.* preterminal axon; *t.n.* terminal node; *t.o.* tendon organ; Ib tendon-organ afferent

Fig. 191. Free ending in association with fat cells

Fig. 192. Free ending in connective tissue. This ending together with that illustrated in Fig. 191 was derived from the same free-ending afferent whose stem diameter was 7.0 μm

Fig. 193. Free endings among connective tissue that contains some fat cells

Fig. 194. Five axon terminals of a free ending associated with small blood vessels. The diameter of the myelinated fibre that supplies the ending is 3.0 μm just prior to the terminal node

Fig. 195. Free-ending terminals in tendon close to a tendon organ. Note the 1.5 μm free-ending afferent travelling with the Ib afferent. From normal flexor digitorum brevis muscle

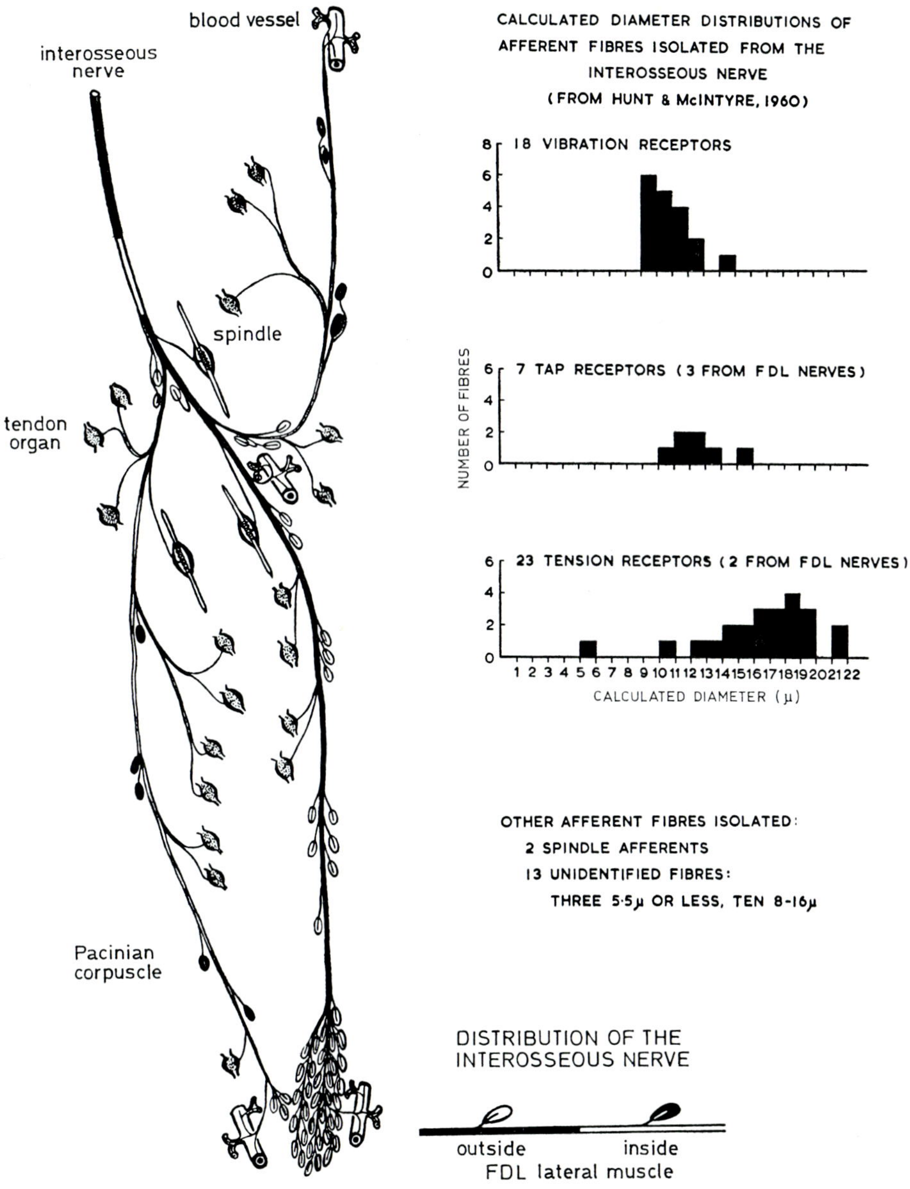

Fig. 196. Schema illustrating the branching of the interosseous nerve of the lower leg of the cat, and the receptors innervated by it, based on a study of fourteen gold chloride preparations. The histograms are those published by HUNT and McINTYRE (1960, Fig. 8) showing the calculated diameter distributions of afferent fibres in the interosseous nerve innervating vibration, tap, and tension receptors. (From BARKER, 1962b)

The incidental association of free-endings with muscle spindles and tendon organs has already been noted (see § II A, 3 and § III).

2. *Pacinian and Paciniform Corpuscles*

Pacinian corpuscles occur on the surfaces of fasciae, aponeuroses and tendons (Stilwell, 1957a, b), but are rarely found within muscle. The FHL muscle is exceptional in containing several that are supplied by branches of the interosseous nerve. In the cat this contains 150–200 afferent fibres distributed in two intramuscular FHL branches and one extramuscular branch that runs between the perimysium of the muscle and the tibial periosteum and interosseous membrane (Barker, 1962b; see Fig. 196). The nerve supplies an average of 20 tendon organs, 3 spindles, a number of free endings, and 60 Pacinian corpuscles, of which an average of 9 are intramuscular. Most of the corpuscles are located in an extramuscular cluster near the ankle (Plate 44, Fig. 200). The majority are large with up to 60 lamellae, but some are less than half this size, though not as small as paciniform corpuscles. The frequency of occurrence of the large and small types suggests that they correspond, respectively, to the vibration and tap receptors identified from single-fibre recordings by Hunt and McIntyre (1960) (see Barker, 1967, p. 19).

The paciniform corpuscle is a much smaller version of the classic Pacinian receptor, the differences being its cylindrical rather than oval shape, the number of capsule lamellae (4–8 instead of 20–60), and the shorter axon terminal (Plate 44). In both receptors the myelinated afferent gives rise to a long, straight terminal, bulbous and occasionally branched at its tip, that is embedded in a finely lamellated, granular inner core surrounded by the concentrically arranged layers of the capsule. Paciniform corpuscles are distributed in a number of different tissues including tendons, aponeuroses, muscle, and the fibrous capsules of joints. They are sparsely distributed in muscle where they occur mostly at the musculotendinous junctions. The average number in the cat soleus muscle is 3, in the Vth interosseous muscle of the foot 8, and in rectus femoris 12, equivalent, respectively, to 4%, 31% and 12% of the number of spindles in the same muscles (Barker, 1962b).

When Miss M. Pang made her analysis of tendon organs (see p. 128), she also examined gold chloride preparations of 253 paciniform corpuscles from cat rectus femoris, semitendinosus and pes interosseous muscles. The mean dimensions of 100 of the receptors were 177.5 μm long by 21.2 μm wide. Of the total sample, 174 (68.8%) were found in clusters of 2 (most frequently) to 9 supplied by the branches of one nerve fibre. The majority (185 or 73%) were situated on or within tendon organs, an association already noted (p. 128). The diameter of 40 stem

---

Plate 44, Figs. 197–200

Pacinian and paciniform corpuscles

Fig. 197. Photograph of one Pacinian corpuscle, and part of another, innervated by the cat's interosseous nerve. (Teased, silver preparation by Dr. M.C. Ip)

Fig. 198. Paciniform corpuscle located in the intramuscular connective tissue of a de-efferented and sympathectomized cat hindlimb muscle. Teased, silver preparation. (From Stacey, 1967)

Fig. 199. Gold chloride preparation of a group of Pacinian corpuscles innervated by the cat's interosseous nerve. (Preparation by Miss M. Pang)

Fig. 200. Transverse section (10 μm) through the distal end of the cat's flexor hallucis longus muscle showing the extramuscular terminal cluster of Pacinian corpuscles innervated by the interosseous nerve. Haematoxylin and eosin stain. (Preparation by Mr. D. Chi)

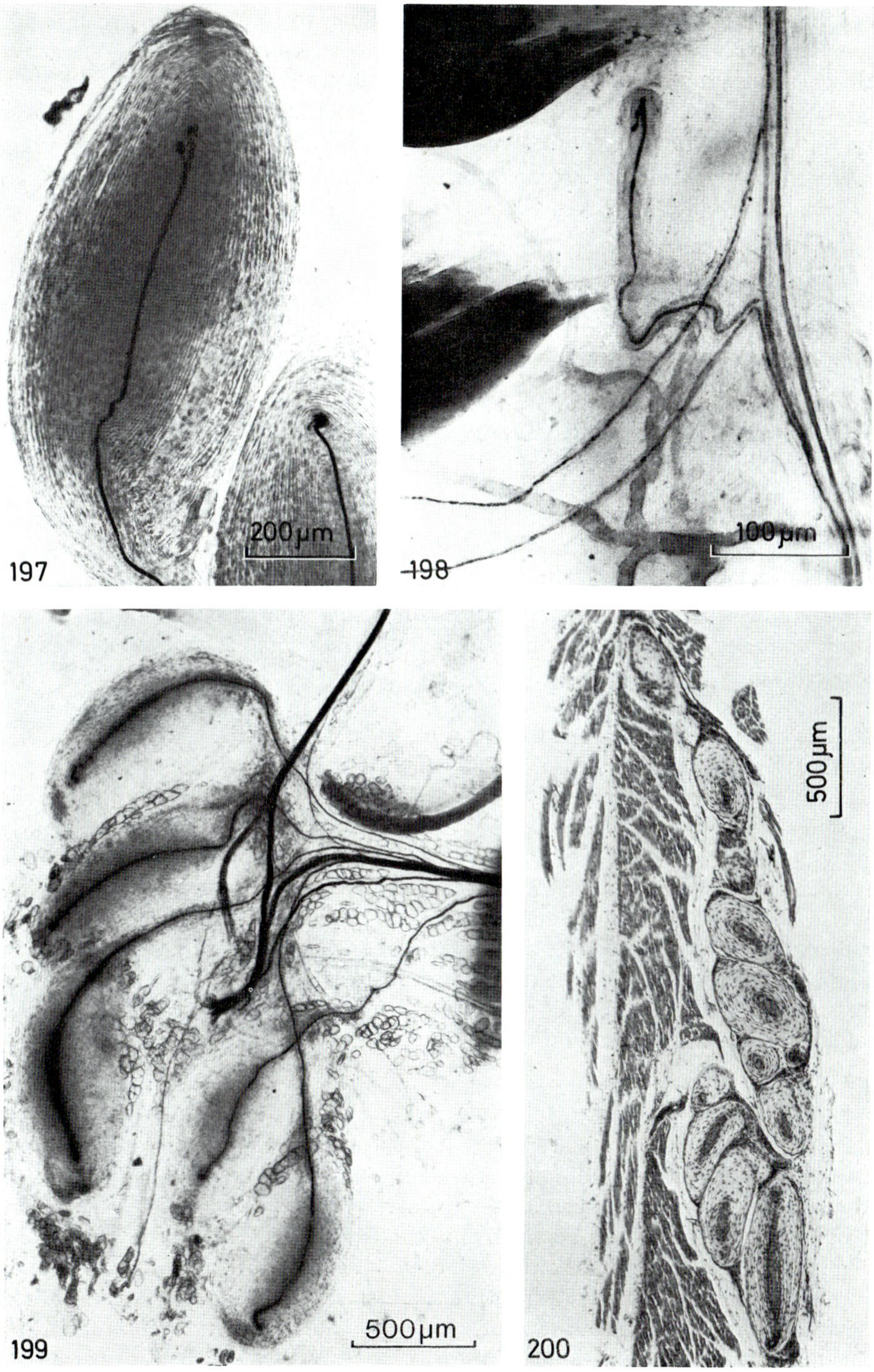

Plate 44

fibres innervating the receptors ranged from 2.8 μm to 11.2 μm, mean 7.2 μm. In a comparable analysis of 65 paciniform corpuscles examined in teased, silver preparations from various cat hindlimb muscles, STACEY (1969) found that 42 (65%) were associated with tendon organs. The diameters of 25 stem fibres ranged from 3.0 μm to 12.0 μm (i.e. belonging to groups I, II, and III).

An arbitrary distinction between paciniform and Pacinian corpuscles is made on the basis of the number of capsule lamellae. Thus HROMADA and POLÁČEK (1958) designate as Pacinian all corpuscles with 15 or more lamellae. The most frequent (69.3%) type of paciniform corpuscle found by them in the joint capsules and periarticular tissue of various mammals had a capsule of 2–5 lamellae, and it is this type that is most common in muscle and tendon. Their Fig. 2 illustrates the various forms of the receptor that they observed in joint capsules.

FITZGERALD (1961, 1962) studied paciniform corpuscles in the epidermis of the pig's snout and concluded that some are in process of degenerating while others are growing, there being a process of 'cyclic growth and decay' among both these receptors and free endings. Similarly TELLO (1932) observed both normal and degenerating paciniform corpuscles in human external genitalia, and concluded that the receptors were continually being renewed.

HROMADA and POLÁČEK (1958) obseved that non-myelinated fibres occasionally form an irregular plexus around paciniform corpuscles terminating on the capsule and among the surrounding blood vessels and connective tissue. These are almost certainly free endings. A thin non-myelinated fibre may also enter paciniform and Pacinian corpuscles and ramify within the central core. Both DOGIEL (1902) and BARKER (1948) describe instances of this type of innervation, and SANTINI has recently demonstrated both by fluorescence (1968, 1969) and electron microscopy (SANTINI, IBATA and PAPPAS, 1971) that it is provided by postganglionic sympathetic fibres. These have a mean conduction velocity of 0.7 m/sec (LOEWENSTEIN, GOTO and NOBACK, 1962), and their stimulation has the effect of facilitating the response of the corpuscle afferent (GOTO and LOEWENSTEIN, 1961), as occurs with sympathetic fibres innervating touch receptors in frog skin (LOEWENSTEIN, 1956). This additional innervation is not evident in the majority of silver preparations of Pacinian and paciniform corpuscles. Likewise GOTO and LOEWENSTEIN (1961) were able to produce the facilitatory effect in only 12 of 91 Pacinian corpuscle preparations.

3. Golgi Endings

In GOLGI's descriptions (1878, 1880) of the tendon organ one end is always muscular and the other tendinous, and he himself called it a 'musculo-tendinous end-organ'. However, a few years later RUFFINI (1894) found a very similar receptor in human subcutaneous connective tissue (Fig. 201A, B), the only difference being that the nerve terminals are distributed among a thinly encapsulated bundle of collagenous fibres in loose connective tissue instead of among a bundle of tendinous collagen attached to a few muscle fibres. Later SFAMENI (1902) found the same receptor in the knee-joint of the dog and named it a 'Ruffini corpuscle'. However, it is apparent both from Sfameni's description and RUFFINI's own later account (1905) that this term was used to include another type of receptor now

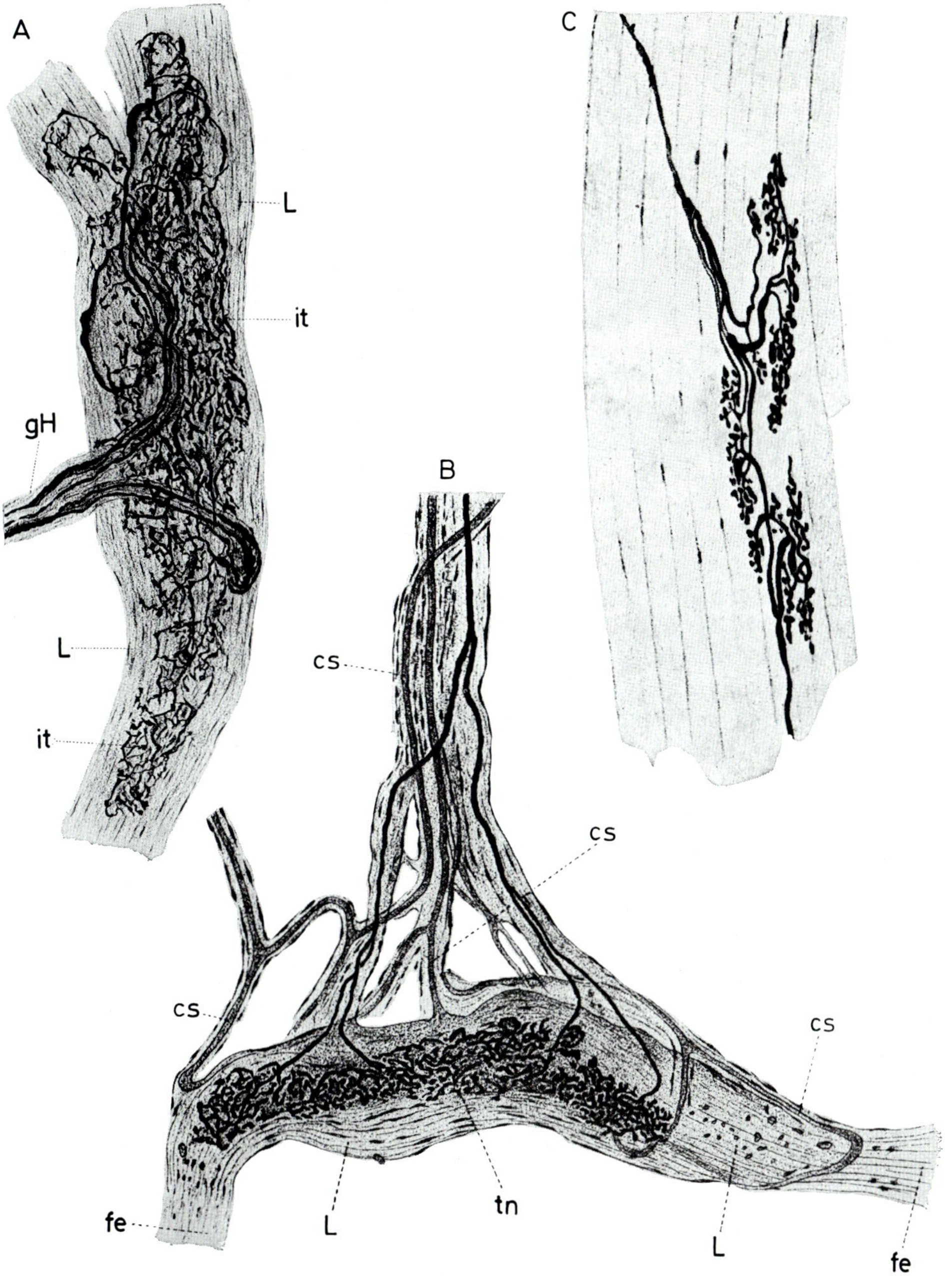

Fig. 201A–C. A, B Golgi endings. Drawings by Ruffini (1894, Figs. 2 and 7) of gold chloride preparations of these receptors in human finger-pad subcutaneous connective tissue. *cs* capillaries; *fe* elastic fibres; *gH* perineural sheath; *it* and *tn* terminal axon branches and axon terminals of the sensory ending; *L* bundle of collagenous fibres in loose connective tissue. C Dogiel endings. Drawing of two afferent fibres terminating in the intramuscular connective tissue of a dog intercostal muscle. Methylene blue preparation. (From Regaud and Favre, 1904, after Dogiel, 1903)

known as the 'Ruffini ending' (see § IV E, 4). The term 'Golgi ending' rather than 'Ruffini corpuscle' is preferred since it serves to emphasize the similarity between the tendon organ, which is so firmly linked with his name, and the receptor that Ruffini (1894) originally described. The similarities include not only the shape and size of the receptor, and the appearance of its terminals, but the fact that it is usually supplied by a single, thick, unbranched afferent.

Golgi endings occur in ligaments, tendons, aponeuroses, fasciae and retinacula (Andrew, 1954; Stilwell, 1957a, b, c, d; Ralston, Miller and Kasahara, 1959; Freeman and Wyke, 1967). In tendons, aponeuroses and fasciae they are supplied by afferents derived extramuscularly from cutaneous nerves (Stilwell, 1957a, b). In cat knee-joint ligaments and related tendons they are supplied by 13–17 μm diameter afferents, and have average dimensions of 600×100 μm (Freeman and Wyke, 1967). Boyd (1954) gives dimensions of 500×125 μm with afferents of 12 μm diameter (cat, cruciate ligament).

Dogiel Endings. Dogiel (1902, 1903) has described the occurrence of non-encapsulated endings in the intramuscular connective tissue of the abdominal and intercostal musculature of various mammals. They are supplied by thick afferents and the terminal arborizations are very similar in appearance to those in tendon organs (see Fig. 201C). Regaud and Favre (1904) describe them as resembling a branch of butchers' broom ('une branche de petit-houx') and refer to them as the 'arborisations en buissons de Dogiel'. Stilwell (1957a) shows an ending of this type present in the tibialis anterior muscle of a monkey in his diagrammatic Fig. 1. He refers to it as a 'flower-spray ending' in the legend, but does not describe such endings in the text. Their terminals are very similar to those of Golgi endings, and they are included here despite their apparent lack of a capsule. Dogiel (1902) stresses their intramuscular location; he rarely found them in musculo-tendinous regions or in tendon.

4. *Ruffini Endings*

Unlike Golgi endings, Ruffini endings (Fig. 202) are not distributed among collagenous bundles, but terminate in a granular amorphous matrix enclosed within a thin capsule. They are also smaller, globular or ovoid rather than cylindrical or fusiform, and occur in clusters supplied by branches of one medium-sized afferent. Alternative descriptive terms are 'spray endings' and 'ball-of-thread endings'; they have also been called 'Golgi-Mazzoni corpuscles'.

Fig. 202A–D. Ruffini endings. A, B Drawings by Ruffini (1894, Figs. 16 and 18) of gold chloride preparations of these receptors in human fingers. The thinly-encapsulated form in A is from subcutaneous finger-pad tissue; the pair of endings with thicker capsules is from the surface of a tendon. C Gold chloride preparation of an afferent fibre supplying five Ruffini endings in the posterior capsule of the cat's knee joint. Camera lucida drawing; receptor capsules not shown. (From Skoglund, 1956, Fig. 9). D Methylene blue preparation of a pair of Ruffini endings from the posterior capsule of the cat's knee joint. *A*, *E* non-myelinated accessory axons, probably of postganglionic sympathetic origin; *B* axon terminals of Ruffini ending; *C* capillary; *D* myelin sheath of *F*, the receptor afferent. Camera lucida drawing; receptor capsule not shown. (From Gardner, 1944, Fig. 14)

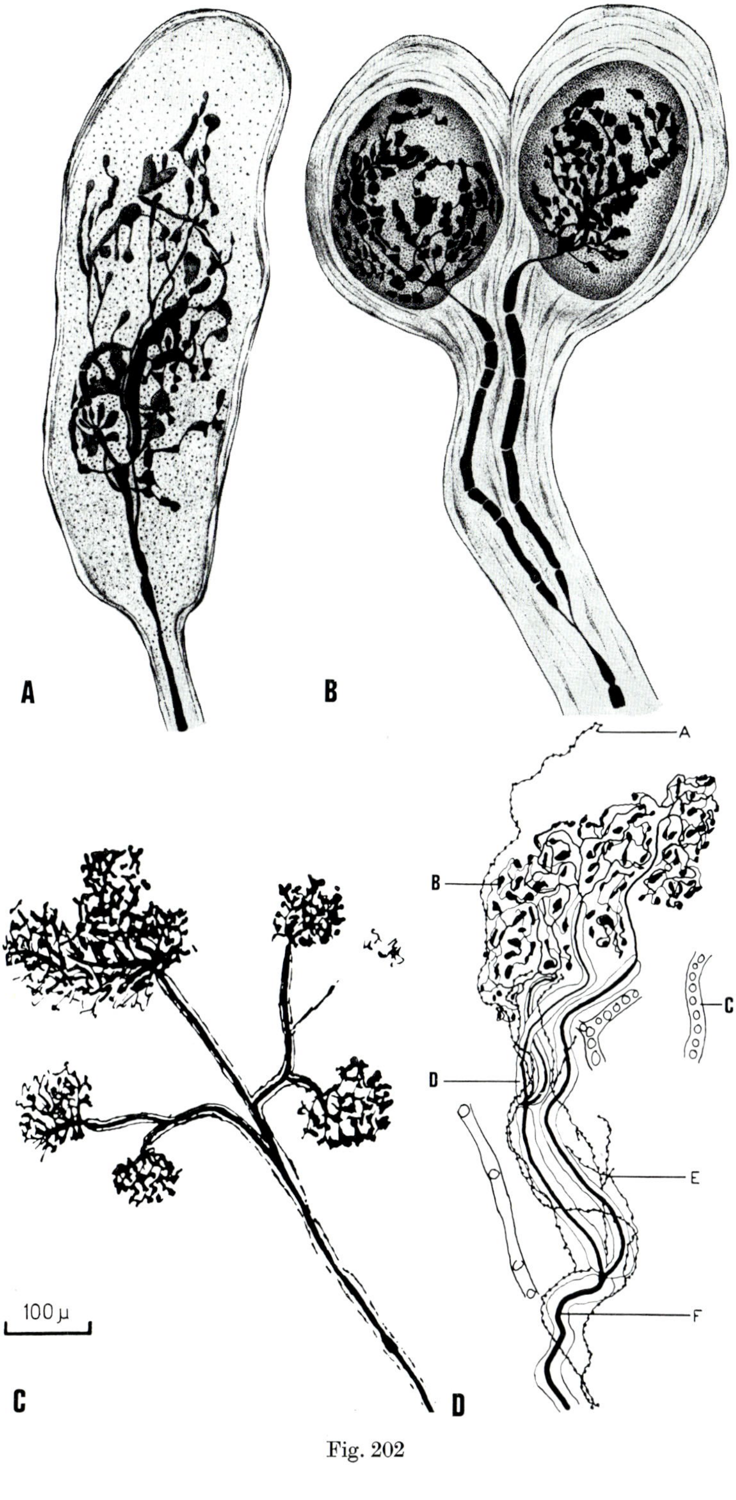

Fig. 202

Stilwell (1957b) has proposed that spray endings, Golgi endings and tendon organs should all be called Ruffini endings! Though it is a pity that Ruffini did not fare better in the eponymous scrambles of his day, this reform is altogether too confusing and has fortunately found no support. Freeman and Wyke (1967) resort to a numerical classification in which Ruffini (spray) endings are Type I, paciniform corpuscles Type II, Golgi endings Type III, and free endings Type IV, but it is doubtful whether many will want to adopt this scheme.

Ruffini endings are present in the fibrous capsules of joints and associated ligaments and tendons (Freeman and Wyke, 1967). Stilwell's figures make it clear that they are also present in fasciae (1957a, Figs. 3, 8, 10), tendon (1957b, Figs. 4, 19; 1957c, Fig. 1), tendon sheaths (1957c, Fig. 14; 1957d, Fig. 2), and retinacula (1957c, Fig. 8). In these locations they are innervated by afferents derived from an extramuscular source. The endings in tendon situated between the collagenous bundles assume an elongated shape (Stilwell, 1957b, Fig. 19).

The mean dimensions of Ruffini endings innervating the cat's knee joint are 100×40 μm (Freeman and Wyke, 1967), and they are found in clusters of up to seven (Boyd, 1954; Skoglund, 1956; Freeman and Wyke, 1967). The diameter range of the supplying afferents is 5–10 μm (Skoglund, 1956); Boyd (1954) found that the mean diameter of 35 afferents was 6.3 μm.

The thin capsule that encloses each terminal arborization is usually composed of only one or two lamellae, and is often not visible in gold chloride preparations, as in Fig. 202C. Thicker capsules of five or six lamellae also occur (Fig. 202B), and both thinly and thickly encapsulated forms of the receptor may be innervated by branches of the same afferent (see Lambertini, 1961, Figs. 15, 17 and 18).

An accessory innervation is occasionally present consisting of one or two thin non-myelinated axons that accompany the afferent into the capsule and end amongst its terminals (Gardner, 1944; Andrew, 1954; Skoglund, 1956). Gardner observed branches of such axons ending among capillaries surrounding the receptors (see Fig. 202D). They are probably of postganglionic sympathetic origin as in Pacinian and paciniform corpuscles.

5. *Eye Muscle Endings*

A variety of receptors may occur in mammalian extraocular muscles other than muscle spindles (see § II A, 7) and tendon organs. There are palisade endings in the musculo-tendinous regions, sensory endings on extrafusal muscle fibres, and various endings in intramuscular connective tissue and tendon.

a) Palisade Endings. In his study of sensory endings in the eye muscles of man, monkey, horse, ox, dog and cat, Dogiel (1906b) describes the occurrence of thick myelinated fibres that travel from muscle into tendon and then turn back to form palisade endings at the ends of the muscle fibres (Fig. 203). The terminals of each receptor surround the end of a single muscle fibre somewhat after the fashion of the basket motor endings in fish and Amphibia (see e.g. Bone, 1964; Barker, 1968a). In some cases branches of the sensory fibre terminate a short distance away from the palisade formation either on the muscle fibre or in adjacent tendon (Fig. 203B). Dogiel observed that the most elaborate palisade endings (Fig. 203A) innervated the ends of the thickest muscle fibres and were supplied by the thickest afferents.

Prior to Dogiel's work, similar endings had been described in the eye muscles of the cat by Huber (1900) and in those of the camel by Crevatin (1902). Huber describes the terminals as mostly lying just distal to the ends of the muscle fibres, each receptor being thinly encapsulated and associated with a tendinous fascicle attached to a single muscle fibre. He found the endings to be numerous in the rectus and oblique muscles, but absent from the retractor bulbi. Tozer and Sherrington (1910) observed palisade endings in the eye muscles of the cat, rabbit, and monkey and confirmed Dogiel's (1906b) findings. Cooper, Daniel and Whitteridge (1955) include typical palisade endings in their schematic diagram of the innervation of a human rectus eye muscle (Fig. 204), but do not describe them in the text. They are absent from sheep eye muscles (D.W. Harker, personal communication).

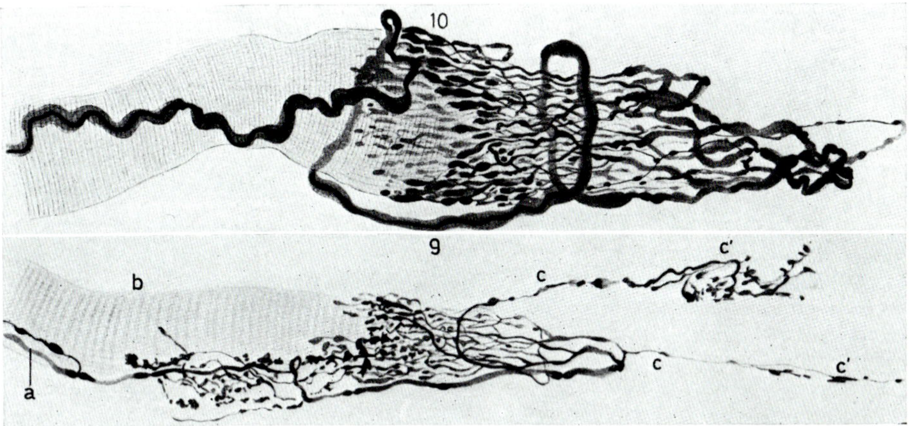

Fig. 203. Drawings by Dogiel (1906b, Figs. 9 and 10) of palisade endings in the rectus eye muscles of the horse. In the lower drawing some of the branches (*c*) of the afferent fibre form endings (c') in nearby tendon. Methylene blue preparations

Sas and Scháb (1952) have investigated the possibility that palisade endings are motor, but found no evidence to support it. Stimulation of one of the nerve branches innervating the cat's inferior oblique muscle, known to consist solely of palisade afferents, produced no visible contraction. Tozer and Sherrington (1910) describe the endings as being "of obviously receptive (sensorial) type", and the fact that branches of the afferents often terminate exclusively in tendon strengthens this impression. The presence of cuffs of cholinesterase at the ends of skeletal muscle fibres has been demonstrated by Couteaux (1953), but there is no evidence to indicate that in extraocular muscles they are specifically related to palisade endings, as suggested by Cheng (1963).

It will be recalled that in Tello's (1917, 1922) description of tendon-organ development, the Ib afferent initially forms terminals around the ends of the myotubes (see p. 133). Later this contact is lost, and the terminals ramify among the bundles of collagen that ultimately constitute the receptor. In the development of palisade endings certain factors evidently operate to ensure that the original contact of the terminals with the myotubes is retained and elaborated.

It is possible that palisade endings are musculo-tendinous stretch receptors associated with non-twitch muscle fibres, the functional relationship being similar to that obtaining between tendon organs and twitch muscle fibres. There are some indications that this may be so. Palisade endings are absent from the cat retractor bulbi muscle (Huber, 1900), which lacks a non-twitch component (Bach-y-Rita and Ito, 1965), and they are also absent from skeletal muscles, which consist

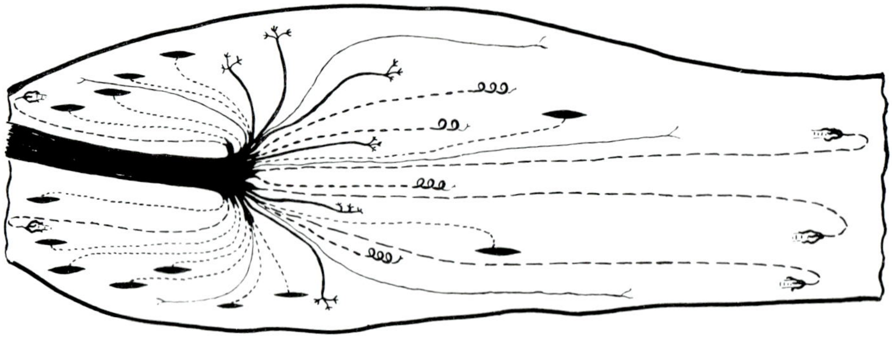

Fig. 204. Much simplified diagram by Cooper, Daniel and Whitteridge (1955) of a human rectus eye muscle, full length (origin, left: insertion, right). The supplying nerve is seen entering the muscle at the junction of the proximal and middle thirds. Muscle spindles are most numerous peripherally and in the proximal third of the muscle; their afferent supply is indicated by dotted lines. In the middle third of the muscle are the motor end-plates whose nerve fibres are shown as solid black lines (grape motor endings and collateral motor supply to spindles not shown). Simple spiral endings are shown lying just distal to the band of motor end-plates. Palisade endings are shown at the musculo-tendinous junctions at each end of the muscle with their afferents indicated by broken lines. The lines ending in small Ys represent nerve fibres supplying some of the other endings that are to be seen in eye muscles

entirely of twitch fibres. The ends of grape-innervated muscle fibres are thick, and it seems more likely that it was these that Dogiel (1906b) observed with palisade endings (as in Fig. 203A) than the ends of thick A fibres. An alternative possibility is that palisade endings are restricted to the orbital rim zone of rectus and oblique muscles, where they innervate the ends of both twitch (C) and non-twitch (G) fibres, while tendon organs are restricted to the central core. The presence of tendon organs in eye muscles has been reported for a number of species (see Dogiel, 1906b; Sampaolo, 1952; Bonavolonta, 1956b, 1958). According to Dogiel (1906b) they are most numerous in the rectus muscles of cattle.

b) Sensory Endings on Extrafusal Muscle Fibres. Belief in the sensory nature of the grape endings in extraocular muscles persisted until the early 'sixties (see e.g. Daniel, 1946; Hosokawa, 1961) when the use of the cholinesterase technique by Kupfer (1960) and Hess (1961b) established that they were motor. Nevertheless there are sensory endings on ordinary muscle fibres in these muscles, and it is understandable that their presence in addition to grape endings should have given rise to some confusion. Most of the information about them has come from studies of human material and is contained in the papers by Daniel (1946) and Sas and Appeltauer (1963). Dogiel (1906b, man and various mammals) includes descriptions of some of the endings in his account of the grape innervation, which he regarded as sensory. The endings occur in the form of simple or complex spirals (Daniel, 1946), and some of the latter may be encapsulated (Sas and Appeltauer, 1963).

Daniel describes the simple spirals as consisting of 3–8 complete turns encircling small muscle fibres, and the more complex and extensive spirals as terminating most frequently on muscle fibres of large or medium diameter. A medium-sized axon may branch to supply simple spirals on three separate muscle fibres, and the terminals of complex spirals may be distributed to two adjacent muscle fibres (Sas and Appeltauer, 1963). Other reports of spiral endings in extraocular muscles are those of Pallot (1934) in cat; Hirano (1941) and Wolter (1955) in man; Cooper and Fillenz (1955) in cat and mangabey monkey; Sampaolo (1952) in cow; and Mahran and Sakla (1965) in the albino mouse.

Most of the complex endings are very irregular with few spirals, as is evident from Dogiel's (1906b) Figs. 2 and 3 reproduced in Fig. 205A, B. Sas and Appeltauer (1963) describe another type of spiral formation present in human extraocular muscles that is enclosed within a thin fusiform capsule approximately 100 μm long by 50 μm wide. One or two axons enter the capsule, branch and encir-

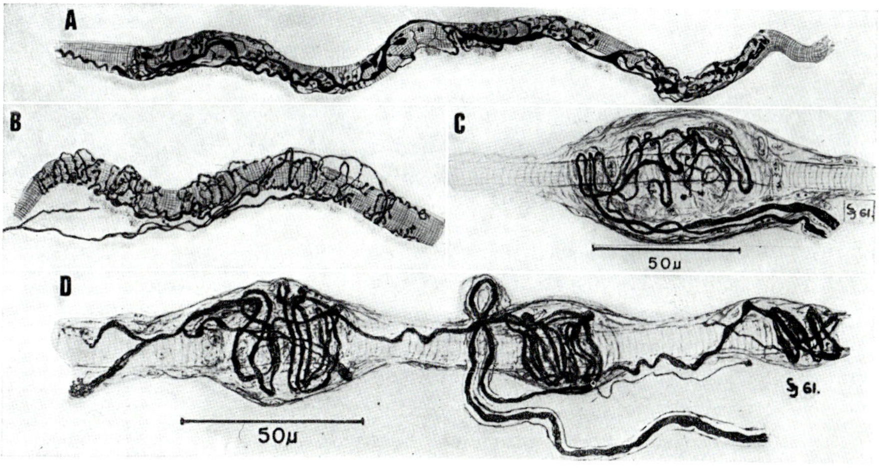

Fig. 205A–D. Extrafusal sensory endings in eye muscles. A, B Drawings by Dogiel (1906b, Figs. 2 and 3) of complex endings in the rectus eye muscles of the horse. Methylene blue preparations. C, D Encapsulated spiral formations in human eye muscles. Drawings of silver impregnated longitudinal sections. (From Sas and Appeltauer, 1963, Figs. 2 and 5)

cle the muscle fibre several times within an intracapsular space, and then end on its surface as little bulbs (Fig. 205C). There is no special nucleation in the region of termination. Three such formations may occur in series on the same muscle fibre supplied by branches of one axon (Fig. 205D), or may be distributed to separate muscle fibres. Some of the spiral formations are more complex, the encircling fibres and terminals being enclosed in a capsule about 800 μm long by 300 μm wide.

In the cat and mangabey monkey the spiral endings are located on small muscle fibres peripherally situated in the orbital rim zone (Cooper and Fillenz, 1955). The simple spirals in human extraocular muscles are also peripherally located on small fibres (Sas and Appeltauer, 1963). In sheep superior rectus the orbital rim zone is composed of A and C twitch fibres as well as G fibres (Harker, 1972a), and though all the fibres are small (diameter 9–15 μm) there is a diameter distribution in which the A fibres are the largest, G fibres the smallest, and C fibres intermediate. The innervation of twitch fibres by spiral endings cannot therefore be excluded, especially in view of Daniel's (1946) observation that the complex spirals most frequently occur on fibres of large and medium diameter. Spindle primary endings are distributed to both G and C-type fibres (see § II A, 7), and it may be that the spiral endings are also distributed to both twitch and non-twitch fibres. Thus in cat eye muscles, which lack spindles, the role of the muscle stretch receptor, as Cooper and Fillenz (1955) showed, is performed by the spiral endings, whereas in sheep eye muscles, which lack spiral endings (D.W. Harker, personal communication), this function is served by the spindles. Alternatively, both systems may operate, as in the eye muscles of man and the albino mouse.

The spiral endings are referred to as 'simple neuromuscular spindles' or 'atypical muscle spindles' by Sas and Appeltauer (1963). Wolter (1955) calls them 'sensory muscle spindles'. This is a confusion best avoided. In that a sensory axon and a muscle fibre are involved, and, in some cases, a capsule, there is obviously some resemblance to the monofibral type of spindle such as occurs in snakes and lizards. But there is no equatorial nucleation, as is characteristic of intrafusal muscle fibres, and evidently no motor innervation specifically distributed on either side of the spiral ending.

c) Endings in Intramuscular Connective Tissue and Tendon. The intramuscular connective tissue of eye muscles is richly supplied with sensory endings. Among these are Ruffini endings (see Dogiel, 1906b, Fig. 15B; Sas and Appeltauer, 1963, Fig. 6). The endings described by Wolter (1955, Fig. 3) as 'flower-bud-like sensory end-plates' are probably receptors of this type; he states that they are very common in human eye muscles. The 'terminaisons en buissons de Dogiel' (see p. 148) are also present (see Dogiel, 1906b, Fig. 13). The 'aboreal sensory terminations' and 'brush-like sensory end-structures' described by Wolter (1955) appear to be of this type, though they are enclosed within a thin capsule. Wolter also describes the occasional occurrence of encapsulated 'sensory spools' and 'sensory end-bulbs'. The spools consist of irregular coils formed en route by an axon; several spools may be formed during the course of a single axon.

Golgi endings and Ruffini endings may occur in the tendons of eye muscles (see Dogiel, 1906b, Figs. 14A, B and 15A).

V. The Composition of Mammalian Muscle Nerves

This chapter began with some remarks about the relative proportions of motor and sensory nerve fibres in cat hindlimb muscle nerves (see p. 2). Having fully described the receptors present in mammalian muscle, it is now possible to examine the composition of the muscle nerves in more detail.

After appropriate fixation and staining for myelin, the number and diameter of the myelinated fibres in a muscle nerve can be ascertained from transverse sections, and a histogram prepared of their size-frequency distribution, i.e. a fibre-diameter spectrum. In a study of twenty-five rabbit muscle nerves FERNAND and YOUNG (1951) found the size-frequency distribution of the fibres to be either bimodal or unimodal. Bimodal spectra were obtained for most limb muscle nerves and nerves to extraocular muscles, unimodal spectra for nerves to muscles of the face, larynx, infrahyoid region and diaphragm (Fig. 206). They suggest that unimodal nerves supply muscles in which proprioceptors are few or absent, and this is certainly true so far as spindle content is concerned (see e.g. BAUM, 1900; BOWDEN and MAHRAN, 1956; COOPER, 1960, Table 1). The presence of the grape motor component in extraocular muscle nerves leads to a much higher percentage of fibres below 8 μm diameter than occurs in limb muscle nerves.

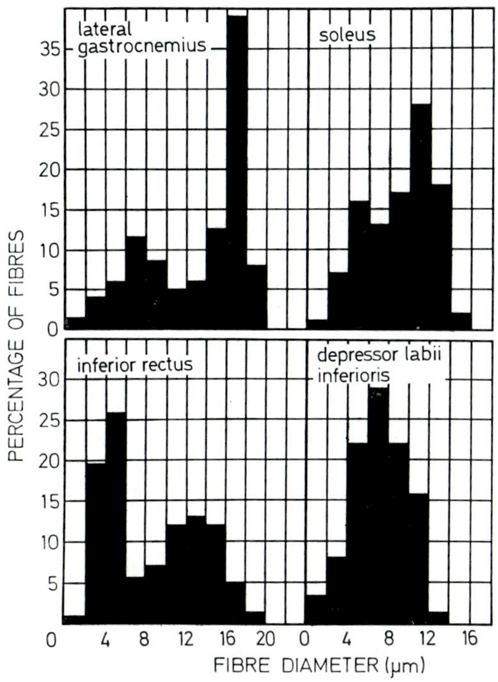

Fig. 206. Myelinated fibre-diameter spectra for two hindlimb muscle nerves, an eye-muscle nerve, and a facial nerve of the rabbit. (From FERNAND and YOUNG, 1951; selected spectra from Figs. 1, 3 and 4 redrawn)

In some distal muscle nerves the fibre-diameter spectrum is only incipiently bimodal. One reason for this may be decrease in diameter owing to the conical tapering of fibres that occurs as they travel distally from the spinal roots. Such tapering may account for the fact that the femoral group of muscles studied by LLOYD and CHANG (1948) received larger afferents than the more distal crural group. Similarly in FERNAND and YOUNG's (1951) study the aggregate of all fibres above 14 μm was 35.5% of all fibres in nerves to the thigh as compared with 29.4 % in the shank. They suggest that tapering may partly account for the incipiently bimodal spectrum they obtained for the soleus nerve in which the large fibres peak at 10–12 μm and there is no fibre larger than 16 μm (Fig. 206). They also suggest that distal muscles such as soleus and the interossei may have small nerves because they arise from small cells at the caudal end of the spinal cord where there are fewer influences stimulating cell size than occur in the more proximal lumbar segments. However, most of the nerve fibres supplying soleus are derived from the same level of the cord (L_7) as those supplying gastrocnemius (SHERRINGTON, 1892; SPRAGUE, 1958), and the nerves to both the lateral (Fig. 206) and medial heads of this muscle have a clearly biomdal spectrum with large fibres peaking at 16–18 μm. If tapering is an important factor affecting fibre size in the soleus nerve, one would expect a similar diameter reduction to affect the gastrocnemius nerves whose origin from the tibial nerve is only slightly more proximal. As we shall see, there are other factors affecting the fibre-diameter spectra of nerves that override tapering, the most important being the proportional distribution of different types of motor unit in the muscles supplied by the nerves concerned.

A. Analysis of Separate Components

1. Methods of Study

It is rarely possible for the histologist to trace individual nerve fibres from the muscle nerve to their endings within the muscle. Most muscle nerves contain too many fibres for their composition to be analysed in this way. It is possible, however, to count and measure the total diameters of the motor or sensory nerve fibres in a muscle nerve after one of these components has degenerated following the removal of portions of the appropriate spinal roots. This information can then be related to what is known about the conduction velocities and functions of the motor and sensory nerve fibres supplying the muscle. Since there is a linear relation between conduction velocity and fibre diameter (HURSH, 1939), an attempt can then be made to analyse the fibre-diameter spectrum in terms of its functional components. Additional information is required in the case of nerve fibres of different function whose conduction velocities and diameters overlap. Thus in the case of sensory fibres data are needed regarding the number and nature of the receptors in the muscle and their ending-to-fibre ratios.

Of the several sources of error inherent in this method of analysis, one of the most serious concerns the consquences of damage inflicted on the intact root during laminectomy when either the dorsal root ganglion or a portion of the ventral root is removed. The nature of the operation is such that more damage is likely to be inflicted on the ventral root when the dorsal root ganglion is removed than vice versa. BOYD and DAVEY (1968) estimate that damage during de-afferentation reduces the motor component of a muscle nerve by 6–7%, and that damage during de-efferentation reduces the sensory component by 3–4%. These estimates are based on their finding that if the same muscle nerve is respectively de-afferented and de-efferented in different cats, the sum of the motor and sensory components thus isolated is 10% less than the total number of myelinated fibres in the normal nerve. At an earlier stage of their work the figure was 15% less, arbitrarily 10% motor, 5% sensory (BOYD, 1962b). The data supporting these estimates are not given, and the source of the normal counts is not specified. Counts of the total number of fibres in a given muscle nerve can vary widely among different cats. For example, the total fibre count for the nerve supplying the medial gastrocnemius muscle is given as 673 by REXED and THERMAN (1948, mean of 2 counts), 778 by O'LEARY, HEINBECKER and BISHOP (1935, mean of 6 counts), and 889 by FORBES (1966, mean of 3 counts). The sum of the motor and sensory counts for a given muscle nerve obtained from different operated cats may thus be higher or lower than the total count obtained from different normal cats, or be about the same (e.g. see Table 3, REXED and THERMAN, 1948). The loss of fibres from de-afferented and de-efferented nerves owing to degeneration following root damage may nevertheless be of the order that BOYD and DAVEY (1968) maintain. Their estimates should be checked in experiments in which both operated and normal counts are made on cats that are litter mates of the same sex.

A critical factor in the preparation of de-afferented or de-efferented nerves is the length of the interval between laminectomy and autopsy. This must be long enough for the appropriate degeneration to take place, and short enough to avoid the possibility of regenerating fibres contaminating the diameter spectra. An interval of not less than 21 days must be allowed for degeneration since short isolated stretches of axon may persist up to this time (YOUNG, 1942). Ellipsoids of degenerating myelin may persist and be mistaken for sound fibres up to 42 days (ECCLES and SHERRINGTON, 1930). Possible sources of regenerating fibres are as follows. (a) Excision of the dorsal root may leave some cell bodies remaining in the central stump resulting in afferent fibres growing across the scar to appear among the intact motor fibres in the muscle nerves. (b) Similarly motor fibres will grow out from the central stump of an excised ventral root, cross the scar, and ultimately appear among the intact sensory fibres in the muscle nerves. (c) Some fibres in the intact root may be severed or crushed during the removal of its partner. The isolated component in the muscle nerves may thus ultimately consist of both normal and regenerating fibres.

Following the severance of the nerve, the growth rate of axons across the scar is 0.25 mm/day (CAJAL, 1928). However, once they reach the peripheral stump they grow much faster, the rate in cats being 2.5 mm/day (BENTLEY and HILL, 1936), in rabbits 3.5 mm/day (GUTMANN, GUTTMANN, MEDAWAR and YOUNG, 1942). In the cat a gap of 6–7 mm is left after the removal of either the dorsal or the ventral root of L_7, representing a scar delay in regeneration of 24–28 days. To this must be added 3 days to allow for the reorganization of the central stump before growth begins (GUTMANN et al., (1942). If no length of nerve is removed, the scar delay is naturally much shorter. After the severance and suture of nerves in rabbits GUTMANN et al. (1942) found that new axon tips appeared in the peripheral stump after 7.3 days. After crushing a nerve the delay was 5.2 days. In view of these considerations it follows that the first regenerating fibres to arrive and contaminate a diameter spectrum of the motor or sensory fibres in a muscle nerve will be those that have regenerated after damage to the intact root, since in their case the scar delay will be minimal.

Almost all the fibre-diameter analyses of muscle nerves that have been undertaken have been concerned with the myelinated component. The possibility of there having been regenerated fibres in the nerves studied is thus a matter for concern only if there was a long enough interval between laminectomy and autopsy for them to have become myelinated. The crucial factor therefore is the rate of advance of functional completion. In rabbits this begins 36 days after suture and 20 days after crushing a nerve, the rates of advance being 2.0 mm/day and 3.0 mm/day, respectively (GUTMANN et al., 1942). Unfortunately the equivalent information is not available for cats, which have been used so extensively for studies of fibre-diameter spectra. However, if it is assumed that the same rates obtain as in rabbits, the maximum interval between laminectomy and autopsy may be determined by calculating the number of days required for fibres crushed in the intact roots to re-establish connexions in the muscles affected and become functionally complete. In practice the critical distance involved for any given muscle nerve is the length between the nearest relevant spinal root and the level at which the nerve is sectioned for microscopical examination. The latter must be a few centimetres proximal to the entry of the nerve into the muscle in order to avoid the fibre-diameter spectrum being affected by fibre branching (see ECCLES and SHERRINGTON, 1930).

In a young adult 2.5 kg cat the distance between the entry of the nerve into the tenuissimus muscle and the nearest spinal root (S_1) contributing to the nerve is about 90 mm. The most proximal level at which it is feasible to cut transverse sections of the nerve is about 20 mm from its entry into the muscle. It is therefore necessary to determine how long it would take for fibres crushed in the intact S_1 root during laminectomy to appear as regenerated myelinated fibres in the muscle nerve 70 mm from the site of injury. Calculations from the data given above show that this would take 43 days. The interval for fibres in the intact S_1 root damaged by severance would be 71 days, and for fibres regenerating from the central stump of an excised S_1 root the interval would be 81 days. These intervals will naturally be shorter for thigh muscles that lie nearer to their spinal roots than tenuissimus, and longer for the muscles of the shank and foot. For example, the distance between the entry of the nerve into tibialis anterior and the nearest spinal root (L_6) contributing to the nerve is about 170 mm, and it is feasible to cut sections of the nerve 30 mm from its entry into the muscle. Regenerated myelinated fibres produced by crush damage to the intact L_6 root would appear at this level after 65 days; those produced by severance after 106 days; and those growing from the central stump of an excised L_6 root would appear after 120 days.

The possibility that fibre-diameter spectra might be contaminated by regeneration has been ignored in some of the studies made. For example, LLOYD and CHANG (1948) prepared afferent fibre-diameter spectra of the hindlimb muscle nerves of two cats kept alive for periods of 122 and 103 days after de-efferentation. Such long survival times would have enabled regenerated myelinated fibres, both motor and sensory, to be present in most of the nerves they studied (see below p.

169). The same criticism applies to the survival times of 45–82 days allowed by ECCLES and SHERRINGTON (1930) in their study of de-afferented cat hindlimb muscle nerves. Contamination by regeneration would have been possible after survival times in the latter half of this range, especially in the case of the thigh muscle nerves studied (semitendinosus and crureus). In assessing the consequences of damage inflicted on the ventral root during de-afferentation, as compared with the incomplete removal of the dorsal root ganglion, they state that: "The former by subsequent degeneration of the injured fibres supplies its own numerical control; the latter not." (p. 328). They evidently regarded the survival time exclusively as a degeneration period, as did Lloyd and Chang.

2. *The Motor Component*

The size-frequency distribution of the motor fibres in most limb muscle nerves is markedly bimodal. Fig. 207 shows the fibre-diameter spectra prepared by BOYD and DAVEY (1962) of de-afferented cat nerves supplying the medial gastrocnemius and soleus muscles, together with a fibre-diameter spectrum of the L_7 ventral root, from which most of their motor fibres are derived. There is electrophysiological evidence (LEKSELL, 1945; KUFFLER, HUNT and QUILLIAM, 1951) that the group of small motor fibres is fusimotor (Leksell's γ efferents), and also direct histological evidence from teasing small motor fibres from muscle nerves to their entry into spindles (ADAL and BARKER, 1965a; ADAL, 1965). The group of large motor fibres qualify on the grounds of diameter, and hence conduction velocity, as those innervating extrafusal muscle fibres, and account for the a wave of the compound action potential (ERLANGER and GASSER, 1937). Apart

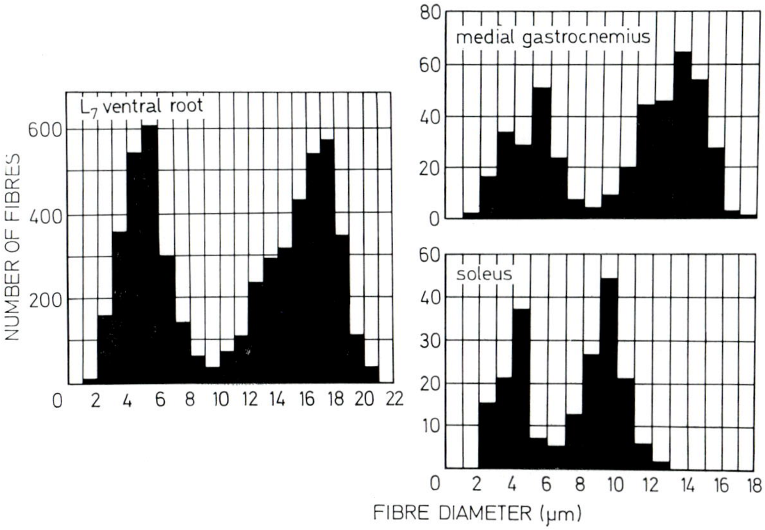

Fig. 207. *Left* Fibre-diameter spectrum of the seventh lumbar ventral spinal root of the cat. *Right* Motor fibre-diameter spectra of the nerves supplying the medial gastrocnemius and soleus muscles of the cat. (From BOYD and DAVEY, 1962; data from Text-figs. 2, 3 and 4 redrawn)

from these skeletomotor α fibres there are others within the large-fibre group that innervate both extra- and intrafusal muscle fibres, namely, the skeletofusimotor β fibres (BESSOU, EMONET-DÉNAND and LAPORTE, 1963, 1965; ADAL and BARKER, 1965a, b). The work of EMONET-DÉNAND, JOFFROY and LAPORTE (1971) and ELLAWAY, EMONET-DÉNAND, JOFFROY and LAPORTE (1972) makes it unlikely that there are other fibres in this group that innervate spindles exclusively (fusimotor α fibres) (see p. 52).

Fig. 207 shows that the diameter range of the group of large fibres is 10–21 μm (peak 17 μm) in the ventral root, dropping to 9–17 μm (peak 13 μm) in the nerve to medial gastrocnemius, and 7–12 μm (peak 9 μm) in the soleus nerve. Reduction in diameter owing to tapering may partly account for the shift to the left of the fibre-diameter spectra for these two muscle nerves, but there is evidently some other factor operating in the case of the soleus nerve. The explanation lies in the nature of the motor units composing the two muscles. Medial gastrocnemius consists of a mixture of types A, B, and C muscle fibres in which those of type A predominate, whereas soleus is exclusively composed of type B fibres (HENNEMAN and OLSON, 1965). A correlation of functional and histochemical studies has shown that A fibres compose large, fast units that contract infrequently and fatigue rapidly, and that the B and C fibres compose small, frequently contracting, fatigue resistant units that are slow and fast, respectively (OLSON and SWETT, 1966; EDSTRÖM and KUGELBERG, 1968; BURKE, ENGEL, LEVINE, TSAIRIS and ZAJAC, 1971). Hence motor units sampled from medial gastrocnemius fall into a conduction velocity range of 55.0–109.9 m/sec (peak at 90.0–94.9 m/sec) with contraction times ranging between 25.5 and 129 msec (peak at 40.2 msec), whereas motor units sampled from soleus have a conduction velocity range of 50.7–81.3 m/sec (peak at 65.0–69.9 m/sec) with contraction times ranging between 90 and 177 msec (peak at 104 msec) (HENNEMAN and OLSON, 1965). Conversion of these conduction velocity ranges into fibre diameters, using a factor of 5.7 for α fibres (ARBUTHNOTT, BOYD and KALU, 1971), gives fibre-diameter ranges of 10–19 μm for medial gastrocnemius and 9–14 μm for soleus.

The comparison between flexor digitorum longus (FDL) and soleus is even more striking. The group of large motor fibres supplying FDL has a diameter range of 11–22.5 μm, and a conduction velocity range of 71.3–114.3 m/sec (peak 90–95 m/sec). The mean percentages of muscle-fibre types are 41.6 type A, 30.9 type B, and 27.9 type C, with motor unit contraction times ranging from 13 to 81 msec (peak 30–35 msec) (OLSON and SWETT, 1966).

FERNAND and YOUNG (1951) found that fibre-diameter spectra for whole pes interosseous nerves of the rabbit were unimodal with a single peak at 8–10 μm (Fig. 208), and suggested, as mentioned above (p. 156), that this might be due partly to tapering, and partly to the fibres originating from small cells at the caudal end of the spinal cord. There is the additional possibility that γ fibres may be few or absent, fusimotor innervation being largely or entirely taken over by β fibres, as among spindles in the distal muscles of the rat's tail (ANDREW, PART and WAIT, 1971). ADAL (1965) has shown that this is so in rabbit hindfoot lumbrical muscles. The diameters of 105 motor fibres measured in teased, osmium preparations of 21 de-afferented lumbrical nerves had a unimodal distribution peaking at 9 μm in a range of 4–11 μm (Fig. 208). Only 20 fibres had diameters less than 8 μm. In nine nerves all the motor fibres fell within the range 8–11 μm, indicating the probability that a γ supply to the spindles was lacking.[7] Intramuscular teasing of the fibres supplied by two of these nerves confirmed that this was so. One of the nerves contained a single 8 μm fibre that provided the entire motor supply for a 1st lumbrical muscle and the only spindle present.

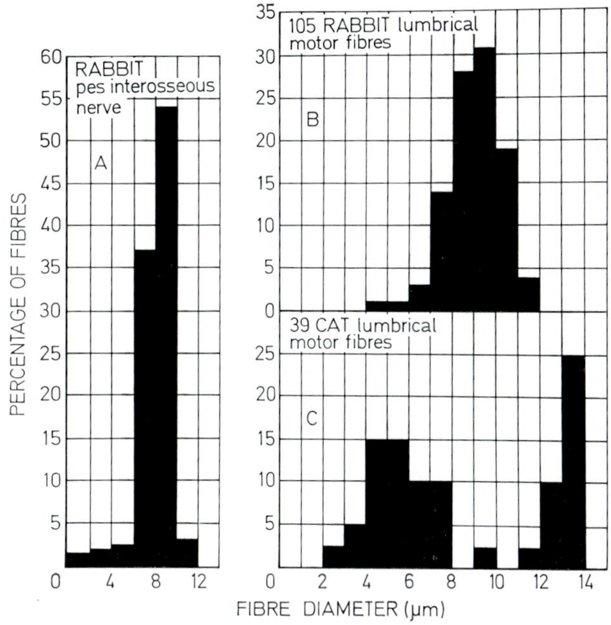

Fig. 208A–C. A Unimodal fibre-diameter spectrum of the rabbit's pes interosseous nerve. Measurements made on Weigert-Pal stained 5 μm thick paraffin section. (From FERNAND and YOUNG, 1951, Fig. 1). B Unimodal fibre-diameter spectrum of 105 rabbit lumbrical motor fibres. Measurements made on teased, osmium preparations of 21 de-afferented nerves from 8 rabbits supplying the three hindfoot lumbrical muscles. (Data from ADAL, 1965, Tables 14–16). C Bimodal fibre-diameter spectrum of 39 cat lumbrical motor fibres. Measurements made on teased, osmium preparations of four de-afferented nerves from three cats supplying three 1st and one 2nd deep lumbrical muscles. (Data from ADAL and BARKER, 1965a, Table 1)

In contrast with the unimodal motor spectrum of these rabbit lumbrical nerves, the diameters given by ADAL and BARKER (1965a) for 39 motor fibres from four cat deep lumbrical nerves have a bimodal distribution (Fig. 208). In these nerves β fibres are scarce, there being only 4 in the total of 39, with diameters of 6.0–12.5 μm, as compared with 20 γ fibres (2.5–7.5 μm) and 15 α fibres (7.5–13.5 μm).

7 The motor endings of spindles innervated solely by β fibres should consist exclusively of p_1 plates (see pp. 53–56). ADAL (1965) found that plate endings were far more common than trail endings in teased, silver preparations of a sample of 92 spindles from 21 rabbit lumbrical muscles. At the time of his study the p_2 plate in rabbit spindles had not been identified, but it is probable that most of the plate endings observed by him were of the p_1 type.

Motor fibres belonging to the group of large fibres begin to branch some distance away from the muscle they eventually enter and innervate. The fibre-diameter spectrum of a de-afferented cat medial gastrocnemius muscle 8 mm from nerve entry, as compared with one 50 mm from nerve entry, shows an increase of fibres below 6 μm and a shift of the large fibre spectrum to the left (ECCLES and SHERRINGTON, 1930). ADAL (1965) found that the distance between nerve entry and the level of initial extramuscular branching was 62 mm for cat medial gastrocnemius, 10 mm for soleus, and 1 mm for the 1st deep lumbrical muscle. It may be that the distance is related to the size of the muscle innervated, or that large fibres begin to branch further away from nerve entry than smaller ones. The majority of the γ fibres traced by ADAL and BARKER (1965a) began branching about 1 mm *after* nerve entry, and the smallest fibres branched only after reaching the spindles (see fibres *a* and *c* in Fig. 209).

As already noted (pp. 52–53), in teased, silver preparations the diameter range of the terminal branches of α fibres is similar to that of the p_1 branches of β fibres, whereas the diameter of the largest terminal branches of γ fibres exceeds the upper limit of this range by about 1 μm (see Fig. 71). Measurements of the intramuscular diameters of fusimotor fibres over distances of 0.5–1.0 mm from spindle entry made by ADAL (1961; cat, gold chloride preparations) had a diameter range of 2.0–8.0 μm (peak 3.5 μm, 13% of fibres above 4 μm) in quadratus femoris, 1.5–6.0 μm (peak 3.5 μm, 7% of fibres above 4 μm) in extensor digitorum brevis, and 1.5–4.0 μm (peak 1.5 μm) in pes interossei. This suggests that γ fibres may taper as they travel to the more distal muscles.

The spatial configurations built up by the repeated intramuscular branching of α and β fibres are described by ADAL (1965) as being either 'pyramidal' (most frequent), 'semipyramidal', or 'monopodial' (least frequent). The latter term describes a configuration in which small branches are given off at intervals from the main axon, as in the monopodial branching of plants. Adal found that branching of parent axons usually produces two daughter fibres, but instances of up to five occur. The diameter of the daughter fibres is usually unequal, markedly so in monopodial branching in which the smaller of two branches may be less than one third as thick as the parent axon and terminate with little further subdivision (see also COOPER, 1929; ECCLES and SHERRINGTON, 1930; VON THIEL, 1959).

The intramuscular branching of γ fibres is of particular interest in view of Boyd's belief that there are distinct groups of thickly-myelinated and thinly-myelinated γ axons in muscle nerves (BOYD, 1962a; BOYD and DAVEY, 1962, 1966, 1968). He was impressed by the fact that measurements made in teased, gold

Fig. 209A–C. Supply of skeletomotor (α) and fusimotor (γ) fibres to a cat's 1st deep lumbrical muscle. A Osmic section of de-efferented nerve cut approximately 12 mm from its entry into the muscle. γt thinly-myelinated γ fibre. B Teased osmic preparation of same nerve approximately 1 mm from its entry into the muscle. C Schematic representation of the intramuscular branching and distribution of γ fibres *a–g* to the six spindles (*Sp I–VI*) in the muscle. Scale indicates intramuscular length (mm) of each fibre and levels at which branching and spindle entry occur. Average total diameters (μm) of the fibres and their branches are indicated by the numbers placed in relation to them. Fibres and fibre branches entering spindles as 'γ_1 fibres' shown in black; 'γ_2 fibres' in white. (From ADAL and BARKER, 1965a)

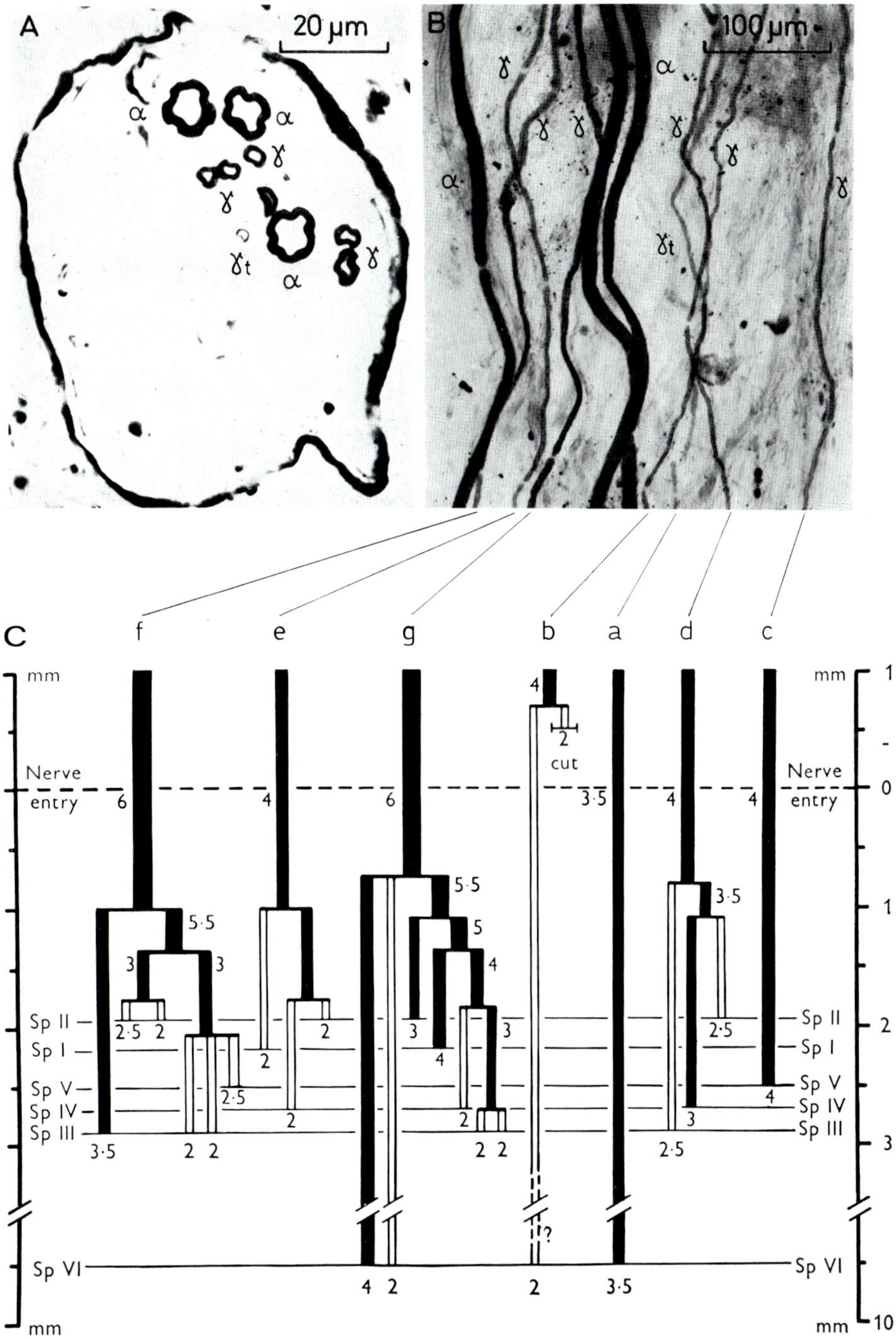
A
20 µm
α
γ
γt
B
100 µm
C
f
e
g
b
a
d
c
mm
Nerve entry
cut
Sp II
Sp I
Sp V
Sp IV
Sp III
Sp VI
1
0
1
2
3
10
mm

Fig. 209

chloride preparations of the axon diameters of fusimotor (presumed γ) fibres about 1 mm from spindle entry had a bimodal distribution with peaks at 1.5–2.0 μm and 2.5–3.0 μm (Boyd, 1962c, Fig. 10). He maintained that axons at this level less than 2 μm diameter ('γ_2 axons') were traceable to diffuse multiterminal endings ('γ_2 network') in the spindles, and that those above 3 μ diameter ('γ_1 axons') terminated in them as plates. Axons measuring 2–3 μm were mostly γ_1 axons, but a few γ_2 axons also occurred within this range. Motor fibre-diameter spectra for muscle nerves, such as those supplying popliteus and tibialis posterior, also showed bimodality in the γ group, there being a thinly-myelinated group peaking at 3–4 μm, and a thickly-myelinated group peaking at 6–7 μm (Boyd and Davey, 1962). Boyd therefore suggested the possibility that his γ_1 axons were derived from large, thickly-myelinated γ-stem fibres, and that the γ_2 axons were derived from small, thinly-myelinated γ-stem fibres (Boyd, 1962c; Boyd and Davey, 1962, 1966). This implied that the small, thinly-myelinated γ fibres would have to branch so as to innervate more than twice as many spindles as the large, thickly-myelinated ones (Boyd, 1962c, calculated the average proportion to be 12 as to 5), since they were found by Boyd and Davey (1962) to be from one half to one third as numerous in muscle nerves.

Later work showed that the γ supply to spindles is not distributed in this way. So far as the nature of the supply close to the spindles is concerned, the most that can be said is that the majority of myelinated fibres less than 2 μm diameter, and all non-myelinated fibres (in de-afferented and sympathectomized muscles), terminate as trail endings and hence qualify as fibres that Boyd designated as γ_2. However, the diameters of most myelinated trail fibres at this level fall between 2 μm and 4 μm, as do those of most fibres terminating as p_2 plates (see pp. 52–53 and Fig. 71).

So far as intramuscular branching is concerned, Adal and Barker (1965a) showed that there is no correlation between the diameters of γ fibres about 1 mm from spindle entry, the most proximal level at which Boyd (1962c, Fig. 10) distinguishes γ_1 and γ_2 fibres, and the diameters and degree of myelination of their stem fibres in the muscle nerve. Of eighteen thickly-myelinated 4.0–7.5 μm γ fibres whose course was traced from muscle nerve to spindle level in teased, osmium preparations, seven divided to produce thick branches of 3.0–4.5 μm diameter (γ_1 calibre), two divided to produce thin branches of 2.0–2.5 μm diameter (γ_2 calibre), and nine divided to produce a mixture of both thick and thin branches. Five thinly-myelinated 2.5–3.5 μm γ-stem fibres did not branch at all in their intramuscular course; each supplied one spindle only, three entering at γ_1 calibre, and two at γ_2 calibre. A schematic representation of the intramuscular branching and distribution of seven γ fibres supplying six spindles in one of the cat lumbrical muscles studied is shown in Fig. 209. In this study Adal and Barker (1965a) also traced the intramuscular course of five β fibres. These had stem diameters ranging from 6.0 to 12.5 μm. Their distribution at these two extremes varied from being predominantly fusimotor to predominantly skeletomotor, respectively (Fig. 210).

The fact remains that some γ fibres in muscle nerves have thick myelin sheaths and others thin. Boyd and Davey (1968) recognize anatomically distinct groups of γ fibres on the basis of difference in myelin thickness, distinguishing 1–5 μm axons with myelin sheaths 0.3–0.4 μm thick (never more than 0.6 μm) from 2–6 μm

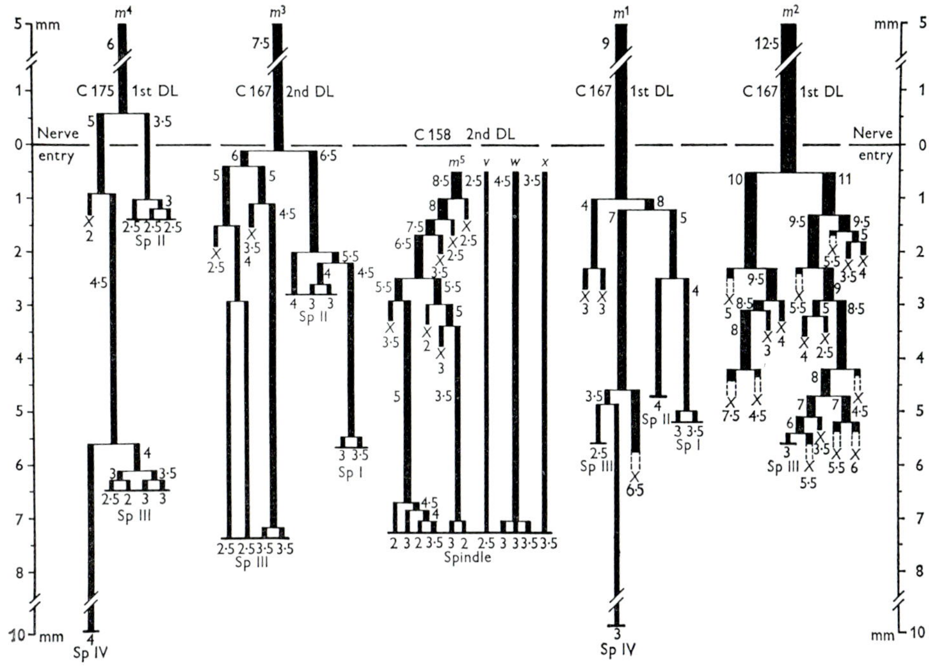

Fig. 210. Schematic representation of the intramuscular branching and distribution of five β skeletofusimotor fibres (m^1–m^5) that contributed to the innervation of four cat deep lumbrical (*DL*) muscles. Scale indicates intramuscular length (mm) of each fibre and the levels at which spindle (*Sp*) entry or extrafusal (*X*) innervation occurs. Average total diameters (μm) of the fibres and their branches are indicated by the numbers placed in relation to them. Fibre m^5 illustrates the monopodial type of branching, fibres m^2 and m^3 the pyramidal type. Fibres *v*, *w* and *x* are γ fibres. (From ADAL and BARKER, 1965a)

axons with myelin sheaths 1.0–1.5 μm thick. In terms of total diameter they separate γ fibres into a thinly-myelinated 2–5 μm group (mean 3.5 μm) and a thickly-myelinated 4.5–8.5 μm group (mean 6.5 μm). It is known, however, that myelin sheath thickness varies widely among axons of the same diameter. This is convincingly demonstrated by SUNDERLAND and ROCHE (1958) in their study of axon-myelin relationships in the median and ulnar nerves of the Australian opossum. Thus in the ulnar nerve they found that axons measuring 1 μm diameter had myelin sheaths ranging in thickness from 0.25 to 1.5 μm, and that axons measuring 6 μm diameter had myelin sheaths ranging in thickness from 0.25 to 2.5 μm. Axons belonging to other diameter classes within the range of 1–6 μm showed a similar range of myelin thickness. Since these data apply to a normal nerve they show that among motor and sensory fibres whose axon diameters range from 1 to 6 μm, the variability of myelin sheath thickness is such as to contribute to populations of thinly-myelinated and thickly-myelinated fibres.

Stacey and I have confirmed that this is true for motor and sensory axons from five cat hindlimb muscle nerves that have been either de-afferented or de-efferented (BARKER, HARKER, STACEY and SMITH, 1972, and unpublished observations). In low-power ($\times$10,000) electron micrographs we measured the diameters and

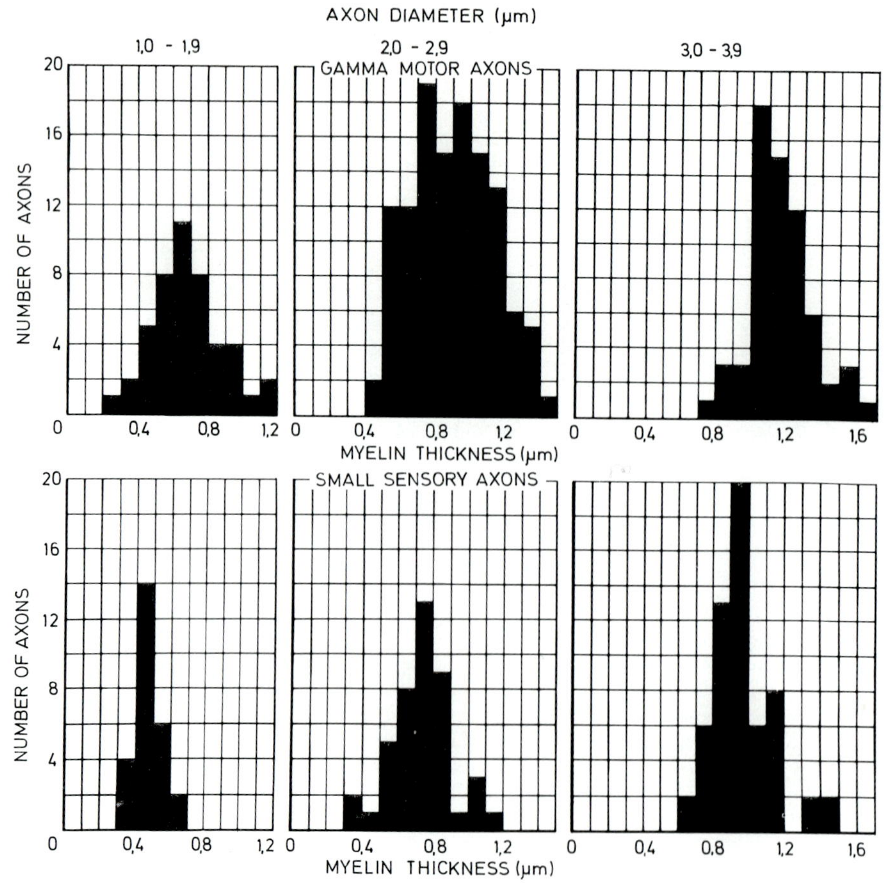

Fig. 211. Histograms of myelin thickness of γ motor axons and small sensory axons with axon diameters falling within the 1 μm, 2 μm and 3 μm classes. (Barker and Stacey, unpublished data relating to various cat hindlimb muscle nerves: see text)

myelin sheath thickness of motor and sensory axons in the 1.0–4.9 μm diameter range. Two hundred and forty-two γ axons were sampled for measurement from one tenuissimus nerve (20), two FHL nerves (44), one EDL nerve (58), two soleus nerves (62), and one common peroneal nerve (58). The afferent sample consisted of 134 axons measured in one FHL nerve (22), one EDL nerve (15), one FDL nerve (13), one soleus nerve (23), and two common peroneal nerves (61). Sampling was random in that the axons measured were those in the 1.0–4.9 μm diameter range that happened to be visible in their entirety between the bars of the grid on which the examined section was mounted. If Boyd and Davey's (1968) classification is correct such sampling of γ fibres should reveal the presence of: (a) γ fibres whose axons fall within the same 1 μm class of diameter, but separate into two populations in terms of myelin thickness ; and (b) few γ axons with myelin measuring 0.5–0.9 μm thick, and none with myelin 0.7–0.9 μm thick. We found, on the contrary, that within any 1 μm class of axon diameter in the 1.0–4.9 μm range, measurements of myelin thickness follow a normal distribution. The same applies to small sensory axons in this diameter range (see Fig. 211). One hundred and eighteen (48.8%) of the γ axons had myelin sheaths 0.5–0.9 μm thick; 75 (31%) had myelin 0.7–0.9 μm thick. Histograms of the total diameters (axon plus myelin) of all the γ fibres and all the small sensory fibres sampled both show a unimodal distribution, the γ fibres peaking at 3.0–3.9 μm, and the sensory fibres at 4.0–4.9 μm (Fig. 212).

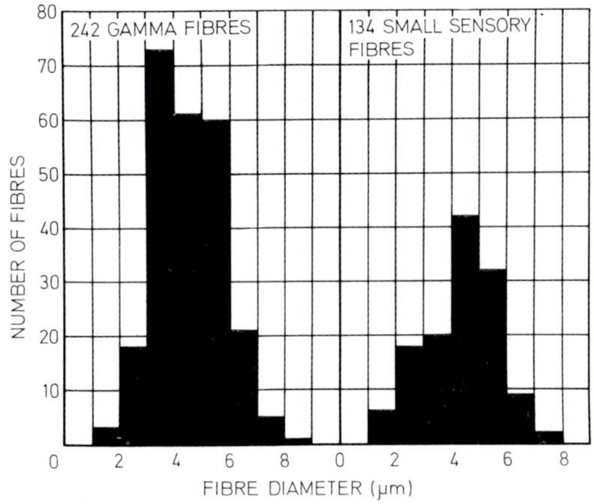

Fig. 212. Fibre-diameter spectra of γ motor fibres and small afferent fibres sampled from various cat hindlimb muscle nerves (see text) show a unimodal distribution. (Barker and Stacey, unpublished data)

Against this background it is scarcely surprising that the diameter spectrum of thinly and thickly-myelinated γ fibres in a cat muscle nerve should sometimes show a bimodal distribution, especially in view of the often relatively small number of fibres involved. Boyd and Davey (1968) found bimodality sometimes evident in the γ fibre-diameter spectra for the tibialis posterior, popliteus, and flexor digitorum brevis nerves, but never evident in the spectra for the nerves supplying extensor digitorum longus, tibialis anterior, and soleus. It may not be without significance that their best example of a bimodal γ fibre-diameter spectrum is that of a popliteus nerve de-afferented 58 days previously (Boyd and Davey, 1962, Fig. 1; 1968, Fig. 12). This is too long a period to be entirely confident about the absence of small thickly-myelinated axons that had regenerated after crush damage to the ventral roots (cf. Sanders, 1948, Fig. 19d). In another popliteus nerve de-afferented 35 days previously the γ spectrum is not obviously bimodal (Boyd and Davey, 1968, Fig. 2d).

In later work from the same laboratory Arbuthnott (1972) reports that in electron micrographs of the de-afferented cat tibialis posterior nerve two groups of γ fibres can be distinguished in terms of number of myelin lamellae, but not in terms of axon area or axon perimeter. The distribution of myelin thickness among γ fibres in the soleus and peroneus longus nerves, however, was unimodal. Becker and Wiesendanger (1967) have shown that this is also true for the cat tenuissimus nerve. They found the mean total diameter of the γ fibres to be 5.1 μm (range 2.6–8.0 μm), the mean axon diameter to be 2.7 μm (range 0.9–5.0 μm), and the mean myelin thickness to be 1.1 μm (range 0.5–1.5 μm). These data were obtained from measurements made on electron micrographs. In studies using the light microscope BOYD and DAVEY (1962, Table 1) report the tenuissimus nerve as containing a mean number of 21 γ fibres among which they distinguish 10 that are thickly myelinated (peak diameter 5.5 μm) and 11 that are thinly myelinated (peak diameter 3.5 μm), later amending these proportions to 15 and 5, respectively (BOYD and DAVEY, 1968, Table 7).[8]

The conclusion reached by BECKER and WIESENDANGER (1967) from their study, which included electrophysiological work, is that there is insufficient evidence to support a distinct classification of γ fibres into separate groups on the basis of total diameter and conduction velocity. There is no correlation between such groups and the functionally distinct dynamic and static γ fibres. Their conduction velocities overlap to a much greater extent than do those of fast γ and slow γ fibres (BROWN et al., 1965; BOYD and DAVEY, 1966). MATTHEWS (1972) summarizes the present position well when he writes: "It may be readily accepted that γ motor fibres do range in size and degree of myelination and that there are interesting differences between different nerves in the relative numbers of different sized γ fibres. But the case has yet to be established that they can be classified on these criteria into two separate populations which are sufficiently distinct to be presumed to have quite different functions" (p. 77).

CSILLIK (1961, 1965) refers to the small motor fibres that supply extrafusal grape endings in frog muscle, and in mammalian extraocular and oesophageal muscles, as γ fibres. In so doing he affords a glimpse of the sorts of problems that arise if the α (skeletomotor), β (skelotofusimotor), γ (fusimotor) terminology is applied outside mammalian skeletal muscle (cf. BARKER, 1968a). It would be as well to leave this scheme to apply within the field for which it was designed. On this narrower basis the classification works well enough, though adjustments of detail are necessary when considering different species.

3. The Sensory Component

Fig. 213 shows the diameter spectrum published by REXED and THERMAN (1948) of the myelinated sensory fibres in the gastrocnemius nerve of a de-efferented cat. The distribution is markedly trimodal and serves to illustrate LLOYD's (1943) classification of muscle afferents into three groups, namely, group I, large afferents of 12–20 μm diameter (peak 15–16 μm); group II, medium-sized afferents of 6–12 μm diameter (peak 8–9 μm); and group III, small afferents of 1–6 μm diameter (peak 3–4 μm). He proposed that the non-myelinated C afferents should be designated as group IV.

---

8 Comparison of BOYD and DAVEY's Table 1 (1962) with their Table 7 (1968) reveals some wide discrepancies. For example, the composition of the motor component of the extensor longus digitorum nerve is given in Table 1 (1962) as 110 α fibres, 65 γ thick fibres, and 25 γ thin fibres. These are mean values from four cats. In Table 7 (1968), however, the motor composition for this nerve is given as 130 α fibres, 40 γ fast (i.e. thick) fibres, and 50 γ slow (i.e. thin) fibres. These values are also from four cats, two of which are common with the 1962 analysis.

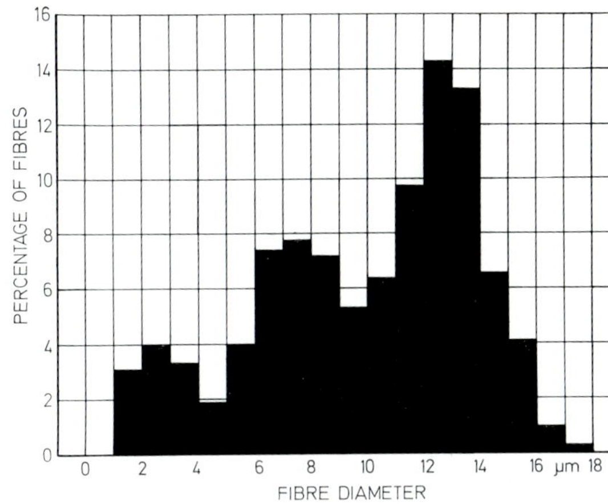

Fig. 213. Myelinated afferent fibre-diameter spectrum of the cat's gastrocnemius muscle nerve (From REXED and THERMAN, 1948, Fig. 4 redrawn)

Lloyd introduced this classification in order to facilitate discussion of reflex responses obtained on stimulating various nerves. The histological work followed later in the form of an analysis of the size/frequency distribution of the myelinated sensory fibres in sixteen cat hindlimb muscle nerves (LLOYD and CHANG, 1948). This established peaks of 15–17 μm for group I fibres, 6–7 μm for group II, and 3 μm for group III; and troughs at 12–13 μm between groups I and II, and at 5 μm between groups II and III, the latter not always well marked.

As these results were being written up, a similar analysis of two cat hindlimb muscle nerves was published by REXED and THERMAN (1948) in which they distinguish between groups of large, medium and small calibre afferents with diameter ranges of 10–17 μm (peak 12–13 μm), 4–10 μm (peak 6–8 μm), and 1–4 μm (peak 2–3 μm), respectively. Lloyd attributes the differences in the peak diameters of the group I fibres in the two studies to differences in shrinkage produced by the two histological methods employed. A much more important difference concerns the survival times of the de-efferented cats, namely, 14–21 days in Rexed and Therman's study as compared with 103 and 122 days in Lloyd and Chang's. The fact that Lloyd and Chang kept their cats for so long after de-efferentation means that regenerated myelinated fibres would have been present in most of the nerves they studied (see p. 158). Consequently most of their fibre-diameter spectra have a prominent 'group III' with a tall peak at 3 μm. Fig. 214 shows their afferent spectrum for a soleus nerve de-efferented 103 days previously, and compares this with a similar spectrum obtained by BARKER, IP and ADAL (1962) for a soleus nerve de-efferented 28 days previously. In the latter there is no clear segregation of groups II and III, and no tall peak at 3 μm. The spectra for three other soleus nerves prepared by BARKER et al. (1962) follow a similar pattern, and the same is true for the soleus spectrum obtained by HAGBARTH and WOHLFART (1952) 25 days after de-efferentation.

Fig. 214 also shows Lloyd and Chang's afferent spectra for the tibialis posterior and peroneus longus nerves de-efferented 103 days previously, and compares these with similar spectra prepared by STACEY (1967, 1969) for the same nerves de-efferented 28 days previously. The marked difference at the small-fibre end of the diameter spectrum is again evident, the longer survival time allowing for the regenerative boost to 'group III'. Subsequent work has proved the usefulness of classifying myelinated muscle afferents into three groups on the basis of diameter as Lloyd proposed, but ironically the spectra prepared by Lloyd and Chang have to be avoided if one wants a valid illustration of afferent trimodality (as in Fig. 213).

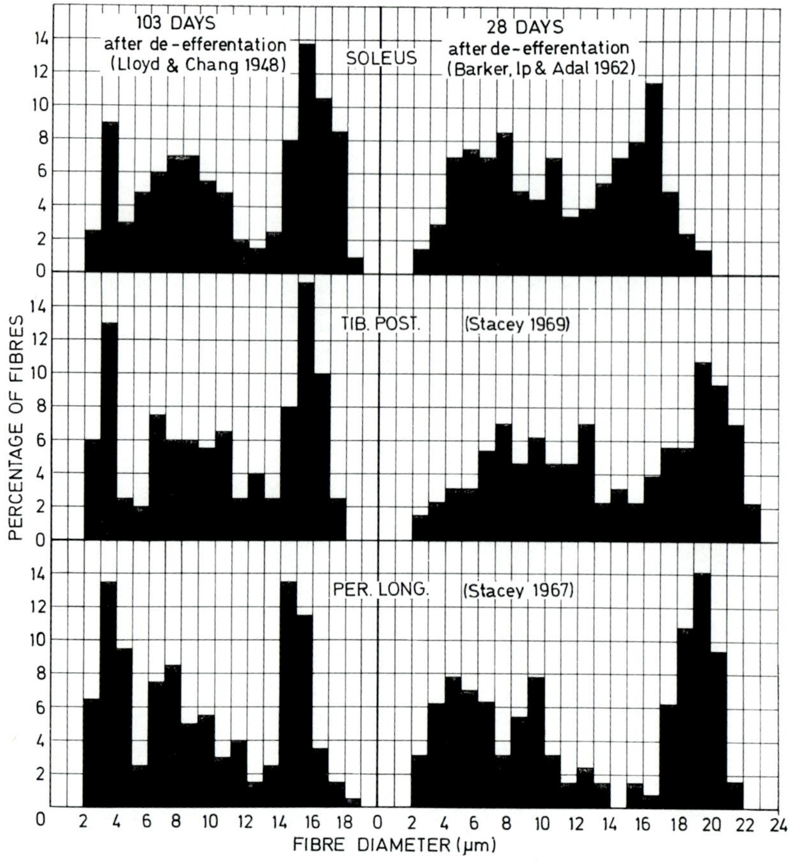

Fig. 214. Myelinated afferent fibre-diameter spectra of the nerves supplying the soleus, tibialis posterior and peroneus longus muscles of the cat. Those on the left were prepared by LLOYD and CHANG (1948) from nerves de-efferented 103 days previously. It is suggested (see text) that the tall peak of fibres at 3 μm is largely due to fibres that have regenerated during this long survival time. The spectra on the right, prepared from nerves de-efferented 28 days previously, show no such peak. (Spectra redrawn from LLOYD and CHANG, 1948, Fig. 2; BARKER, IP and ADAL, 1962, Fig. 2; STACEY, 1967, part of data for Fig. 10B; and STACEY 1969, Fig. 4. In order to make all the spectra comparable the diameter of the nerve fibres in each column is given at the bottom left of the column)

HUNT (1954) carried Lloyd's classification a stage further by determining the function and conduction velocity of isolated single afferents sampled from the nerves supplying cat soleus and medial gastrocnemius muscles. Converting conduction velocities into diameters showed that the afferents innervating muscle spindles had a bimodal distribution partly within group I (12–20 μm) as primary afferents (designated by Hunt as group Ia fibres), and partly within group II (4–12 μm) as secondary afferents. The large afferents innervating tendon organs had a unimodal distribution accounting for the rest of group I (hence designated by Hunt as Ib fibres) and for part of the upper diameter range of group II. It thus comes about that an afferent can be designated with one Roman numeral in the physiological sense, while belonging anatomically to a diameter range designated by another.

Similar investigations of group III afferents were carried out by PAINTAL (1960, 1962) and BESSOU and LAPORTE (1960, 1961). Most of these proved to be activated by pressure receptors located at the surface or within the body of the muscle or at musculo-tendinous junctions. They vary in threshold, rate of adaptation, and size of receptive field. Their functional characteristics and location suggest that they may be identified as paciniform corpuscles and free endings. Moreover PAINTAL (1960) found conduction velocities above 24 m/sec in some pressure afferents that would place some in group II and a few in group I. This correlates histologically with the fact that the diameters of both free-ending and paciniform afferents place them in groups I (few), II (some) and III (many) (see § IV E, 1 and § IV E, 2). Some of the group III afferents studied by Paintal and Bessou and Laporte could not be activated by any of the mechanical stimuli tested. A few others proved to be stretch afferents supplying either tendon organs or spindle secondary endings.

PAINTAL's (1960) study led him to conclude that: "The Group II band of afferent fibres can no longer be regarded as a homogeneous group of stretch fibres, owing to the presence of a fair number of fibres connected to pressure receptors" (p. 268). An attempt to assess the number and diameter distribution of myelinated free-ending afferents in a muscle nerve was made by BARKER et al. (1962), who correlated the receptor populations of four cat soleus muscles with the myelinated afferent fibre-diameter spectra of the nerves supplying them. After allowing for the innervation of spindles, tendon organs and paciniform corpuscles teased out in each case, they found an average of 29.2% of the total number of afferents in the nerve remained unallocated. They suggested that these provided a supply of free endings to the muscle. ZELENÁ and HNIK (1963b) arrive at a similar estimate (28.4%) of the proportion of free-ending afferents in the soleus nerve of the rat.

The average percentages of free-ending afferents in groups I, II, and III were calculated by BARKER et al. (1962) to be 10.6, 35.6, and 76.0, respectively. That about one-third of group II afferents should be engaged in free-ending innervation is a surprisingly high proportion, even allowing for PAINTAL's (1960) observations. It may be that we underestimated the number of secondary endings in our gold chloride preparations, though the presence of 25% additional undetected secondary endings was allowed for in our calculations. A check on the number of spindle capsules obtained from teasing each soleus muscle was provided by counting the number present in serial sections of the contralateral soleus and applying the prediction formulae of CHIN, COPE and PANG (1962). This indicated the probability that very few spindle capsules, if any, were lost in the teasing process.

BOYD and DAVEY (1968) tackle this problem in another way by correlating the spindle and tendon-organ counts given in the literature for a number of cat hindlimb muscles with the number of group I and II afferents found by various workers in the corresponding de-efferented nerves. On this basis they show that nearly all the group I and II afferents in the nerves to medial gastrocnemius, soleus, semitendinosus, tenuissimus, flexor digitorum longus, and flexor hallucis longus can be accounted for. However, the nerves to tibialis posterior, popliteus, tibialis anterior and extensor digitorum longus show an excess of group I and II afferents, which they allocate to free endings and extramuscular receptors. With so many variables and approximations involved it is doubtful whether such correlations are very meaningful. For example, the number of tendon organs used in the correlations for two of the muscles represent single counts, and for five other muscles the number is calculated (in the absence of actual counts) by assuming a spindle to tendon-organ ratio of 1.5:1. However, the number of tendon organs in a given muscle varies widely in different cats, e.g. in seven rectus femoris muscles BARKER and IP (1961) found it ranged between 47 and 104. The spindle to tendon-organ ratio also varies among different muscles, e.g. among six muscles BARKER (1962b) found ratios varying from 1.3:1 to 3.0:1. In four of the nerves Boyd and Davey use a single afferent fibre spectrum as the basis for correlation; in three others they use only two spectra. Yet it is known that there may be considerable variability among afferent counts for the same nerve in different cats, as is illustrated, for example, by the following counts for the nerve to tibialis anterior: 248, 291, 361 (REXED and THERMAN, 1948); 428 (HAGBARTH and WOHLFART, 1952).

Some muscle nerves certainly contain afferents arising from joint receptors, and are thus unsuitable for muscle receptor/muscle afferent correlations. Examples are the femoral, obturator, and, rarely, the medial gastrocnemius nerves in the cat, which contain knee-joint afferents (FREEMAN and WYKE, 1967), as also does the popliteus nerve (SKOGLUND, 1956). The flexor hallucis longus nerve is unsuitable because some of the spindles and tendon organs in the FHL muscle are supplied by afferents in the interosseous nerve (BARKER, 1962b; see Fig. 196); and the nerve to tibialis anterior is unsuitable because one of its intramuscular branches leaves the muscle to innervate dorsal foot muscles (BARKER and CHIN, 1960).

As already noted (§IV E, 1) all the non-myelinated group IV afferents terminate as free endings. In the tibialis posterior nerve of the cat these afferents have a diameter range of 0.15–1.25 μm (peak at 0.35 μm) and make up two thirds of the total sensory component (STACEY, 1969). These data were obtained using an electron microscope. Counts of non-myelinated afferents made from electron micrographs were about four times greater than those made using the light microscope. Previous assessments of the proportion of non-myelinated to myelinated afferents in light microscope studies of cat nerves must therefore be regarded as underestimates. They are 1.0:1.0 for the vastus medialis and 3.6:1.0 for the saphenous nerves (RANSON and DAVENPORT, 1931), and 3.0:1.0 for the phrenic nerve (HINSEY, HARE and PHILLIPS, 1939).

The composition of the afferent component of a muscle nerve may therefore be summarized as follows. (i) Primary or I a afferents supplying annulospiral endings to spindles. Those innervating spindles that also receive secondary endings have diameters falling within the upper part of the group I diameter range, whereas those innervating spindles without secondaries have diameters falling within the lower part (ADAL and BARKER, 1962) with perhaps a few included within group II. (ii) Tendon-organ or I b afferents included mostly in group I, with some in group II, and a few occasionally in group III. (iii) Secondary afferents supplying secondary endings to spindles; included mostly in group II with some in group III. (iv) Paciniform and myelinated free-ending afferents included mostly in group III, with some in group II, and a few in group I. (v) Non-myelinated free-ending afferents comprising the whole of group IV and about two thirds of the total sensory component.

Nerves that contain large α skeletomotor fibres also contain large group I afferents. Thus in BOYD and DAVEY's (1968) study the medial gastrocnemius nerve contained group I afferents up to 20 μm in diameter (mean 15–16 μm), whereas in the soleus nerve the maximum diameter of such afferents was 15 μm (mean 11 μm). LLOYD and CHANG's (1948) analysis showed that extensor muscles are innervated by larger afferents than their flexor antagonists. In addition to such differentials there is also a proximal/distal decrease in the diameter of afferent fibres in limb muscle nerves owing to tapering, as occurs in motor fibres. Branching of afferent fibres in muscle nerves is restricted almost exclusively to fibres not exceeding 10 μm (ECCLES and SHERRINGTON, 1930).

4. The Sympathetic Component

Postganglionic vasomotor fibres are both myelinated and non-myelinated, the former having diameters of up to 6 μm. Those contributing to the perivascular plexus of muscle arteries are either derived directly from ganglia in the sympathetic chain, or are distributed together with the motor and sensory fibres in muscle nerves. The distribution to limb muscles is via their supplying nerves, except in the upper thigh region where the plexus is derived directly from lumbar sympathetic ganglia (see LE GROS CLARK, 1958).

SHERRINGTON (1894) and ECCLES and SHERRINGTON (1930) have shown that the number of myelinated sympathetic fibres in muscle nerves is negligible. In teased, osmium preparations of cat medial gastrocnemius and semitendinosus nerves that had been both de-afferented and de-efferented Eccles and Sherrington found only two myelinated fibres in each nerve that had not degenerated. Their diameters were 2–3 μm. Apparently no quantitative assessment has been made of the non-myelinated sympathetic component of a muscle nerve.

Acknowledgements

I am greatly indebted to Dr. Michael Stacey, who has not only provided most of the electron micrographs that illustrate this chapter, but has also made up the plates, redrawn the histograms, and been largely responsible for the bibliography. I also wish to thank David Hutchinson for photographic assistance; Jacqui Morley and Joan Nathan for secretarial help; Sue Grocott for help with illustrations and electron microscopy; and Pat Grenville for much help and encouragement.

The assistance of the following authors in providing original photographs or figures is gratefully acknowledged: A. Barets; A. Crowe; M. Gladden; D.W. Harker; C.C. Hunt; N.C.R. Merrillees; A. Milburn; D.J. Pallot; O. Pompeiano; U. Proske; T.W. Schoultz; and R.J. Sterling.

The following publishers have kindly given permission for the reproduction of figures from works acknowledged by reference in the figure legends; Academic Press (*Journal of Ultrastructure Research*); American Physiological Society (*Journal of Neurophysiology*); Cambridge University Press (*Journal of Anatomy, Journal of Cell Science, Journal of Physiology*); Chapman and Hall Ltd. (*Journal of Neurocytology*); C. V. Mosby Co. (*Investigative Ophthalmology*); Excerpta Medica Foundation; Gauthier-Villars, Paris (*Comptes Rendus des séances de l'Académie des Sciences*); Hong Kong University Press; J. A. Churchill Ltd.; Macmillan Journals Ltd. (*Nature, Brain*); Masson et Cie, Paris (*Actualités Neurophysiologiques, Archive d'Anatomie Microscopique et de Morphologie*); Springer-Verlag (*Pflügers Archiv*); The Editor, *Acta Anatomica*; The Editor, *Acta Physiologica scandinavica*; The National Research Council of Canada (*Canadian Journal of Physiology and Pharmacology*); The Royal Society (*Proceedings, Philosophical Transactions*); Williams and Wilkins Co. (*American Journal of Physical Medicine*); Wistar Press (*American Journal of Anatomy, Journal of Comparative Neurology, Anatomical Record*).

References

Adal, M.N.: The proprioceptive component of certain muscle nerves in the cat. M. Sc. Thesis: University of Hong Kong 1961.

Adal, M.N.: A study of the motor supply to mammalian skeletal muscle. Ph. D. Thesis: University of Durham 1965.

Adal, M.N.: The fine structure of the sensory region of cat muscle spindles. J. Ultrastruct. Res. **26**, 332–354 (1969).

Adal, M.N., Barker, D.: Intramuscular diameters of afferent nerve fibres in the rectus femoris muscle of the cat. In: Symposium on Muscle Receptors. Barker, D., Ed., pp. 249–256. Hong Kong: Hong Kong Univ. Press 1962.

Adal, M.N., Barker, D.: Intramuscular branching of fusimotor fibres. J. Physiol. (Lond.) **177**, 288–299 (1965a).

Adal, M.N., Barker, D.: Motor supply to hindlimb muscles of the cat and rabbit J. Anat. (Lond.) **99**, 918–919 (1965b).

Adal, M.N., Barker, D.: The fine structure of cat fusimotor endings. J. Physiol. (Lond.) **192**, 50–52 (1967).

Allen, W.F.: Distribution of the spinal nerves in *Polistotrema* and some special studies on the development of spinal nerves. J. comp. Neurol. **28**, 137–213 (1917).

de Anda, G., Rebollo, M.A.: The neuromuscular spindles in the adult chicken: I. Morphology. Acta anat. (Basel) **67**, 437–451 (1967).

Andrew, B.L.: The sensory innervation of the medial ligament of the knee joint. J. Physiol. (Lond.) **123**, 241–250 (1954).

Andrew, B.L. (Ed.): Symposium on Control and Innervation of Skeletal Muscle. Dundee: Thomson (Edinburgh: Livingstone) 1966.

Andrew, B.L., Part, N.J.: Properties of fast and slow motor units in hindlimb and tail muscles of the rat. Quart. J. exp. Physiol. **57**, 213–225 (1972).

Andrew, B.L., Part, N.J.: The distribution of fusimotor fibres to distal caudal muscles in the rat (in preparation, 1973).

Andrew, B.L., Part, N.J., Wait, F.: Muscle spindles without γ-efferents. J. Physiol. (Lond.) **219**, 28–29 (1971).

Appelberg, B., Bessou, P., Laporte, Y.: Action of static and dynamic fusimotor fibres on secondary endings of the cat's spindles. J. Physiol. (Lond.) **185**, 160–171 (1966).

Arbuthnott, E.R.: Three groups of motor nerve fibres in cat peripheral nerves differentiated by the number of lamellae in the myelin sheath. J. Physiol. (Lond.) **227**, 44–45 (1972).

Arbuthnott, E. R., Boyd, I. A., Kalu, K. U.: The somatosensory system. Proc. of Sat. Symp. XXV Int. Cong. Phys. Sci. (in press, 1971).

Bach-y-Rita, P., Ito, F.: *In vivo* microelectrode studies of the cat retractor bulbi fibres. Invest. Ophthal. **4**, 338 (1965).

Banker, B.Q., Girvin, J.P.: The ultrastructural features of the mammalian muscle spindle. J. Neuropath. exp. Neurol. **30**, 155—195 (1971).

Banker, B.Q., Przybylski, R.J., van der Meulen, J.P., Victor, M., Eds.: Research in Muscle Development and the Muscle Spindle. Amsterdam: Excerpta Medica 1972.

Banks, R.W.: Histochemical studies on rabbit intrafusal fibres. J. Anat. (Lond.) **108**, 613 (1971).

Banks, R.W., James, N.T., Meek, G.A.: Morphometric studies on intrafusal and extrafusal muscle fibres. J. Anat. (Lond.) **111**, 489 (1972).

Barets, A.: Les récepteurs intra-musculaires des nageoires chez les sélaciens. Arch. Anat. micr. Morph. exp. **45**, 254–260 (1956).

Barker, D.: The innervation of the muscle-spindle. Quart. J. micr. Sci. **89**, 143–186 (1948).

Barker, D., Ed.: Symposium on Muscle Receptors. Hong Kong: Hong Kong Univ. Press 1962a.

Barker, D.: The structure and distribution of muscle receptors. In: Symposium on Muscle Receptors. Barker, D., Ed., pp. 227–240. Hong Kong: Hong Kong Univ. Press 1962b.

Barker, D.: Three types of motor ending in cat spindles. J. Physiol. (Lond.) **186**, 27–28 (1966).

Barker, D.: The innervation of mammalian skeletal muscle. In: Ciba Foundation Symposium on Myotatic, Kinesthetic and Vestibular Mechanisms. de Reuck, A.V.S., Knight, J., Eds., pp. 3–15. London: Churchill 1967.

Barker, D.: L'innervation motrice du muscle strié des vertébrés. Actualités neurophysiol. **8**, 23–71 (1968a).

Barker, D.: The disposition of trail endings in mammalian spindles. J. Physiol. (Lond.) **196**, 51–52 (1968b).

Barker, D., Bessou, P., Jankowska, E., Pagès, B., Stacey, M.: Distribution des axones fusimoteurs statiques et dynamiques aux fibres musculaires intrafusales, chez le chat. C. R. Acad. Sci. (Paris) **275**, 2527–2529 (1972).

Barker, D., Chin, N.K.: The number and distribution of muscle-spindles in certain muscles of the cat. J. Anat. (Lond.) **94**, 473–486 (1960).

Barker, D., Cope, M.: Tandem muscle-spindles in the frog. J. Anat. (Lond.) **96**, 49–57 (1962a).

Barker, D., Cope, M.: The innervation of individual intrafusal muscle fibres. In: Symposium on Muscle Receptors. Barker, D., Ed., pp. 263–269. Hong Kong: Hong Kong Univ. Press 1962b.

Barker, D., Emonet-Dénand, F., Laporte, Y., Proske, U., Stacey, M.: Identification des terminaisons motrices des fibres fusimotrices statiques chez la chat. C. R. Acad. Sci. (Paris) **271**, 1203–1206 (1970).

Barker, D., Emonet-Dénand, F., Laporte, Y., Proske, U., Stacey, M.: Identification of the endings and function of cat fusimotor fibres. J. Physiol. (Lond.) **216**, 51–52 (1971).

Barker, D., Emonet-Dénand, F., Laporte, Y., Proske, U., Stacey, M.: Morphological identification and intrafusal distribution of the endings of static fusimotor axons in the cat. J. Physiol. (Lond.) **230**, 405–427 (1973).

Barker, D., Gidumal, J.L.: The morphology of intrafusal muscle fibres in the cat. J. Physiol. (Lond.) **157**, 513–528 (1961).

Barker, D., Harker, D.W.: Two types of multiply-innervated muscle fibre in the superior rectus muscle of the sheep. J. Physiol. (Lond.) **222**, 74–75 (1972).

Barker, D., Harker, D.W., Stacey, M.J., Smith, C.R.: Fusimotor innervation. In: Research Concepts in Muscle Development and the Muscle Spindle. Banker, B.Q. et al., Eds., pp. 227–250. Amsterdam: Excerpta Medica 1972.

Barker, D., Hunt, J.P.: Mammalian intrafusal muscle fibres. Nature (Lond.) **203**, 1193 (1964).

Barker, D., Ip, M.C.: The primary and secondary endings of the mammalian muscle spindle. J. Physiol. (Lond.) **153**, 8–10 (1960).

Barker, D., Ip, M.C.: A study of single and tandem types of muscle-spindle in the cat. Proc. roy. Soc. B. **154**, 377–397 (1961).

Barker, D., Ip, M.C.: A silver method for demonstrating the innervation of mammalian muscle in teased preparations. J. Physiol. (Lond.) **169**, 73–74 (1963).

Barker, D., Ip, M.C.: The motor innervation of cat and rabbit muscle spindles. J. Physiol. (Lond.) **177**, 27–28 (1965).

Barker, D., Ip, M.C.: Sprouting and degeneration of mammalian motor axons in normal and de-afferentated skeletal muscle. Proc. roy. Soc. B. **163**, 538–554 (1966).

Barker, D., Ip, M.C., Adal, M.N.: A correlation between the receptor population of the cat's soleus muscle and the afferent fibre diameter spectrum of the nerve supplying it. In: Symposium on Muscle Receptors. Barker, D., Ed., pp. 257–261. Hong Kong: Hong Kong Univ. Press 1962.

Barker, D., Milburn, A.: Increase in number of intrafusal muscle fibres during the development of muscle spindles in the rat. J. Physiol. (Lond.) **222**, 159–160 (1972).

Barker, D., Stacey, M.J.: Somatic non-myelinated axons in cat hindlimb muscles. Proc. int. Union Physiol. Sci. **7**, 30 (1968).

Barker, D., Stacey, M.J.: Rabbit intrafusal muscle fibres. J. Physiol. (Lond.) **210**, 70–72 (1970).

Barker, D., Stacey, M.J., Adal, M.N.: Fusimotor innervation in the cat. Phil. Trans. B. **258**, 315–346 (1970).

Barrios, P., Haase, J., Heinrich, W.: Fusimotorische Alpha-reflexe an prätibialen Flexorenspindeln der Katze. Pflügers Arch. ges. Physiol. **296**, 49–69 (1967).

Batten, F.E.: The muscle-spindle under pathological conditions. Brain **20**, 138–179 (1897).

Baum, J.: Beiträge zur Kenntnis der Muskelspindeln. Anat. Hefte **13**, 250–305 (1900).

Becker, H.W., Weisendanger, M.: Elektrophysiologische und elektronmikroskopische Charakterisierung der afferenten Innervation des M. tenuissimus der Katze. Helv. physiol. pharmacol. Acta **25**, 262–286 (1967).

Bentley, F.H., Hill, N.: Nerve grafting. Brt. J. Surg. **24**, 368–387 (1936).

Berndt, J.M., Oswaldo-Cruz, E., Rocha-Miiranda, C.E.: Identification of beta fibres at spindle entry. J. comp. Neurol. **136**, 419–422 (1969).

Bessou, P., Emonet-Dénand, F., Laporte, Y.: Occurence of intrafusal muscle fibre innervation by branches of slow a motor fibres in the cat. Nature (Lond.) **198**, 594–595 (1963).

Bessou, P., Emonet-Dénand, F., Laporte, Y.: Motor fibres innervating extrafusal and intrafusal muscle fibres in the cat. J. Physiol. (Lond.) **180**, 649–672 (1965).

Bessou, P., Laporte, Y.: Activation des fibres afférentes myelinisées de petit calibre, d'origine musculaire (fibres du groupe III). C. R. Acad. Sci. (Paris) **154**, 1093–1096 (1960).

Bessou, P., Laporte, Y.: Étude des récepteurs musculaires innervés par les fibres afférentes due groupe III (fibre myelinisées fines), chez le chat. Arch. ital. Biol. **99**, 293–321 (1961).

Bessou, P., Pagès, B.: Spindle secondary ending responses elicited by stimulation of static fusimotor axons. J. Physiol. (Lond.) **202**, 569–584 (1969).

Bloom, F.E., Barrnett, R.J.: Fine structural localization of noradrenaline in vesicles of autonomic nerve endings. Nature (Lond.) **210**, 599 (1966).

Bonavolonta, A.: Ricerche comparative sulle espansioni nervose sensitive nei muscoli estrinseci dell'occhio dell'uomo e di altri mammiferi. 1. I fusi neuro-muscolari. Quad. Anat. prat. **11**, 48–73 (1956a).

Bonavolonta, A.: Ricerche comparative sulle espansioni nervose sensitive nei muscoli estrinseci dell'occhio dell'uomo e di altri mammiferi. 2. Gli organi muscolo-tendinei di Golgi. Quad. Anat. prat. **11**, 151–166 (1956b).

Bonavolonta, A.: Ulteriori ricerche comparative sulla innervazione propriocettiva dei muscoli oculo-estrinseci. Quad. Anat. prat. **14**, 48–55 (1958).

Bone, Q.: Some observations upon the peripheral nervous system of the hagfish, *Myxine glutinosa*. J. mar. biol. Ass. U. K. **43**, 31–47 (1963).

Bowden, R.E.M.: Muscle spindles in the human foetus. Acta Biol. Szeged. **9**, 35–39 (1963).

Bowden, R.E.M., Mahran, Z.Y.: The functional significance of the pattern of innervation of the muscle quadratus labii superioris of the rabbit, cat and rat. J. Anat. (Lond.) **90**, 217–227 (1956).

Boyd, I.A.: The histological structure of the receptors in the knee joint of the cat correlated with their physiological response. J. Physiol. (Lond.) **124**, 476–488 (1954).

BOYD, I.A.: The tenuissimus muscle of the cat. J. Physiol. (Lond.) **133**, 35–36 (1956).

BOYD, I.A.: Simple and compound mammalian muscle spindles. J. Physiol. (Lond.) **145**, 55–56 (1959).

BOYD, I.A.: The diameter and distribution of the nuclear bag and nuclear chain muscle fibres in the muscle spindles of the cat. J. Physiol. (Lond.) **153**, 23–24 (1960).

BOYD, I.A.: The nuclear-bag fibre and nuclear-chain fibre systems in the muscle spindles of the cat. In: Symposium on Muscle Receptors. BARKER, D., Ed., pp. 185–190. Hong Kong: Hong Kong Univ. Press 1962a.

BOYD, I.A.: Final discussion. In: Symposium on Muscle Receptors. BARKER, D., Ed., pp. 271–273. Hong Kong: Hong Kong Univ. Press. 1962b.

BOYD, I.A.: The structure and innervation of the nuclear bag muscle fibre system and the nuclear chain muscle fibre system in mammalian muscle spindles. Phil. Trans. B. **245**, 81–136 (1962c).

BOYD, I.A.: The mammalian muscle spindle – an advanced study (film). J. Physiol. (Lond.) **214**, 1 (1971a).

BOYD, I.A.: Specific fusimotor control of nuclear bag and nuclear chain fibres in cat muscle spindles. J. Physiol. (Lond.) **214**, 30–31 (1971b).

BOYD, I.A., DAVEY, M.R.: The groups of origin in the nerves to skeletal muscle of the γ_1 and γ_2 fusimotor fibres present close to, and within, mammalian muscle spindles. In: Symposium on Muscle Receptors. BARKER, D., Ed., pp. 191–198. Hong Kong: Hong Kong Univ. Press 1962.

BOYD, I.A., DAVEY, M.R.: The distribution of two types of small motor nerve fibre to different muscles in the hind-limb of the cat. In: Muscular Afferents and Motor Control, Nobel Symposium I. GRANIT, R., Ed., pp. 59–68. Stockholm: Almqvist & Wiksell 1966.

BOYD, I.A., DAVEY, M.R.: Composition of Peripheral Nerves. Edinburgh: Livingstone 1968.

BOYD, I.A., GLADDEN, M.H., MCWILLIAM, P.N.: Static and dynamic fusimotor action in isolated cat muscle spindles with intact nerve and blood supply. J. Physiol. (Lond.) **230**, 29–30 (1973).

BRIDGMAN, C.F.: The structure of tendon organs in the cat: a proposed mechanism for responding to muscle tension. Anat. Rec. **162**, 209–220 (1968).

BRIDGMAN, C.F.: Comparisons in structure of tendon organs in the rat, cat and man. J. comp. Neurol. **138**, 369–372 (1970).

BRIDGMAN, C.F., ELDRED, E.: Hypothesis for a pressure-sensitive mechanism in muscle spindles. Science **143**, 481–482 (1964).

BRIDGMAN, C.F., SHUMPERT, E.E., ELDRED, E.: Insertions of intrafusal fibres in muscle spindles of the cat and other mammals. Anat. Rec. **164**, 391–402 (1969).

BROOKE, M.H., KAISER, K.K.: Three "myosin adenosine triphosphatase" systems: The nature of their pH lability and sulfhydryl dependence. J. Histochem. Cytochem. **18**, 670–672 (1970).

BROWN, M.C.: A comparison of the spindles in two different muscles of the frog. J. Physiol. (Lond.) **216**, 553–563 (1971).

BROWN, M.C., BUTLER, R.G.: Depletion of intrafusal muscle fibres glycogen by stimulation of fusimotor fibres. J. Physiol. (Lond.) **229**, 25–26 (1973).

BROWN, M.C., CROWE, A., MATTHEWS, P.B.C.: Observations of the fusimotor fibres of the tibialis posterior muscle of the cat. J. Physiol. (Lond.) **177**, 140–159 (1965).

BRZEZINSKI, D.K., VON: Untersuchungen zur Histochemie der Muskelspindeln. I. Mitteilung: Topochemie der Polysaccharide. Acta histochem. (Jena) **12**, 75–79 (1961a).

BRZEZINSKI, D.K., VON: Untersuchungen zur Histochemie der Muskelspindeln. II. Mitteilung: Zur Topochemie und Funktion des Spindelraumes und der Spindelkapsel. Acta histochem. (Jena) **12**, 277–288 (1961b).

BULLER, A.J., LEWIS, D.M.: Further observations on mammalian cross-innervated skeletal muscle. J. Physiol. (Lond.) **178**, 343–358 (1965).

BURKE, R.E., ENGEL, W.K., LEVINE, D.N., TSAIRIS, P., ZAJAC, F.E.: Mammalian motor units: physiological-histochemical correlation in three types in cat gastrocnemius. Science **174**, 709–712 (1971).

Cajal, S. Ramón y: Degeneration and regeneration of the nervous system. (Transl. and ed. by May, R.M.). Oxford: University Press 1928.

Carli, G., Diete-Spiff, K., Pompeiano, O.: Skeletomotor and fusimotor control of gastrocnemius muscle from Deiter's nucleus. Experientia (Basel) **22**, 583–584 (1966).

Carli, G., Diete-Spiff, K., Pompeiano, O.: Responses of the muscle spindles and of the extrafusal fibres in an extensor muscle to stimulation of the lateral vestibular nucleus in the cat. Arch. ital. Biol. **105**, 209–242 (1967a).

Carli, G., Diete-Spiff, K., Pompeiano, O.: Mechanisms of muscle spindle excitation. Arch. ital. Biol. **105**, 273–289 (1967b).

de Castro, F.: Technique pour la coloration du systeme nerveux quand il est pourvu de ses etuis osseux. Trab. Lab. Invest. biol. Univ. Madr. **23**, 427–446 (1925).

Cattaneo, A.: Organes nerveux terminaux muscolo-tendineux, leurs conditions normales et leur manière de se comporter après la section des racines nerveuses et des nerfs spinaux. Arch. ital. Biol. **10**, 337–357 (1888).

Cauna, N.: Fine structure of the receptor organs and its probable functional significance. In: Ciba Foundation Symposium on Touch, Heat and Pain. de Reuck, A.V.S., Knight, J., Eds., pp. 117–127. London: Churchill 1966.

Cazzato, G., Walton, J.N.: The pathology of the muscle spindle. A study of biopsy material in various muscular and neuromuscular diseases. J. neurol. Sci. **7**, 15–70 (1968).

Ceccherelli, G.: Sulle "terminazioni nervose a paniere" del Giacomini, nei muscoli dorsali degli anfibi anuri adulti. Anat. Anz. **24**, 428–435 (1904).

Cheng, K.: Cholinesterase activity in human extraocular muscles. Jap. J. Ophthalmology **7**, 174–183 (1963).

Chin, N.K.: The structure and innervation of the muscle spindle in the latissimus dorsi anterior and posterior muscles of the domestic fowl. Ph. D. Thesis: University of Hong Kong 1970.

Chin, N.K., Cope, M., Pang, M.: Number and distribution of spindle capsules in seven hindlimb muscles of the cat. In: Symposium on Muscle Receptors. Barker, D., Ed., pp. 24–248. Hong Kong: Hong Kong Univ. Press 1962.

Ciaccio, G.V.: Sur les plaque nerveuses finales dans les tendons des vertébrés. Arch. ital. Biol. **14**, 31–57 (1890).

Cilimbaris, P.A.: Histologische Untersuchungen über die Muskelspindeln der Augenmuskeln. Arch. mikr. Anat. **75**, 692–747 (1910).

Cipollone, L.T.: Ricerche sull'anatomia normale e patologica delle terminazioni nervose nei muscoli striati. Suppl. Ann. Med. Nav. **3**, 222–282 (1897).

Cliff, G.F., Ridge, R.M.A.P.: Innervation of extrafusal and intrafusal fibres in snake muscle. J. Physiol. (Lond.) **233**, 1–18 (1973).

Coërs, C.: Histochemical identification of motor nerve endings in muscle spindles. In: Symposium on Muscle Receptors. Barker, D., Ed., pp. 221–226. Hong Kong: Hong Kong Univ. Press 1962.

Coërs, C., Durand, J.: Données morphologiques nouvelles sur l'innervation des fuseaux neuromusculaires. Arch. Biol. (Liège) **67**, 685–715 (1956).

Cole, W.V.: Motor endings in the striated muscle of vertebrates. J. comp. Neurol. **102**, 671–716 (1955).

Cooper, S.: The relation of active to inactive fibres in fractional contraction of muscle. J. Physiol. (Lond.) **69**, 1–13 (1929).

Cooper, S.: Muscle spindles in the intrinsic muscles of the human tongue. J. Physiol. (Lond.) **122**, 193–202 (1953).

Cooper, S.: Muscle spindles and other muscle receptors. In: The Structure and Function of Muscle. Bourne, G.H., Ed., pp. 381–420. New York: Academic Press 1960.

Cooper, S.: Muscle spindles and motor units. In: Symposium on Control and Innervation of Skeletal Muscle. Andrew, B.L., Ed., pp. 9–17. Dundee: Thompson (Edinburgh: Livingstone) 1966.

Cooper, S., Daniel, P.M.: Muscle spindles in human extrinsic eye muscles. Brain **72**, 1–24 (1949).

Cooper, S., Daniel, P.M.: Human muscle spindles. J. Physiol. (Lond.) **133**, 1–3 (1956).

Cooper, S., Daniel, P.M.: Muscle spindles in man: their morphology in the lumbricals and the deep muscles of the neck. Brain **86**, 563–586 (1963).

Cooper, S., Daniel, P.M.: Elastic tissue in muscle spindles of man and the rat. J. Physiol. (Lond.) **192**, 10–11 (1967).

Cooper, S., Daniel, P.M., Whitteridge, D.: Muscle spindles and other sensory endings in the extrinsic eye muscles: the physiology and anatomy of these receptors and of their connexions with the brainstem. Brain **78**, 564–583 (1955).

Cooper, S., Fillenz, M.: Afferent discharges in response to stretch from the extraocular muscles of the cat and monkey and the innervation of these muscles. J. Physiol. (Lond.) **127**, 400–413 (1955).

Corvaja, N., Magherini, P.C., Pompeiano, O.: Ultrastructure of glycogen-membrane complexes in sensory nerve fibres of cat muscle spindles. Z. Zellforsch. **121**, 199–217 (1971).

Corvaja, N., Marinozzi, V., Pompeiano, O.: Close appositions and junctions of plasma membranes of intrafusal muscle fibres in mammalian muscle spindles. Pflügers Arch. ges. Physiol. **296**, 337–345 (1967).

Corvaja, N. Marinozzi, V., Pompeiano, O.: Muscle spindles in the lumbrical muscle of the cat. Arch. ital. Biol. **107**, 365–543 (1969).

Corvaja, N., Pompeiano, O.: The differentiation of two types of intrafusal fibres in rabbit muscle spindles. Pflügers Arch. **317**, 187–197 (1970).

Couteaux, R.: Particularités histochimiques des zones d'insertion du muscle strié. C. R. Soc. Biol. (Paris) **147**, 1974–1976 (1953).

Crevatin, F.: Sopra le terminazioni nervose nei tendini dei pipistrelli. Rend. Sess. Roy. Accad. Sci. Ist. Bologna **5**, 31–34 (1901).

Crevatin, F.: Su di alcune forma di terminazioni nervose nei muscoli dell'occhio del dromedario. Rend. Sess. Roy. Accad. Sci. Ist. Bologna **6**, 57–61 (1902).

Crowe, A., Ragab, A.H.M.F.: Some preliminary studies on spindle-like structures in tortoise muscle. J. Physiol. (Lond.) **192**, 22–23 (1967).

Crowe, A., Ragab, A.H.M.F.: The innervation and capsular structure of the tortoise muscle spindle. J. Physiol. (Lond.) **201**, 5–6 (1969).

Crowe, A., Ragab, A.H.M.F.: The structure, distribution and innervation of spindles in the extensor digitorum brevis 1 muscle of the tortoise *Testudo graeca*. J. Anat. (Lond.) **106**, 521–538 (1970a).

Crowe, A., Ragab, A.H.M.F.: Studies on the fine structure of the capsular region of tortoise muscle spindles. J. Anat. (Lond.) **107**, 257–269 (1970b).

Crowe, A., Ragab, A.H.M.F.: A histochemical investigation of intrafusal fibres in tortoise muscle spindles. J. Histochem. Cytochem. **20**, 200–204 (1972).

Csillik, B.: Cholinesterase-active myoneural structures of alpha and gamma efferent fibres. Histochemie der Cholinesterase. Bibl. anat. (Basel) **2**, 161–173 (1961).

Csillik, B.: Functional structure of the post-synaptic membrane in the myoneural junction. Budapest: Akadémiai Kiadó 1965.

Cuajunco, F.: Embryology of the neuromuscular spindle. Contr. Embryol. Carneg. Instn. **19**, 45–72 (1927).

Cuajunco, F.: Development of the neuromuscular spindle in human fetuses. Contr. Embryol. Carneg. Instn. **28**, 95–128 (1940).

Daniel, P.M.: Spiral nerve endings in the extrinsic eye muscles of man. J. Anat. (Lond.) **80**, 189–192 (1946).

Dietert, S.E.: The demonstration of different types of muscle fibres in human extraocular muscle by electron microscopy and cholinesterase staining. Invest. Ophthal. **4**, 51–63 (1965).

Dogiel, A.S.: Methylenblautinktion der motorischen Nervenendigungen in den Muskeln der Amphibien und Reptilien. Arch. mikr. Anat. **35**, 305–320 (1890).

Dogiel, A.S.: Die Nervenendigungen im Bauchfell, in den Sehnen, den Muskelspindeln und dem Centrum tendineum des Diaphragmas beim Menschen und bei Säugetieren. Arch. mikr. Anat. **59**, 1–31 (1902).

Dogiel, A.S.: Nervenendigungen in der Pleura des Menschen und der Säugetiere. Arch. mikr. Anat. **62**, 244–250 (1903).

DOGIEL, A.S.: Zur Frage über den fibrillären Bau der Sehnenspindeln oder der Golgi'schen Körperchen (organo nervoso terminale musculo-tendineo). Arch. mikr. Anat. **67**, 638–646 (1906a).

DOGIEL, A.S.: Die Endigungen der sensiblen Nerven in den Augenmuskeln und deren Sehnen beim Menschen und den Säugetieren. Arch. mikr. Anat. **68**, 501–526 (1906b).

DÜRING, M., ANDRES, K.H.: Die Feinstruktur der Muskelspindeln von Mammalia. Anat. Anz. **124**, 566–573 (1969).

ECCLES, J.C.: The Physiology of nerve cells. Baltimore: The Johns Hopkins Press 1957.

ECCLES, J.C., SHERRINGTON, C.S.: Numbers and contraction-values of individual motor-units examined in some muscles of the limb. Proc. roy. Soc. B. **106**, 326–357 (1930).

EDSTRÖM, L., KUGELBERG, E.: Histochemical composition, distribution of fibres and fatiguability of single motor units. J. Neurol. Neurosurg. Psychiat. **31**, 424–433 (1968).

ELDRED, E., BRIDGMAN, C.F., SWETT, J.E., ELDRED, B.: Quantitative comparisons of muscle receptors of the cat's medial gastrocnemius, soleus and extensor digitorum brevis muscles. In: Symposium on Muscle Receptors. BARKER, D., Ed., pp. 207–213. Hong Kong: Hong Kong Univ. Press 1962.

ELDRED, E., YELLIN, H., GADBOIS, L., SWEENEY, S.: Bibliography on Muscle Receptors; their Morphology, Pathology and Physiology. Exp. Neurol. Suppl. **3** (1967).

ELLAWAY, P.H., EMONET-DÉNAND, F., JAMI, L., JOFFROY, M.: Proportion des fibres fusimotrices statiques et dynamiques dans les muscles peroneus longus et flexor hallucis longus du chat. C.R. Acad. Sci. (Paris) **274**, 3597–3600 (1972).

ELLAWAY, P.H., EMONET-DÉNAND, F., JOFFROY, M.: Mise en évidence d'axones squeletto-fusimoteurs (axones β) dans le muscle premier lombrical superficiel du Chat. J. Physiol. (Paris) **68**, 617–623 (1971).

ELLAWAY, P.H., EMONET-DÉNAND, F., JOFFROY, M., LAPORTE, Y.: Lack of exclusively fusimotor α axons in flexor and extensor leg muscles of the cat. J. Neurophysiol. **35**, 149–153 (1972).

EMONET-DÉNAND, F., JANKOWSKA, F., LAPORTE, Y.: Skeleto-fusimotor fibres in the rabbit. J. Physiol. (Lond.) **210**, 669–680 (1970).

EMONET-DÉNAND, F., JOFFROY, M., LAPORTE, Y.: Absence d'axones α exclusivement fusimoteurs dans les muscles prétibiaux du Chat. J. Physiol. (Paris) **63**, 46 (1971).

EMONET-DÉNAND, F., JOFFROY, M., LAPORTE, Y.: Fibres fusimotrices dont l'action sur la sensibilité phasique des terminaisons primaires dépend de leur fréquence de stimulation. C. R. Acad. Sci. (Paris) **275**, 89–91 (1972).

EMONET-DÉNAND, F., LAPORTE, Y.: Frequencygrams of rabbit spindle primary endings elicited by stimulation of fusimotor fibres. J. Physiol. (Lond.) **201**, 673–684 (1969).

ENGEL, W.K., IRWIN, R.L.: A histochemical-physiological correlation of frog skeletal muscle fibres. Amer. J. Physiol. **213**, 511–518 (1967).

ERLANGER, J., GASSER, H.S.: Electrical Signs of Nervous Activity. Philadelphia: University of Pennsylvania Press 1937.

ERULKAR, S.D., SHELANSKI, M.L., WHITSEL, B.L., OGLE, P.: Studies of muscle fibres of the tensor tympani of the cat. Anat. Rec. **149**, 279–297 (1964).

FARRELL, P.R., FEDDE, M.R.: Uniformity of structural characteristics throughout the length of skeletal muscle-fibres. Anat. Rec. **164**, 219–230 (1969).

FEINDEL, W.H., WEDDELL, G., SINCLAIR, D.C.: Pain sensibility in deep somatic structures. J. Neurol. Neurosurg. Psychiat. **11**, 113–117 (1948).

FERNAND, V.S.V., HESS, A.: The occurence, structure and innervation of slow and twitch muscle fibres in the tensor tympani and stapedius of the cat. J. Physiol. (Lond.) **200**, 547–554 (1969).

FERNAND, V.S.V., YOUNG, J.Z.: The sizes of the nerve fibres of muscle nerves. Proc. roy. Soc. B. **139**, 38–58 (1951).

FESSARD, A., SAND, A.: Stretch receptors in the muscles of fishes. J. exp. Biol. **14**, 383–404 (1937).

FITZGERALD, M.J.T.: Developmental changes in epidermal innervation. J. Anat. (Lond.) **95**, 495–514 (1961).

FITZGERALD, M.J.T.: On the structure and life history of bulbous corpuscles (*Corpuscula Nervorum Terminalia Bulboidea*). J. Anat. (Lond.) **96**, 189–208 (1962).

FLOYD, K.: Small multiple nerve endings in sheep oesophageal muscle. J. Physiol. (Lond). **216**, 37–38 (1971).

FORBES, W.: Experimental and histological studies of function in peripheral nerves in the cat. Ph. D. Thesis: University of Glasgow 1966.

FORSTER, L.: Note on foetal muscle spindles. J. Physiol. (Lond.) **28**, 201–203 (1894).

FREEMAN, M. A. R., WYKE, B.: The innervation of the knee joint. An anatomical and histological study in the cat. J. Anat. (Lond.) **101**, 505–532 (1967).

FUKAMI, Y.: Tonic and phasic muscle spindles in the snake. J. Neurophysiol. **33**, 28–35 (1970).

FUKAMI, Y., HUNT, C. C.: Structure of snake muscle spindles. J. Neurophysiol. **33**, 9–27 (1970).

GAIRNS, F. W.: A modified gold chloride method for the demonstration of nerve endings. Quart. J. micr. Sci. **74**, 151–153 (1930).

GARDNER, E.: The distribution and termination of nerves in the knee-joint of the cat. J. comp. Neurol. **80**, 11–32 (1944).

GARVEN, H. S. D.: The nerve-endings in the panniculus carnosus of the hedgehog, with special reference to the sympathetic innervation of striated muscle. Brain **48**, 380–441 (1925).

GERMINO, N. I., D'ALBORA, H.: Succinic-dehydrogenase activity in the neuromuscular spindles of the chick. Experientia (Basel) **21**, 45–46 (1965).

GIACOMINI, E.: Sui fusi neuro-musculari dei Sauropsidi. Atti Accad. Fisiocr. Siena Sez. med.-fis. **9**, 215–230 (1898).

GILL, H. I.: Neuromuscular spindles in human lateral pterygoid muscles. J. Anat. (Lond.) **109**, 157–167 (1971).

GINSBORG, B. L., MACKAY, B.: A histochemical demonstration of two types of motor innervation in avian skeletal muscle. Bibl. anat. (Basel) **2**, 174–181 (1961).

GLADDEN, M. H.: Muscle spindle innervation in the intertransverse caudal muscles of the rat. Experientia (Basel) **25**, 604–606 (1969).

GLADDEN, M. H.: The elastic fibres of muscle spindles. In: Abstracts of Communications given at Sixth Symposium on Current Research in Muscular Dystrophy and Related Diseases. Abstr. 1. 1972a.

GLADDEN, M. H.: Elastic fibres in muscle spindles of the cat. J. Physiol. (Lond.) **227**, 45–46 (1972b).

GOGLIA, G.: Ricerche ultrastrutturali sopra i fusi neuromuscolari del gatto, del coniglio e del ratto. Acta med. romana **8**, 7–53 (1970).

GOLGI, C.: Intorno alla distribuzione e terminazione dei nervi nei tendini dell'uomo e di altri vertebrati. Rend. Ist. Lombardo Sci. Lettere B **11**, 445–453 (1878).

GOLGI, C.: Sui nervi dei tendini dell'uomo e di altri vertebrati e di un nuovo organo nervoso terminale musculo-tendineo. Mem. Roy. Accad. Sci. Torino **32**, 359–385 (1880).

GOTO, K., LOEWENSTEIN, W. R.: An accessory small nerve fibre in a mechanoreceptor. Biol. Bull. **121(2)**, 391 (1961).

GRANIT, R. (Ed.): Muscular Afferents and Motor Control. Nobel Symposium I. New York: J. Wiley 1966.

GRAY, E. G.: The spindle and extrafusal innervation of a frog muscle. Proc. roy. Soc. B. **146**, 416–430 (1957).

GRAY, E. G.: The structure of fast and slow muscle fibres in the frog. J. Anat. (Lond.). **92**, 559–562 (1958).

GREENE, T., JAMPEL, R.: Muscle spindles in the extraocular muscles of the Macaque. J. comp. Neurol. **126**, 547–550 (1966).

GREGOR, A.: Über die Vertheilung der Muskelspindeln in der Musculatur des menschlichen Fötus. Arch. Anat. Physiol., Anat. Abt., 112–196 (1904).

GRUNER, JEAN-E.: La structure fine du fuseau neuromusculaire humain. Rev. neurol. **104**, 490–507 (1961).

GUTH, L., SAMAHA, F. J.: Qualitative differences between actomyosin ATPase of slow and fast mammalian muscle. Exp. Neurol. **25**, 138–152 (1969).

GUTMANN, E., GUTTMANN, L., MEDAWAR, P. B., YOUNG, J. Z.: The rate of regeneration of nerve. J. exp. Biol. **19**, 14–44 (1942).

HAGBARTH, K. E., WOHLFART, G.: The number of muscle-spindles in certain muscles in cat in relation to the composition of the muscle nerves. Acta anat. (Basel) **15**, 85–104 (1952).

Hamada, I.: Morphological study of nerve endings in the ocular muscles (In Japanese). J. Kyoto Prefect. med. Univ. **2**, 1641–1678 (1928); cited by Hosokawa, 1961.

Harker, D.W.: The structure and innervation of sheep superior rectus and levator palpebrae extraocular muscles. I. Extrafusal muscle fibres. Invest. Ophthal. **11**, 956–969 (1972a).

Harker, D.W.: The structure and innervation of sheep superior rectus and levator palpebrae extraocular muscles. II. Muscle spindles. Invest. Ophthal. **11**, 970–979 (1972b).

Harker, D.W.: The structure and innervation of sheep extraocular and foot muscles. Ph. D. Thesis: University of Durham 1974.

Hassall, A.H.: The Microscopic Anatomy of the Human Body, Vol. 1. London: Samuel Highley 1849.

Henneman, E., Olson, C.B.: Relations between structure and function in the design of skeletal muscles. J. Neurophysiol. **28**, 581–598 (1965).

Hennig, G.: Die Nervendigungen der Rattenmuskelspindel im elektronen und phasekontrastmikroskopischen Bild. Z. Zellforsch. **96**, 275–294 (1969).

Hess, A.: The structure of extrafusal muscle fibres in the frog and their innervation studied by the cholinesterase technique. Amer. J. Anat. **107**, 129–152 (1960).

Hess, A.: Two kinds of motor nerve endings on mammalian intrafusal muscle fibres revealed by the cholinesterase technique. Anat. Rec. **139**, 173–184 (1961a).

Hess, A.: The structure of slow and fast extrafusal muscle fibres in the extraocular muscles and their nerve endings in guinea pig. J. cell. comp. Physiol. **58**, 63–80 (1961b).

Hess, A.: Structural differences of fast and slow extrafusal muscle fibres and their nerve endings in chickens. J. Physiol. (Lond.) **157**, 221–231 (1961c).

Hess, A.: Two kinds of extrafusal muscle fibres and their nerve endings in the garter snake. Amer. J. Anat. **113**, 347–364 (1963).

Hess, A.: The structure of vertebrate slow and twitch muscle fibres. Invest. Ophthal. **6**, 217–228 (1967).

Hess, A.: Vertebrate slow muscle fibres. Physiol. Rev. **50**, 40–62 (1970).

Hewer, E.E.: The development of nerve endings in the human foetus. J. Anat. (Lond.) **69**, 369–379 (1935).

Hines, M.: The innervation of the muscle spindle. Proc. Res. nerv. ment. Dis. **9**, 124–137 (1930).

Hines, M., Tower, S.S.: Studies on the innervation of skeletal muscles. II. Of muscle spindles in certain muscles of the kitten. Bull. Johns Hopk. Hosp. **42**, 264–307 (1928).

Hinsey, J.C.: Some observations on the innervation of skeletal muscle of the cat. J. comp. Neurol. **44**, 87–195 (1927).

Hinsey, J.C.: The innervation of skeletal muscle. Physiol. Rev. **14**, 514–585 (1934).

Hinsey, J.C., Hare, K., Phillips, R.A.: Sensory components of the phrenic nerve of the cat. Proc. Soc. exp. Biol. (N.Y.) **41**, 411–414 (1939).

Hirano, N.: Histologische Untersuchungen über die nervöse Innervation der menschlichen äußeren Augenmuskeln. Albrecht v. Graefes Arch. klin. exp. Ophthal. **142**, 560–575 (1941).

Holmes, W.: Holmes's silver method. In: Carleton's Histological Technique. 4th edit. Drury, R.A.B., Wallington, E.A., Eds. London: Oxford University Press 1967.

Homma, S., Seki, Y.: Muscle spindles in phasic and tonic muscle. Tohoku J. exp. Med. **83**, 391–397 (1964).

Honee, G.L.J.M.: An investigation on the presence of muscle spindles in the human lateral pterygoid muscle. Ned. T. Tandheelk. **73**, 43–48 (1966).

Hosokawa, H.: Proprioceptive innervation of striated muscles in the territory of cranial nerves. Tex. Rep. Biol. Med. **19**, 405–464 (1961).

Houk, J., Henneman, E.: Responses of Golgi tendon organs to active contractions of the soleus muscle of the cat. J. Neurophysiol. **30**, 466–481 (1967).

Hromada, J., Poláček, P.: A contribution to the morphology of encapsulated nerve endings in the joint capsule and the periarticular tissue. Acta anat. (Basel) **33**, 187–202 (1958).

Hubbard, S.J., Hess, A.: The fine structure of the primary sensory zone of cat muscle spindles. Anat. Rec. **157**, 262 (1967).

HUBER, G.C.: Sensory nerve terminations in the tendons of the extrinsic eye-muscles of the cat. J. comp. Neurol. **10**, 152–158 (1900).

HUBER, G. C., DE WITT, L. M.: A contribution on the motor nerve-endings and on the nerve-endings in muscle spindles. J. comp. Neurol. **7**, 169–230 (1897).

HUBER, G. C., DE WITT, L. M.: A contribution on the nerve terminations in neuro-tendinous and end-organs. J. comp. Neurol. **10**, 159–208 (1900).

HUNT, C.C.: Relation of function to diameter in afferent fibres of muscle nerves. J. gen. Physiol. **38**, 117–131 (1954).

HUNT, C.C., MCINTYRE, A.K.: Characteristics of responses from receptors from the flexor longus digitorum muscle and the adjoining interosseous region of the cat. J. Physiol. (Lond.) **153**, 74–87 (1960).

HUNT, C.C., WYLIE, R.M.: Responses of snake muscle spindles to stretch and intrafusal muscle fibre contractions. J. Neurophysiol. **33**, 1–8 (1970).

HUNTER, J.I.: Lectures on the sympathetic innervation of striated muscle. Brit. med. J. **1**, 197–201 (1925).

HURSH, J.B.: Conduction velocity and diameter of nerve fibres. Amer. J. Physiol. **127**, 131–139 (1939).

IP, M.C.: The number and variety of proprioceptors in certain muscles of the cat. M. Sc. Thesis: University of Hong Kong 1961.

ITO, F.: Abortive spike of leaf-like receptors in the frog muscle. J. Physiol. Soc. Jap. **30**, 411–412 (1968a).

ITO, F.: Functional properties of leaf-like muscle receptors in frog sartorius muscle. Jap. J. Physiol. **18**, 590–600 (1968b).

ITO, F.: Responses in the sensory nerve terminal of leaf-like receptors in the frog sartorius muscle. Jap. J. Physiol. **21**, 431–441 (1971).

ITO, F., TOYAMA, K., ITO, R.: A comparative study on structure and function between the extrafusal receptor and the spindle receptor in the frog. Jap. J. Physiol. **14**, 12–33 (1964).

JAHN, U.: Morphological observations on living neuromuscular spindles. Acta anat. (Basel) **39**, 341–350 (1959).

JAMES, N.T.: The histochemical demonstration of myoglobin in muscle spindles. Histochemical Journal **3**, 333–338 (1971a).

JAMES, N.T.: The histochemical demonstration of three types of intrafusal fibre in rat muscle spindles. Histochemical Journal **3**, 457–462 (1971b).

JAMES, N.T., MEEK, G.A.: The ultrastructure of the avian muscle spindle. J. Anat. (Lond.) **111**, 489–491 (1972).

JONES, E.G.: The innervation of muscle spindles in the Australian opposum, *Trichosurus vulpecula*, with special reference to the motor nerve endings. J. Anat. (Lond.) **100**, 733–759 (1966).

KARLSEN, K.: The location of motor end plates and the distribution and histological structure of muscle spindles in jaw muscles of the rat. Acta odont. scand. **23**, 521–547 (1965).

KARLSSON, U.L.: The frog muscle spindle: Ultrastructure and intrafusal stretch characteristics. In: Research in Muscle Development and the Muscle Spindle. BANKER, B.Q. et al., Eds., pp. 299–332. Amsterdam: Excerpta Medica 1972.

KARLSSON, U.L., ANDERSSON-CEDERGREN, E.: Motor myoneural junctions on frog intrafusal muscle fibres. J. Ultrastruct. Res. **14**, 191–211 (1966).

KARLSSON, U.L., ANDERSSON-CEDERGREN, E.: Small leptomeric organelles in intrafusal muscle fibres of the frog as revealed by electron microscopy. J. Ultrastruct. Res. **23**, 417–426 (1968).

KARLSSON, U.L., ANDERSSON-CEDERGREN, E.: Satellite cells of the frog muscle spindle as revealed by electron microscopy. J. Ultrastruct. Res. **34**, 426–438 (1970).

KARLSSON, U.L., ANDERSSON-CEDERGREN, E., OTTOSON, D.: Cellular organization of the frog muscle spindle as revealed by serial sections for electron microscopy. J. Ultrastruct. Res. **14**, 1–35 (1966).

KARNOVSKY, M.J., ROOTS, L.: A 'direct-coloring' thiocholine method for cholinesterase. J. Histochem. Cytochem. **12**, 219–221 (1964).

KATZ, B.: The terminations of the afferent nerve fibre in the muscle spindle of the frog. Phil. Trans. B. **243**, 221–240 (1961).

KELLY, A.M., ZACKS, S.I.: The histogenesis of rat intercostal muscle. J. Cell Biol. **42**, 135–152 (1969).

KENNEDY, W.R.: Innervation of normal human muscle spindles. Neurology (Minneap.) **20**, 463–475 (1970).

KERSCHNER, L.: Bemerkungen über ein besonderes Muskelsystem im willkürlichen Muskel. Anat. Anz. **3**, 126–132 (1888a).

KERSCHNER, L.: Beitrag zur Kenntnis der sensiblen Endorgane. Anat. Anz. **3**, 288–296 (1888b).

KERSCHNER, L.: Über Muskelspindeln. Anat. Anz. **7**, 85–89 (1892).

KERSCHNER, L.: Über die Fortschritte in der Erkenntnis der Muskelspindeln. Anat. Anz. **8**, 449–458 (1893).

KIDD, G.L.: Excitation of primary muscle spindle endings by β-axon stimulation. Nature (Lond.) **203**, 1248–1251 (1964).

KÖLLIKER, A.: On the termination of nerves in muscles, as observed in the frog, and on the disposition of the nerves in the frog's heart. Proc. roy. Soc. B. **12**, 65–84 (1862).

KUBOTA, K., MASEGI, T.: Muscle spindle distribution in snout musculature of the Japanese shrew-mole. Anat. Rec. **172**, 703–710 (1972).

KUFFLER, S.W.: The two skeletal nerve-muscle systems in frog. Naunyn-Schmiedebergs Arch. Pharmak. exp. Path. **220**, 116–135 (1953).

KUFFLER, S.W., GERARD, R.W.: The small-nerve motor system to skeletal muscle. J. Neurophysiol. **10**, 383–394 (1947).

KUFFLER, S.W., HUNT, C.C., QUILLIAM, J.P.: Function of medullated small-nerve fibres in mammalian ventral roots: efferent muscle spindle innervation. J. Neurophysiol. **14**, 29–54 (1951).

KÜHNE, W.: Über die Endigung der Nerven in den Nervenhügeln der Muskeln. Virchows Arch. path. Anat. **30**, 187–220 (1864).

KULCHITSKY, N.: Nerve endings in muscles. J. Anat. (Lond.) **58**, 152–168 (1924a).

KULCHITSKY, N.: Nerve endings in muscles of the frog. J. Anat. (Lond.) **59**, 1–17 (1924b).

KUPFER, C.: Motor innervation of extra-ocular muscle. J. Physiol. (Lond.) **153**, 522–526 (1960).

LAMBERTINI, G.: Nouvelles recherches sur les récepteurs nerveux de l'homme et du singe practiquées a l'aide de la méthode de Ruffini. Bull. Ass. Anat. (Nancy) **47**, 3–30 (1961).

LANDON, D.N.: Electron microscopy of muscle spindles. In: Control and innervation of skeletal muscle. ANDREW, B.L., Ed., pp. 96–111. Edinburgh-London: Livingstone 1966.

LANDON, D.N.: Observations on the morphogenesis of rat skeletal muscle. J. Anat. (Lond.) **107**, 385–387 (1970).

LANDON, D.N.: A quantitative study of some fine structural features of developing myotubes in the rat. J. Anat. (Lond.) **110**, 170–171 (1971).

LANDON, D.N.: The fine structure of developing muscle spindles in the rat. J. Anat. (Lond.) **111**, 512–513 (1972a).

LANDON, D.N.: The fine structure of the equatorial regions of developing muscle spindles in the rat. J. Neurocytol. **1**, 189–210 (1972b).

LANGLEY, J.N.: The nerve fibre constitution of peripheral nerves and of nerve roots. J. Physiol. (Lond.) **56**, 382–395 (1922).

LÄNNERGREN, J., SMITH, R.S.: Types of muscle fibres in toad skeletal muscle. Acta physiol. scand. **68**, 263–274 (1966).

LE GROS CLARK, W.E.: The tissues of the body, 5th Ed. Oxford: Clarendon Press 1965.

LEKSELL, L.: The action potential and excitatory effects of the small ventral root fibres to skeletal muscle. Acta physiol. scand. **10**, (Suppl. 31), 1–84 (1945).

LLOYD, D.P.C.: Neuron patterns controlling transmission of ipsilateral hind limb reflexes in cat. J. Neurophysiol. **6**, 293–315 (1943).

LLOYD, D.P.C., CHANG, H.T.: Afferent fibres in muscle nerves. J. Neurophysiol. **11**, 199–208 (1948).

LOEWENSTEIN, W.R.: Modulation of cutaneous mechanoreceptors by sympathetic stimulation. J. Physiol. (Lond.) **132**, 40–60 (1956).

LOEWENSTEIN, W.R., GOTO, K., NOBACK, C.: C fibre innervation of a mechanoreceptor. Experientia (Basel) **18**, 460 (1962).

LOWENSTEIN, O.: Pressure receptors in the fins of the dogfish *Scylliorhinus canicula*. J. exp. Biol. **33**, 417–421 (1956).

MAHRAN, Z.Y., SAKLA, F.B.: The pattern of innervation of the extrinsic ocular muscles and the intra-orbital ganglia of the albino mouse. Anat. Rec. **152**, 173–184 (1965).

MAIER, A., ELDRED, E.: Comparisons in the structure of avian muscle spindles. J. comp. Neurol. **143**, 25–40 (1971).

MAIER, A., ELDRED, E., EDGERTON, V.G.: The effects on spindles of muscle atrophy and hypertrophy. Exp. Neurol. **37**, 100–123 (1972).

MAIER, A., DE SANTIS, M., ELDRED, E.: Absence of muscle spindles in avian extraocular muscles. Exp. Eye Res. **12**, 251–253 (1971).

MARCHAND, R., BRIDGMAN, C.F., SHUMPERT, E., ELDRED, E.: Association of tendon organs with spindles in muscles of the cat's leg. Anat. Rec. **169**, 23–32 (1971).

MARCHAND, E.R., ELDRED, E.: Postnatal increase of intrafusal fibres in the rat muscle spindle. Exp. Neurol. **25**, 655–676 (1969).

MATHER, V., HINES, M.: Studies in the innervation of skeletal muscle. V. The limb muscles of the newt, *Triturus torosus*. Amer. J. Anat. **54**, 177–201 (1934).

MATTHEWS, P.B.C.: Muscles spindles and their motor control. Physiol. Rev. **44**, 219–288 (1964).

MATTHEWS, P.B.C.: Mammalian muscle receptors and their central actions. London: Edward Arnold 1972.

MATTHEWS, P.B.C., WESTBURY, D.R.: Some effects of fast and slow motor fibres on muscle spindles of the frog. J. Physiol. (Lond.) **178**, 178–192 (1965).

MAVRINSKAIA, L.F.: On the relationship between the development of the nerve endings of the skeletal muscles and the appearance of movement activity in the human foetus. Arkh. Anat. Gistol. Embriol. **38**, 61–68 (1960).

MAYNARD, J.A., TIPTON, C.M.: The effects of exercise training and denervation on the morphology of intrafusal muscle fibres. Int. Z. angew. Physiol. **30**, 1–9 (1971).

MAYR, R.: Untersuchungen an isolierten Muskelspindeln der Ratte nach Cholinesterasedarstellung und Sudanschwarz-Färbung. Z. Zellforsch. **93**, 594–606 (1969).

MAYR, R.: Zwei elektronenmikroskopisch unterscheidbare Formen sekundärer sensorischer Endigungen in einer Muskelspindel der Ratte. Z. Zellforsch. **110**, 97–107 (1970).

MERRILLEES, N.C.R.: Some observations on the fine structure of a Golgi tendon organ of a rat. In: Symposium on Muscle Receptors. BARKER, D., Ed., pp. 199–206. Hong Kong: Hong Kong Univ. Press 1962.

MERRILLEES, N.C.R., SUNDERLAND, S., HAYHOW, W.: Neuromuscular spindles in the extraocular muscles in man. Anat. Rec. **108**, 23–30 (1950).

MIESCHER, J. F.: Über eigenthümliche Schläuche in den Muskeln einer Hausmaus. Bericht Ver. Nat. Ges., Basel. **5**, 198–202 (1842).

MILBURN, A.: The early development of muscle spindles in the rat. J. Cell Sci. **12**, 175–195 (1973).

MUNK, H.: Zur Anatomie und Physiologie der Quergestreiften Muskelfasern der Wirbelthiere, mit Anschluss von Beobachtungen über die electrischen Organe der Fische. Göttingen Nachrichten No. **1**, 1–14 (1858).

OGATA, T., MORI, M.: Histochemical demonstration of the three types of intrafusal fibres of muscle spindles, a study on oxidative enzymes. Acta Med. Okayama **16**, 347–350 (1962).

OGATA, T., MORI, M.: Histochemical study of exidative enzymes in vertebrate muscles. J. Histochem. Cytochem. **12**, 171–182 (1964).

O'LEARY, J., HEINBECKER, P., BISHOP, G.H.: Analysis of function of a nerve to a muscle. Amer. J. Physiol. **110**, 636–658 (1935).

OLSON, C.B., SWETT, C.P.: A functional and histochemical characterization of motor units in a heterogeneous muscle (flexor digitorum longus) of the cat. J. comp. Neurol. **128**, 475–498 (1966).

ONANOFF, J. G.: Sur la nature des faisceaux neuro-musculaires. C. R. Acad. Sci. (Paris) **42**, 432–433 (1890).

ORKLAND, R.K.: A further study of electrical responses in slow and twitch muscle fibres of the frog. J. Physiol. (Lond.) **167**, 181–191 (1963).

Oshima, T.: Anatomical studies of the muscle spindle. J. Nagasaki med. Ass. **16**, 525–548 (1938); cited by Hosokawa 1961.

Ovalle, W.K.: Fine structure of rat intrafusal muscle fibres. The polar region. J. Cell Biol. **51**, 83–103 (1971).

Ovalle, W.K.: Fine structure of rat intrafusal muscle fibres. The equatorial region. J. Cell Biol. **52**, 382–396 (1972a).

Ovalle, W.K.: Motor nerve terminals on rat intrafusal muscle fibres; a correlated light and electron microscopic study. J. Anat. (Lond.) **111**, 239–253 (1972b).

Ovalle, W.K., Smith, R.S.: Histochemical identification of three types of intrafusal muscle fibres in the cat and monkey based on the myosin ATPase reaction. Canad. J. Physiol. Pharmacol. **50**, 195–202 (1972).

Page, S.G.: Intrafusal muscle fibres in the frog. J. Microscopie **5**, 101–104 (1966).

Page, S.G.: Fine structure of tortoise skeletal muscle. J. Physiol. (Lond.) **197**, 707–715 (1968).

Paintal, A.S.: Functional analysis of Group III afferent fibres of mammalian muscles. J. Physiol. (Lond.) **152**, 250–270 (1960).

Paintal, A.S.: Responses and reflex effects of pressure-pain receptors of mammalian muscles. In: Symposium on Muscle Receptors. Barker, D., Ed., pp. 133–142. Hong Kong: Hong Kong Univ. Press 1962.

Pallot, D.J.: Structural and functional studies on snake muscle spindles. Ph. D. Thesis: University of Bristol 1973.

Pallot, D.J., Ridge, R.M.A.P.: Ultrastructure of snake muscle spindles. J. Physiol. (Lond.) **218**, 17–18 (1971).

Pallot, D.J., Ridge, R.M.A.P.: The fine structure of the long-capsule muscle spindles in the snake *Natrix* sp. J. Anat. (Lond.) **113**, 61–74 (1972).

Pallot, D.J., Ridge, R.M.A.P.: The fine structure of the short-capsule muscle spindles of *Natrix* sp. J. Anat. (Lond.) **114**, 13–24 (1973).

Pallot, G.: Contribution à l'étude des terminaisons nerveuses dans le muscle striée: Recherches cytologiques sur les fuseaux de type Kühne. Bull. Histol. Techn. micr. **11**, 337–364 (1934).

Palmer, J.M., Stilwell, D.L.: Analysis of innervation of tenuissimus muscle of cat. Anat Rec. **130**, 434 (1958).

Perroncito, A.: Sur la terminaison des nerfs dans les fibres musculaires striés. Arch. ital. Biol. **36**, 245–254 (1901).

Perroncito, A.: Études ultérieures sur la terminaisons des nerfs dans les muscles a fibres striés. Arch. ital. Biol. **38**, 392–411 (1902).

Pezard, A., May, R.M.: Les terminaisons nerveuses du muscle couturier de la grenouille et la question de sa partie aneurale. Ann. Physiol. Physicochim. biol. **13**, 460–473 (1937).

Poloumordwinoff, D.: Recherches sur les terminaisons nerveuses sensitives dans les muscles striés volontaires. Soc. Sci. Arcachon Sta. Biol. Bull. **3**, 73–79 (1898).

Porayko, O., Smith, R.S.: Morphology of muscle spindles in the rat. Experientia (Basel) **24**, 588–589 (1968).

Proske, U.: Responses of muscle spindles in the lizard. Nature (Lond.) **213**, 1144–1146 (1967).

Proske, U.: The innervation of muscle spindles in the lizard *Tiliqua nigrolutea*. J. Anat. (Lond.) **105**, 217–230 (1969).

Proske, U.: The muscle spindles in slow and twitch skeletal muscle of the lizard. J. Physiol. (Lond.) **230**, 429–448 (1973).

Proske, U., Vaughan, P.: Histological and electrophysiological identification of lizard muscle fibres. J. Physiol. (Lond.) **199**, 495–509 (1968).

Ragab, A.H.M.F.: The structure and innervation of tortoise muscle spindles. Ph. D. Thesis: University of Durham 1970.

Rakhawy, M.T., Shehata, S.H., Badawy, Z.H.: The proprioceptive innervation of the lateral pterygoid muscle in man and some other mammals. Acta anat. (Basel) **79**, 581–598 (1971).

Ralston, H.J., Miller, M.R., Kasahara, M.: Nerve endings in human fasciae, tendons, ligaments, periosteum, and joint synovial membrane. Anat. Rec. **136**, 137–147 (1959).

RANSON, S.W., DAVENPORT, H.K.: Sensory unmyelinated fibres in the spinal nerves. Amer. J. Anat. **48**, 331–353 (1931).

REBOLLO, M.A., DE ANDA, G.: The neuromuscular spindles in the adult chicken. II. Histochemistry. Acta anat. (Basel) **67**, 595–608 (1967).

REGAUD, C.: Les terminaisons nerveuses et les organes nerveux sensitifs de l'appareil locomoteur. II. Les terminaisons nerveuses et les organes nerveux sensitifs des tendons, de leurs gaines, des enveloppes conjonctives musculo-tendineuses du perioste et des tissus articulaires. Rev. gén. Histol. **1**, 587–689 (1907).

REGAUD, C., FAVRE, M.: Les terminaisons nerveuses et les organes nerveux sensitifs de l'appareil locomoteur. I. Les terminaisons nerveuses et les organes nerveux sensitifs des muscles striés squelettaux. Rev. gén. Histol. **1**, 1–140 (1904).

REXED, B., THERMAN, P.O.: Calibre spectra of motor and sensory nerve fibres to flexor and extensor muscles. J. Neurophysiol. **11**, 133–140 (1948).

RIDGE, R.M.A.P.: Different types of extrafusal muscle fibres in snake costocutaneous muscles. J. Physiol. (Lond.) **217**, 393–418 (1971).

ROBERTSON, J.D.: Electron microscopy of the motor end-plate and the neuromuscular spindle. Amer. J. phys. Med. **39**, 1–43 (1960).

RUFFINI, A.: Sur un reticule nerveux special, et sur quelques corpuscules de Pacini qui se trouvent en connexion avec les organes musculo-tendineux du chat. Arch. ital. Biol. **18**, 101–105 (1893a).

RUFFINI, A.: Sur la terminaison nerveuse dans les faisceaux musculaires et leur signification physiologique. Arch. ital. Biol. **18**, 106–114 (1893b).

RUFFINI, A.: Sur un nouvel organe nerveux terminal et sur la présence des corpuscules Golgi-Mazzoni dans le conjonctif sous-cutané de la pulpe des doigts de l'homme. Arch. ital. Biol. **21**, 249–265 (1894).

RUFFINI, A.: Sopra due speciali modi d'innervazione degli organi di Golgi. Ric. Lab. Anat. Norm. Roy. Univ. Roma **6**, 231–253 (1897).

RUFFINI, A.: On the minute anatomy of the neuromuscular spindles of the cat, and on their physiological significance. J. Physiol. (Lond.) **23**, 190–208 (1898).

RUFFINI, A.: Sur les expansions nerveuses de la peau chez l'homme et quelques autres mammifères. Rev. gén. Histol. **1**, 421–540 (1905).

RUMPELT, H.J., SCHMALBRUCH, H.: Zur Morphologie der Bauelemente von Muskelspindeln bei Mensch und Ratte. Z. Zellforsch. **102**, 601–630 (1969).

SAĞLAM, M.: Morphologische und quantitative Untersuchungen über die Muskelspindeln in der Nackenmuskulatur (M. biventer cervicis, M. rectus capitis dorsalis, and M. rectus capitis lateralis) des Bunt- und Blutspechtes. Acta anat. (Basel) **69**, 87–104 (1968).

SAMPAOLO, C.L.: Le espansioni sensitive nei muscoli oculari del vitello. Boll. Soc. ital. Biol. sper. **28**, 1305–1308 (1952).

SANDERS, F.K.: The thickness of the myelin sheaths of normal and regenerating peripheral nerve fibres. Proc. roy. Soc. B. **134**, 323–357 (1948).

SANTINI, M.: Noradrenergic fibres in Pacinian corpuscles. Anat. Rec. **160**, 494 (1968).

SANTINI, M.: New fibres of sympathetic nature in the inner core region of Pacinian corpuscles. Brain Res. **16**, 53–538 (1969).

SANTINI, M., IBATA, Y.: The fine structure of thin unmyelinated axons within muscle spindles. Brain Res. **33**, 289–302 (1971).

SANTINI, M., IBATA, Y., PAPAS, G.D.: The fine structure of the sympathetic axons within the Pacinian corpuscle. Brain Res. **33**, 279–287 (1971).

SAS, J., APPELTAUER, C.: Atypical muscle spindles in the extrinsic eye muscles of man. Acta anat. (Basel) **55**, 311–322 (1963).

SAS, J., SCHÁB, R.: Die Sogenannten »Palisaden-Endigungen« der Augenmuskeln. Acta morph. Acad. Sci. hung. **2**, 259–266 (1952).

SCALZI, H.A., PRICE, H.M.: The arrangement and sensory innervation of the intrafusal fibres in the feline muscle spindle. J. Ultrastruct. Res. **36**, 375–390 (1971).

SCALZI, H.A., PRICE, H.M.: Electron-microscopic observations of the sensory region of the mammalian muscle spindle. In: Research in muscle development and the muscle spindle. BANKER et al., Eds., pp. 254–263. Amsterdam: Excerpta Medica 1972.

Schoultz, T.W., Swett, J.E.: The fine structure of the Golgi tendon organ. J. Neurocytol. **1**, 1–26 (1972).

Sfameni, A.: Recherches anatomiques sur l'existence des nerfs et sur leur mode de se terminer dans le tissu adipeux, dans le périoste, dans le périchondre et dans les tissus qui renforcent les articulations. Arch. ital. Biol. **38**, 49–101 (1902).

Shantha, T.R., Golarz, M.N., Bourne, G.H.: Histological and histochemical observations on the capsule of the muscle spindle in normal and denervated muscle. Acta anat. (Basel) **69**, 632–646 (1968).

Sherrington, C.S.: Notes on the arrangements of some motor fibres in the lumbosacral plexus. J. Physiol. (Lond.) **13**, 621–772 (1892).

Sherrington, C.S.: On the anatomical constitution of nerves to skeletal muscles; with remarks on recurrent fibres in the ventral spinal nerve-root. J. Physiol. (Lond.) **17**, 211–258 (1894).

Sherrington, C.S.: Further note on the sensory nerves of muscles. Proc. roy. Soc. B. **61**, 247–249 (1897).

Shimizu, M.: Ultrastructure of Golgi's tendon spindle of mouse (In Japanese). J. Yonago Med. Ass. **22**, 285–296 (1971).

Siemmerling, C.: Ein Fall von gummöser Erkrankung der Hirnbasis mit Beteiligung des Chiasma Nervorum Opticorum. Arch. Psychiat. Nervenkr. **19**, 423 (1888).

Skoglund, S.: Anatomical and physiological studies of knee joint innervation in the cat. Acta physiol. scand. **36**, Suppl. 124, 1–101 (1956).

Slawik, F.F.: Über das Vorkommen von Muskelspindeln in der Muscularis propria des menschlichen Ösophagus. Anat. Anz. **93**, 133–137 (1942).

Smirnoff, A.E.: Sur les nerfs des tendons chez *Rana temporaria*, *Rana esculenta* and *Bufo viridis*. C. r. Séanc. Soc. nat. Univ. Kasan (In Russian) (1890); cited by Regaud, 1907.

Smirnoff, A.E.: Sur les nerfs des tendons chez *Rana temporaria*, *Rana esculenta* and *Bufo vulgaris*. Mem. Acad. Imp. Sci. Saint-Petersbourg (In Russian) **73**, Suppl. 4 (1893); cited by Regaud, 1907.

Smith, R.S.: Intrafusal musculature in *Xenopus laevis*. Acta physiol. scand. **59**, Suppl. 213 (1963).

Sommerkamp, H.: Das Substrat der Dauerverkürzung am Froschmuskel (Physiologische und pharmakologische Sonderstellung bestimmter Muskelfasern). Naunyn-Schmiedebergs Arch. Pharmak. exp. Path. **128**, 99–115 (1928).

Spiro, A.J., Beilin, R.L.: Histochemical duality of rabbit intrafusal fibres. J. Histochem. Cytochem. **17**, 348–349 (1969).

Sprague, J.M.: The distribution of dorsal root fibres on motor cells in the lumbosacral spinal cord of the cat, and the site of excitatory and inhibitory terminals in monosynaptic pathways. Proc. roy. Soc. B. **149**, 534–556 (1958).

Stacey, M.J.: The innervation of mammalian skeletal muscle by medium and small diameter afferent nerve fibres. Ph. D. Thesis: University of Durham 1967.

Stacey, M.J.: Free endings in skeletal muscle of the cat. J. Anat. (Lond.) **105**, 231–254 (1969).

Steer, J.C., Horney, F.D.: Evidence for passage of cerebrospinal fluid among spinal nerves. Canad. med. Ass. J. **98**, 71–74 (1968).

Stefanelli, A.: Terminazioni nervose di moto e di senso nei petromizonti. Arch. Zool. ital. **17**, 85–101 (1932).

Sterling, R.J.: Extrafusal and intrafusal end-plate morphology in frog muscle (in preparation 1973).

Stilwell, D.L., jr.: Regional variations in the innervation of deep fasciae and aponeuroses. Anat. Rec. **127**, 635–654 (1957a).

Stilwell, D.L., jr.: The innervation of tendons and aponeuroses. Amer. J. Anat. **100**, 289–317 (1957b).

Stilwell, D.L., jr.: The innervation of the deep structures of the hand. Amer. J. Anat. **101**, 75–99 (1957c).

Stilwell, D.L., jr.: The innervation of deep structures of the foot. Amer. J. Anat. **101**, 59–74 (1957d).

Sunderland, S., Roche, A.F.: Axon-myelin relationships in peripheral nerve fibres. Acta anat. (Basel) **33**, 1–37 (1958).

SUTTON, A.C.: On the development of the neuro-muscular spindle in the extrinsic eye muscles of the pig. Amer. J. Anat. **18**, 117–144 (1915).

SWASH, M., FOX, K.P.: Muscle spindle innervation in man. J. Anat. (Lond.) **112**, 61–80 (1972).

SWETT, J.E., ELDRED, E.: Distribution and numbers of stretch receptors in medial gastrocnemius and soleus muscles of the cat. Anat. Rec. **137**, 453–460 (1960a).

SWETT, J.E., ELDRED, E.: Comparison in structure of stretch receptors in medial gastrocnemius and soleus muscles of the cat. Anat. Rec. **137**, 461–473 (1960b).

SZEPSENWOL, J.: The neuromuscular spindle in the lizard *Anolis cristatellus*. Cellule **61**, 21–37 (1960).

TELLO, J.F.: Genesis de las terminaciones nerviosas motrices y sensitivas. I. En el sistema locomotor de los vertebrados superiores. Histogenesis muscular. Trab. Lab. Invest. biol. Univ. Madrid **15**, 101–199 (1917).

TELLO, J.F.: Die Entstehung der motorischen und sensiblen Nervenendigungen. Z. ges. Anat. **64**, 348–440 (1922).

TELLO, J.F.: Contribution à la connaissance des terminaisons sensitives dans les organes génitaux externes et de leur développement. Trav. rech. biol. Univ. Madrid **28**, 1–58 (1932).

VON THIEL, W.: Bau und funktionelle Bedeutung einiger isolierter markhaltiger Nervenfasern des Musculus piriformis der Katze. Acta anat. (Basel) **37**, 137–153 (1959).

THOMPSON, J.: Parallel spindle systems in the small muscles of the rat tail. J. Physiol. (Lond.) **211**, 781–799 (1970).

TIEGS, O.W.: Innervation of voluntary muscle. Physiol. Rev. **33**, 90–144 (1953).

TOWER, S.S.: Atrophy and degeneration in the muscle spindle. Brain **55**, 77–90 (1932).

TOZER, F.M., SHERRINGTON, C.S.: Receptors and afferents of the third, fourth, and sixth cranial nerves. Proc. roy. Soc. B. **82**, 450–457 (1910).

TRANZER, J.P., THOENEN, H.: Electronmicroscopic localization of 5-hydroxydopamine (3, 4, 5-trihydroxy-phenyl-ethylamine) as new 'false' sympathetic transmitter. Experientia (Basel) **23**, 743–744 (1967).

TUFFERY, A.R.: Growth and degeneration of motor end-plates in normal cat hind limb muscles. J. Anat. (Lond.) **110**, 221–247 (1971).

VOSS, H.: Untersuchungen über Zahl, Anordnung und Länge der Muskelspindeln in den Lumbricalmuskeln des Menschen und einiger Tiere. Z. mikr.-anat. Forsch. **42**, 509–524 (1937).

VOSS, H.: Ein ungewöhnliches und reichliches Vorkommen von freien Zellen im Kapselraum der Muskelspindeln beim Neugeborenen. Anat. Anz. **110**, 417–419 (1962a).

VOSS, H.: Sonderformen der Muskelspindeln des Menschen. Anat. Anz. **110**, 420–424 (1962b).

VOSS, H.: Untersuchungen über die absolute und relative Zahl der Muskelspindeln in weiteren Muskelgruppen (Mm. Scaleni und Rückenmuskeln) des Menschen. Anat. Anz. **112**, 276–279 (1963).

WALKER, L.B., JR.: Diameter spectrum of intrafusal muscle fibres in muscle spindles of the dog. Anat. Rec. **133**, 385 (1958).

WALKER, L.B., JR., RAJAGOPAL, M.D.: Neuromuscular spindles in the human tongue. Anat. Rec. **133**, 438 (1959).

WEDDELL, G., HARPMAN, J.A.: The neurohistological basis for the sensation of pain provoked from deep fascia, tendon, and periosteum. J. Neurol. Psychiat. **3**, 319–328 (1940).

WEISMANN, A.: Über das Wachsen der quergestreiften Muskeln nach Beobachtungen am Frosch. Z. Rat. Med. **10**, 263–284 (1961).

WERNER, J.K.: Development of the neuromuscular spindle. Amer. J. phys. Med. **51**, 192–207 (1972).

WINCKLER, G., FOROGLOU, C.: Caractéristiques des 'nuclear bag' et des 'nuclear chain' fibres des fuseaux neuro-musculaires. Brain Res. **42**, 289–296 (1972).

WHITTERIDGE, D.: The effect of stimulation of intrafusal muscle fibres on sensitivity to stretch of extraocular muscle spindles. Quart. J. exp. Physiol. **44**, 385–393 (1959).

WIRSEN, C., LARSSON, K.S.: Histochemical differentiation of skeletal muscle in foetal and newborn mice. J. Embryol. exp. Morph. **12**, 759–767 (1964).

Wohlfart, G., Henriksson, K. G.: Observations on the distribution, number and innervation of Golgi musculo-tendinous organs. Acta anat. (Basel) **41**, 192–204 (1960).

Wolter, J. R.: Morphology of the sensory nerve apparatus in striated muscle of the human eye. Arch. Ophthal. **53**, 201–207 (1955).

Yellin, H.: A histochemical study of *Muscle spindles* and their relationship to extrafusal fibre types in the rat. Amer. J. Anat. **125**, 31–46 (1969).

Yellin, H., Guth, L.: The histochemical classification of muscle fibers. Exp. Neurol. **26**, 424–432 (1970).

Yokomatsu, K.: Studies of nerves and nerve endings in the visual organs. VI. Nerve endings in the extrinsic eye muscles (In Japanese). Acta anat. nippon. **4**, 1331–1354 (1932); cited by Hosokawa, 1961.

Young, J. Z.: The functional repair of nervous tissue. Physiol. Rev. **22**, 318–374 (1942).

Zelená, J.: The morphogenetic influence of innervation on the ontogenetic development of muscle-spindles. J. Embryol. exp. Morph. **5**, 283–292 (1957).

Zelená, J.: Effect of innervation on the development of skeletal muscle. Babakova Sbirka (Praha) **12**, 1–102 (1959).

Zelená, J.: Development of muscle receptors after tenotomy. Physiol. bohemoslov. **12**, 30–36 (1963).

Zelená, J.: Development, degeneration and regeneration of receptor organs. In: Progress in Brain Research, Vol. 12. Amsterdam: Elsevier 1964.

Zelená, J., Hnik, P.: Absence of spindles in muscles of rats reinnervated during development. Physiol. bohemoslov. **9**, 373–381 (1960a).

Zelená, J., Hnik, P.: Irreversible elimination of muscle receptors. Nature (Lond.) **188**, 946–947 (1960b).

Zelená, J., Hnik, P.: Effect of innervation on the development of muscle receptors. In: The Effect of Use and Disuse on Neuromuscular Functions. Gutmann, E., Hnik, P., Eds., pp. 95–105. Prague: Publishing House of the Czechoslovak Academy of Sciences 1963a.

Zelená, J., Hnik, P.: Motor and receptor units in the soleus muscle after nerve regeneration in very young rats. Physiol. bohemoslov. **12**, 277–290 (1963b).

Zenker, W., Krammer, E.: Untersuchungen über Feinstruktur und Innervation der inneren Augenmuskulatur des Huhnes. Z. Zellforsch. **83**, 147–168 (1967).

Chapter 2

The Physiology of Muscle Receptors

By

CARLTON C. HUNT, St. Louis, Missouri (USA)

With 21 Figures

Contents

I. Introduction

In the preceding section BARKER has described in detail the structure of sensory receptors in skeletal muscle of various vertebrates. A variety of receptor structures are present, ranging from the highly organized, complex sense organ, the muscle spindle to free nerve endings.

Although certain aspects of structure and reflex function of muscle receptors were known by the turn of the century, it was not until the development of recording devices capable of detecting the action potential of nerve fibres that the receptors themselves could be studied physiologically. The introduction of single fibre recording by ADRIAN and his coworkers in the 1920's allowed the responses of individual sense organs to natural stimuli to be studied and opened a whole era of receptor physiology. The frog muscle spindle was one of the first receptors to be studied and it provided much of the early information on the general properties of sense organs. In particular it showed how stimulus intensity was transduced into impulse frequency and how factors such as adaptation modified the response.

The present discussion will be largely limited to certain vertebrates, frogs, reptiles, birds and mammals, with particular emphasis on the latter. Also, most attention will be given to the specialized mechanoreceptors of muscle which signal muscle length and tension. Two aspects are of importance: one is a description of the behavior of the various receptors to the appropriate natural stimuli in terms of impulse frequency in their afferent axons. Another concerns the mechanisms underlying the receptor behavior.

Muscle receptors responding to changes in muscle length or tension have properties seen in mechanoreceptors generally:

A mechanical deformation of the sensory nerve terminal leads to an alteration in the permeability of the nerve membrane which depolarizes the ending. The depolarization thus produced (the receptor or generator potential) at some point initiates impulse activity which is then propagated along the axon. The sequence of events may be outlined as follows:

Mechanical stimulus
↓
Deformation of receptor terminals
↓
Conductance change in terminals
↓
Receptor (Generator) potential
↓
Impulse initiation

While many of the steps involved are not clearly understood, the overall transduction from stimulus input to impulse frequency output has been carefully determined. MATTHEWS (1972) recent book gives an excellent and detailed account of such studies on mammalian receptors. In view of this, the present discussion will emphasize recent attempts to study the mechanisms underlying the behavior of receptors. Articles on certain aspects of this problem have already appeared in volume I of this handbook (OTTOSON and SHEPHERD, TERZUOLO and KNOX; they will therefore not be treated in detail here.

II. Mammalian Receptors

A. The Mammalian Muscle Spindle

1. Background

The physiology of the mammalian muscle spindle has been studied extensively. Its input to the central nervous system, signalling muscle length and rate of change of length, plays a major role in the reflex regulation of muscle movement, both at spinal and supraspinal levels. The mammalian spindle is subject to efferent control, mostly by neurons other than those innervating extrafusal muscle fibres. The reflex regulation of this spindle efferent system is an important component of the control of motor behavior by the central nervous system.

As in other vertebrate spindles, as well as in some invertebrate muscle receptors, the mammalian spindle has afferent fibres which terminate in sensory endings applied to specialized muscle fibres. BARKER's chapter has considered in detail the structural characteristics of the mammalian muscle spindle and its location in relation to the other elements within skeletal muscle. The physiological features to be considered include a description of the discharge in afferent fibres from muscle spindles in response to mechanical stimuli such as stretch or muscle contraction, the modulation of this behavior by activity in the motor axons to muscle spindles and the mechanisms underlying these responses.

There have been many studies of the afferent discharge from mammalian muscle spindles in response to change in muscle length, during muscular contraction and on stimulation of the motor innervation to spindles. Since this article was begun, an excellent detailed account of such studies has appeared in MATTHEWS' (1972) book on mammalian muscle receptors. In view of this, the present account will include a briefer review of such material and will emphasize some of the more recent direct attempts to investigate mechanisms of muscle spindle behavior.

The pioneering study of the physiological behavior of mammalian spindles was carried out by B.H.C. MATTHEWS (1933) who analyzed the afferent discharge in response to stretch and to motor nerve stimulation. In his classical paper the relation of discharge frequency in spindle afferent fibers to amplitude and velocity of stretch was clearly shown as well as the pause or reduction in discharge which accompanies contraction of the "in parallel" extrafusal muscle fibres. The subsequent elucidation of the γ motor innervation to the spindle (LEKSELL, 1945; KUFFLER et al., 1951) was followed by many investigations on the spindle afferent discharge in response to various changes in muscle length and to activation of its motor supply.

Discovery of many details of the sensory and motor innervation of mammalian spindles has proceeded hand in hand. However, the present state of knowledge can most readily be presented by considering these two aspects separately. Since most of the work on mammalian spindles has been done in cat, the following discussion will refer to studies on that species unless noted otherwise.

2. Sensory Innervation

a) Fibre Diameters. Myelinated afferent fibres in nerves to limb muscles show a trimodal distribution of external fibre diameters (LLOYD and CHANG, 1948). The fibres range in size from 1 μ to 22 μ with some variation among different muscles. The minima in the distribution at approximately 4 and 12 μ has been used to divide the population into three groups: group I (12–22 μ), group II (4–12 μ) and group III (1–4 μ). Since the external diameter of myelinated fibres is directly related to conduction velocity (conduction velocity in m/sec approximately equals 6 $\times$ diameter in μ), the diameter of an individual afferent fibre whose conduction velocity is known may be calculated. Individual fibres from muscle spindle receptors may be identified by their response to stretch and to muscle contraction (see below). The calculated diameter of such fibres indicates that they fall mainly into groups I and II of the diameter spectrum (MERTON, 1953; Hunt, 1954). Group I contains fibres which terminate in tendon organs (Ib fibres) as well as in muscle spindles (Ia fibres) as may be seen in Fig. 1. As discussed by BARKER, the diameter

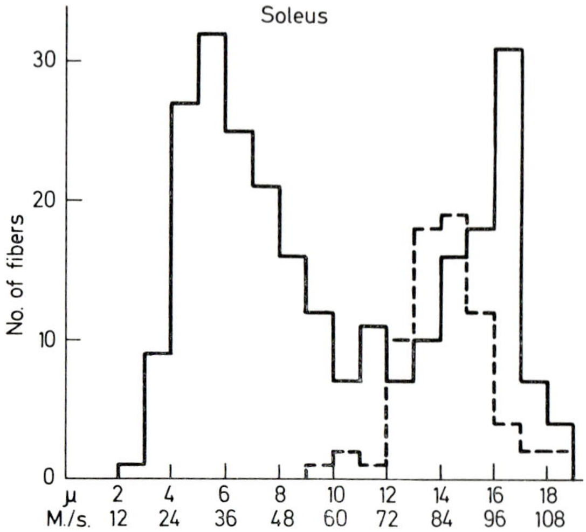

Fig. 1. Diameter distribution of afferent fibres from muscle spindles —— *A* and tendon organs ---- *B*. Soleus nerve. From HUNT (1954)

of afferent fibres terminating in primary endings is greater than those terminating in secondary endings. The reasonable assumption has been made that most Ia fibres in muscle nerves arise from primary endings and most group II from secondary endings. This is further suggested by differences in the responses to stretch recorded from Ia or group II fibres (see below). The division between group Ia and group II fibres at 12 μ, corresponding to a conduction velocity of 72 m/sec, does not necessarily indicate a clear-cut separation of fibres from primary and secondary endings. However, the assumption that spindle afferent fibres conducting above 80 m/sec are from primary endings and those conducting below 60 m/sec are from secondary endings (MATTHEWS, 1963) seems correct (see below).

b) Responses to Maintained Stretch. Muscle-spindle afferent axons conducting at velocities characteristic of either group I or group II fibres, exhibit a regular sustained discharge when the muscle is maintained at greater than some threshold length. Axons conducting at group II velocity (presumably from secondary endings) show, on the average, only a slightly higher threshold for sustained discharge to static stretch than the group Ia axons (presumably from primary endings). Soleus and medial gastrocnemius axons showed an average threshold for sustained discharge of 3.3 gm tension for Ia and 19 gm for group II fibres, although many Ia and some group II fibres discharged at zero initial tension (HUNT, 1954). Both type receptors are slowly adapting in that discharge continues indefinitely to maintained suprathreshold stretch.

An important relationship is that between static muscle length and impulse frequency. Several investigators have found that, over a certain range of length, discharge frequency is related to length in an approximately linear fashion (ELDRED et al., 1953; HARVEY and MATTHEWS, 1961b). This is true for both primary and secondary endings in the absence of activity in motor fibres to the spindles. The slope of the relation between impulse frequency and length gives a measure of the position sensitivity of the ending. BESSOU and LAPORTE (1962),

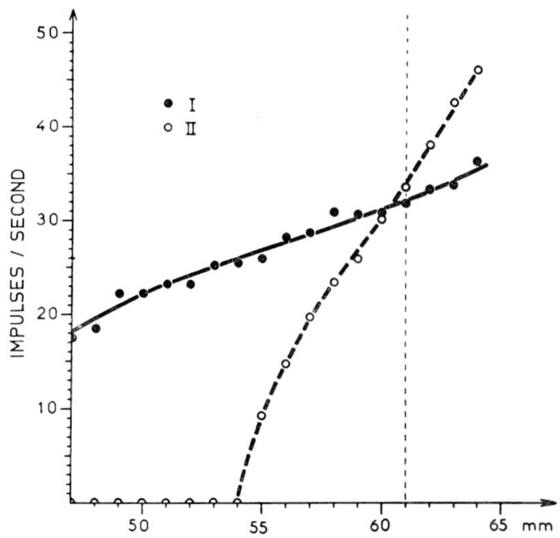

Fig. 2. Relation between muscle length and static discharge frequency from a primary and a secondary ending. From BESSOU and LAPORTE (1962)

recording from primary and secondary endings of the same spindle of the tenuissimus muscle found that the secondary ending usually had a higher threshold to maintained stretch than the primary, but that the slope of the relation between discharge frequency and muscle length was usually steeper in the secondary ending (see Fig. 2). There appears to be much variation in the relative position sensitivities of primary and secondary endings. In deefferented soleus spindles

HARVEY and MATTHEWS (1961b) found the slope to be approximately equal in primary and secondary endings. The slope of the relationship also was found to vary in spindles from other muscles (see below).

c) Responses During Ramp-and-Hold Stretch. The most common method used to study the dynamic and static responses to stretch has been to increase muscle length from an initial value, at a constant velocity, and then to maintain the new length (ramp-and-hold extension). Both velocity and amplitude of stretch may be varied. Various mechanical or electromechanical devices have been used to achieve this.

B.H.C. MATTHEWS (1933) found that most of the spindle afferent fibres he examined had a conspicuous dynamic response. That is, the frequency of discharge was greater while extension was increasing than during its maintenance. (He recorded from axons in partly sectioned muscle nerves and had no measure of diameter or conduction velocity. In retrospect it appears that he recorded largely from primary endings.) COOPER (1961) made the important observation that spindle afferent fibres conducting at group II velocities (from secondary endings) showed a different response during extension than those conducting at group I velocities (from primary endings). Fibres from secondary endings showed much less dynamic response than those from primary endings. In these experiments the spindles were under the influence of background activity in fusimotor fibres. These findings were confirmed and extended in deefferented spindles by P.B.C. MATTHEWS and colleagues and by LAPORTE, BESSOU and their coworkers (see below).

In isolated mammalian spindles, responses from anatomically identified primary and secondary endings have been found to show the same differences in behavior to stretch as seen in group Ia and group II fibres respectively (HUNT and OTTOSON, 1973). This confirms the well accepted association between spindle afferent fibre conduction velocity and type of ending.

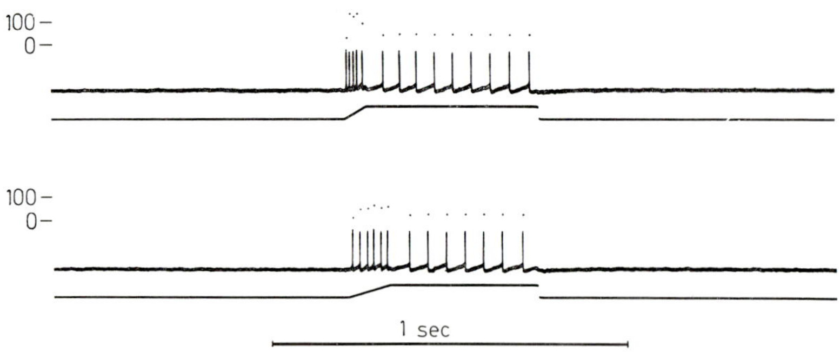

Fig. 3. Response of a primary ending of an isolated mammalian spindle to ramp-and-hold stretch. Frequency of discharge shown by dots above impulse discharge. Lower trace displacement. From HUNT and OTTOSON, unpublished records

The response of a primary ending to a ramp-and-hold stretch may be seen in Fig. 3. Impulse discharge is shown in the middle trace and the record of displacement in the lower trace. The upper trace is a plot of instantaneous frequency, each point representing an impulse with a position on the ordinate proportional to 1/interval between that and the preceding impulse. The record is from an isolated spindle; the behavior is similar to that seen in Ia fibres from intact muscle. Shortly after the onset of stretch several impulses occur at high frequency. This is followed by a further high frequency discharge during the later part of the ramp. The discharge then falls during the static phase to a lower frequency than during the dynamic phase. The discharge occurring early in extension usually consists of a brief high frequency burst known as the *initial burst.* It varies with velocity of stretch, as does the subsequent later dynamic phase of the discharge.

An example of the response of a secondary ending of an isolated spindle is shown in Fig. 4. There is no initial burst and the frequency of discharge during the dynamic phase is only slightly higher than during the static phase.

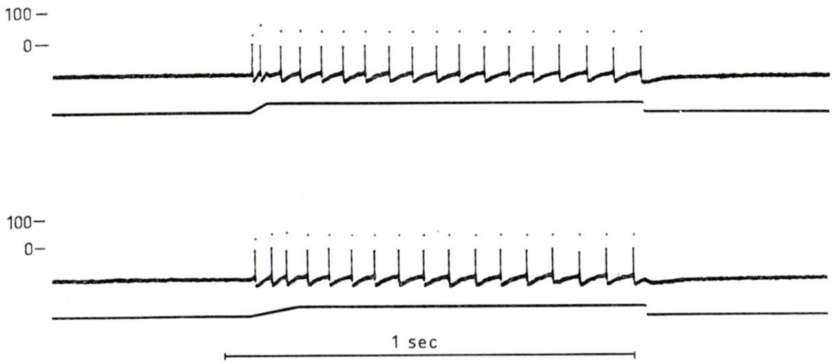

Fig. 4. Response of a secondary ending of an isolated mammalian spindle to ramp-and-hold stretch. Frequency of discharge shown by dots above impulse discharge. Lower trace displacement. From Hunt and Ottoson, unpublished records

The discharge of primary and secondary endings may be readily compared from the impulse frequency plots during applied stretch (Figs. 3 and 4). It may again be noted that the initial burst is evident in the responses from primary endings and not from secondary endings. Also, the high dynamic sensitivity of the primary endings is apparent on comparing the frequency during the dynamic phase with that during the static phase.

The transition in the frequency response of primary endings from the end of the dynamic phase to the static level varies. At the end of a ramp stretch the frequency may fall abruptly to the static level; there may be a brief pause before the discharge resumes, or, more commonly the frequency may drop suddenly at first and then slowly fall to the static level.

Primary endings show a greater fall in frequency on release from stretch than do secondary endings. This often takes the form of a pause in the discharge followed by a gradual return to baseline frequency. A brief pause in the discharge of primary endings may also be seen during ramp and hold stretch at the end of the dynamic phase prior to the maintained discharge during the static phase.

The initial burst, commonly seen in primary but not in secondary endings, increases in frequency as velocity of stretch is increased. It is most conspicuous when the spindle has been at rest. Repetitive stretch abolishes it, as does fusimotor activity. The initial burst has been attributed to cross linkages set up between thick and thin filaments of the intrafusal fibres which increase the stiffness over a short range (see Matthews, 1972).

The difference in the frequency of discharge near the end of the ramp and the static frequency 0.5 sec later has been called the dynamic index (Crowe and Matthews, 1964). This has been used as a measure of the dynamic sensitivity of the ending. It is characteristically high in fibres from primary endings and lower in fibres from secondary endings.

In a study of deefferented spindles Matthews (1963) found that with few exceptions the dynamic indices of axons conducting above 80 m/sec were higher than those of fibres conducting below 60 m/sec. In both groups of axons the dynamic index increased as velocity of stretch was increased, but the dynamic index from presumed primary endings generally remained higher. Of 14 fibres whose conduction velocities ranged from 60–80 m/sec, 5 behaved as typical primary endings, 3 as the usual secondary endings and the remaining 6 showed intermediate behavior. The slope of the relation between dynamic index and velocity of stretching (3–30 mm/sec) tended to be much steeper for fast (80 m/sec) than for slow (60 m/sec) conducting spindle afferent fibres. The evidence indicates that conduction velocity of the afferent fibre is useful in indicating whether it comes from a primary or secondary ending especially if the conduction velocity is greater than 80 m/sec or less than 60 m/sec. Identification is further aided by determination of the amount of dynamic response to stretch. For some fibres in the 60–80 m/sec conduction velocity range identification remains uncertain.

d) Responses to Vibration. As might be expected from their high dynamic sensitivity to stretch, primary endings are readily excited by longitudinal vibration applied to the muscle tendon (Brown et al., 1967). Vibration applied to the muscle directly also evokes a discharge from primary endings at the frequency of vibration (Kuffler et al., 1951; Granit and Henatsch, 1956, Bianconi and van der Meulen, 1963; Crowe and Matthews, 1964b).

Brown, Engberg and Matthews (1967) found that the most sensitive primary endings in soleus could be driven by vibration of less than 10 μ applied to the tendon. The discharge from such endings followed in a 1:1 relationship vibration frequencies of 200 to 500 Hz at about the same threshold. The threshold varied with initial tension, becoming lower as the muscle was stretched slightly from resting length and then constant as further stretch was applied. Fig. 5 shows, in contrast, that secondary endings appeared relatively insensitive to vibration as did tendon organs. Only at high amplitudes and low frequencies of vibration could an occasional secondary ending be driven.

e) Responses to other Waveforms of Stretch. Lennerstrand and Thoden (1968) studied the dynamic responses of primary and secondary endings of cat spindles by utilizing various waveforms of extension in addition to ramp-and-hold stretch such as, triangular length changes of constant velocity, sinusoidal length changes and a parabolic waveform. Analysis of the responses to sinusoidal

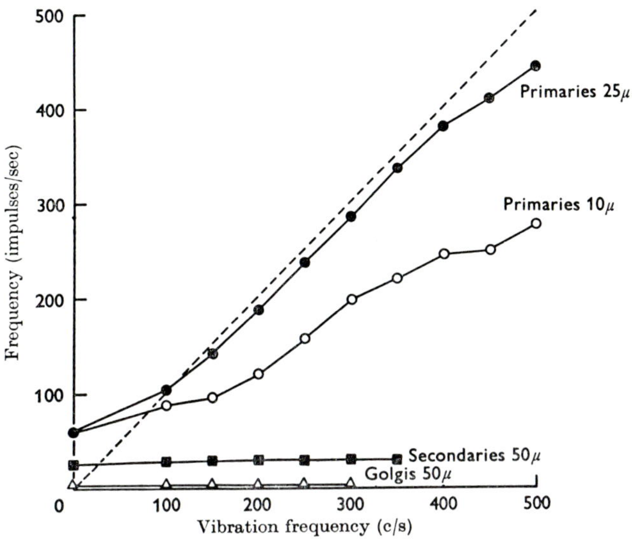

Fig. 5. Effect of vibration of indicated amplitude on primary and secondary endings of muscle spindles and on Golgi tendon organs. Plot shows relation between discharge frequency and vibration frequency. Dashed line shows relationship expected if each cycle of vibration produced one impulse. From BROWN, ENGBERG and MATTHEWS (1967)

stretch of both primary and secondary endings suggested that the input-output relations did not obey a power function. Using triangular waveforms the discharge frequency did not follow the velocity of stretch but showed two transients: one had a slow rise time and was called an integrative or i-transient. The other showed a rapid rise and quick decay at onset of extension and relaxation and was termed derivative or d-transient. Primary endings showed much more conspicuous initial bursts and d-transients to ramp extension than to triangular waveforms. These responses did not seem related to acceleration as had been suggested by SCHÄFER (1967), SCHÄFER and HENATSCH (1966, 1967). Parabolic stretch, which imposes a step change in acceleration, showed no component of the response frequency, after subtraction of the position response, which could be attributed to acceleration.

The question of whether or not the initial burst signals information concerning acceleration has been extensively discussed. Under a restricted set of conditions it may: when there is no activity in fusimotor fibres and there has been a period of quiescence during which the spindle has not been stretched. In normal function such conditions would seldom be expected to occur.

LENNERSTRAND (1968) determined the position sensitivity of primary and secondary endings in several ankle flexors and extensors. Values were obtained 30 sec after a length change. The position sensitivity of secondary endings was greater than for primary endings in the same muscle. For example, in soleus the mean value for primary endings was 3.9 impulses/sec/mm and for secondary endings 5.9.

Velocity sensitivity was determined by applying a triangular waveform of extension. After ignoring initial bursts and other transients, and subtracting the position response, the resultant dynamic response frequency was plotted, for a given velocity, as a function of stretch amplitude. The slope of this relation was called the "slow" velocity response and its intercept with the ordinate the "quick" velocity response. The "quick" velocity component differed in primary and secondary endings. In primary endings its magnitude was nearly linearly related to velocity in the 2–32 mm/sec range. The "slow" velocity component in primary endings reached a maximum at very low velocities and then remained constant as velocity was increased. When length was decreased, primary endings tended to cease firing.

Secondary endings showed less dynamic response (cf. Cooper, 1961; Harvey and Matthews, 1961; Bessou and Laporte, 1962), with triangular stretch the frequency-length curve was only slightly higher during extension than during relaxation, the difference increasing as velocity was increased. The "quick" velocity component was small at low velocities, often did not increase until velocity exceeded 2 mm/sec and then increased linearly with velocity. In secondary endings the "slow" velocity component was not seen at velocities below 1–2 mm/sec. It then increased as velocity was increased but not linearly.

3. Mechanisms Determining Responses to Stretch

a) Receptor Potential and Impulse Initiation. Lippold, Nicholls and Redfearn (1960) recorded from a fine nerve branch close to the tenuissimus muscle, permitting detection of receptor (generator) potential as well as discharge from a muscle spindle afferent fibre. Whether this was from a primary or secondary ending was not determined. Judging from the responses to stretch they must have recorded from primary endings; the discharge showed a conspicuous dynamic phase. The receptor potential also showed a dynamic component in its response to stretch. Through electrodes placed on the muscle, current steps were applied which affected the afferent discharge frequency, evidently by influencing the impulse generating site. Depolarizing current augmented and hyperpolarizing current diminished discharge frequency with little or no evident accommodation. The effect of applied current suggested that the dynamic response to stretch depended upon characteristics of the receptor potential and not some property of the impulse initiating site. A recent study by Emonet-Dénand and Houk (1969) indicates that primary endings do show a small degree of accommodation to current applied to the tenuissimus muscle.

Recently, Hunt and Ottoson (1973) have recorded the receptor potential from primary and secondary endings of isolated mammalian spindles after block of impulse activity by tetrodotoxin. In response to ramp and hold stretch the receptor potential showed a dynamic and static phase (see Fig. 6). In primary endings the dynamic phase exhibited an early component beginning at the onset of stretch and characterized by a rapid rise in the receptor potential. Its slope and magnitude depended on velocity of stretch. This early component was labile, being reduced or abolished with repetitive stretch. In its timing, velocity-dependence and lability it appeared related to the initial burst seen in the discharge from primary endings. Following this early component there was a subsequent rise in the receptor potential during the remainder of the ramp called the late component

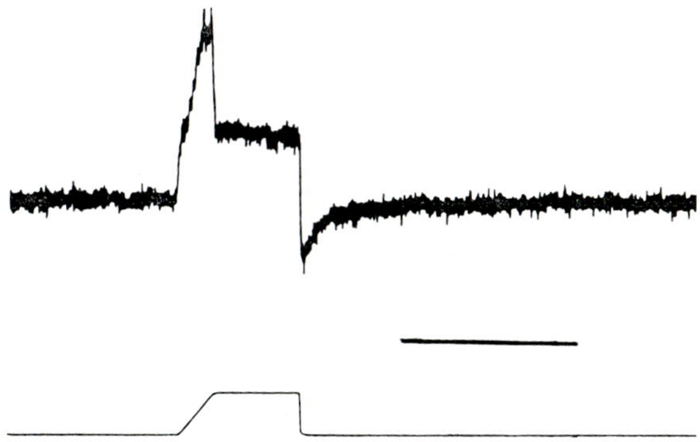

Fig. 6. Receptor potential of a primary ending of an isolated mammalian spindle. Lower trace displacement. Time 1 sec. From HUNT and OTTOSON, unpublished records

of the dynamic phase. This also varied in rate of rise and peak amplitude with the velocity of stretch. At the end of the ramp the receptor potential fell to a steady level maintained for the duration of the static phase. On release of stretch the potential decayed to the base line level. The amplitude of the receptor potential at the end of the dynamic phase is considerably above the static level, in keeping with the high dynamic sensitivity of the primary ending.

The receptor potential of secondary endings was similar but with important quantitative differences. The early component of the dynamic phase was smaller or absent (Fig. 7). The later component did not usually continue to rise during the

Fig. 7. Receptor potential of a secondary ending of an isolated mammalian spindle. Lower trace displacement. Time 1 sec. From HUNT and OTTOSON, unpublished records

ramp to the same extent as in primary endings so that the peak amplitude of the dynamic phase was often only slightly greater than the static level. There was considerable variation in this regard among secondary endings as might be expected from the variability in their dynamic sensitivity.

In both primary and secondary endings, the amplitude of the static phase was linearly related to amplitude of stretch over the range examined.

Comparison of the discharge frequency in response to a ramp and hold stretch, recorded prior to the application of TTX, with the receptor potential recorded after block of impulse activity suggested that the impulse frequency in primary endings might depend on rate of rise as well as amplitude of receptor potential. This was also suggested by a fall in the apparent firing level of impulses in primary endings when the velocity of stretch was high. With high velocities of stretch there was an undershoot of the receptor potential in primary endings followed by a rise to the static level. On release of stretch a brief hyperpolarization was seen under some conditions. Also, following brief stretch a transient hyperpolarization was seen.

b) Mechanical Factors. The finding that primary endings are disposed about both nuclear bag and nuclear chain fibres, while secondary endings are mainly about chain fibres, led to the plausible argument that the endings on the bag fibres are responsible for the dynamic behavior of the primary ending (see (Matthews, 1972). It has also been assumed that mechanical properties of the bag fibres cause the deformation of the primary endings to be greater during application of stretch and dependent on rate of stretch. The bag fibre has been considered to change in length on extension, not as a purely elastic element but with an added viscous element in the polar regions that would provide the rate-dependent dynamic component.

Boyd (1970, 1971) indicates that he has observed, in isolated visualized spindles, opening of the spirals of the primary ending during ramp stretch with a subsequent slippage causing a reduction in the displacement during the early portion of the static phase of stretch. However, high speed photographs of isolated spindles, visualized by differential interference microscopy, have shown no overshoot in displacement of annulospiral endings on bag fibres during an applied ramp and hold stretch (Beacham et al., 1971; Hunt and Beacham, unpublished observations). This makes it seem unlikely that the overall mechanical properties of the bag fibres, such as a higher viscosity in the polar regions, account for the dynamic response. This does not exclude the possible role of mechanical factors in the region of the sensory terminations as a cause of the dynamic behavior of primary endings. Similar studies on frog spindles showed no displacement of the overall sensory region which would account for their dynamic response (Ottoson and Shepherd, 1970; see below).

The nature of the dimensional changes in the sensory terminals following stretch is not known. It seems likely that an increment in the longitudinal dimension of the terminal is responsible for conductance changes leading to the generator potential. With stretch there is also a decrease in intrafusal fibre diameter causing a diminution in the radial dimension of the nerve terminal. This might also play a role in altering the conductance of the terminals.

There is little direct information about the mechanical properties of the intrafusal fibres. These have been assumed to be similar to certain properties of extrafusal fibres. The polar regions, especially of the bag fibres, have been considered to have a higher viscosity than the sensory region because the myofilaments occupy a higher fraction of the cross sectional area of the fibre towards the poles.

c) Summary. Primary endings, with terminations on both bag and chain fibres, exhibit a high dynamic sensitivity to stretch as well as a well maintained static discharge. The receptor potential also shows a prominent dynamic as well as a static component. The basis for the dynamic behavior is not yet clear. Mechanical factors resulting in displacement of the primary terminations with a dynamic overshoot to ramp-and-hold stretch have been suggested but direct measurement of displacements of the sensory region have failed to support this. Other possibilities such as conductance changes or potential changes in the terminals due to electrogenic pump effects have not been explored.

Dynamic sensitivity is seen in receptors other than mechanoreceptors, for example in the barnacle photoreceptor where conductance changes appear to be responsible for the fall from the dynamic to static level of the receptor potential.

Secondary endings whose terminations are largely on chain fibres exhibit less dynamic sensitivity both in their discharge in response to stretch and in their receptor potentials.

Because of the different effects of stimulating static and dynamic fusimotor axons on the response of primary endings to applied stretch (see below) there is much attraction to the idea that the terminals on bag fibres impart the dynamic sensitivity and that the terminals on chain fibres determine mainly static sensitivity. Such an argument suggests that the difference between terminals on the 2 types of fibre would most likely have a mechanical basis. It would be difficult, *a priori*, to assume that the terminals themselves, derived from the same parent axon, had different properties depending on which type intrafusal fibre they ended upon. With the information now available it does not seem possible to provide a definite explanation for the greater dynamic receptor response of primary endings as compared to secondaries.

The characteristics of the impulse initiating sites of primary and secondary endings are also not well understood. More detailed studies will be needed to determine whether or not impulse frequency is related to amplitude of receptor potential in a linear manner and whether there is any significant degree of accommodation. It is possible that sensitivity to rate of change of receptor potential may play a role in the dynamic behavior of the discharge in primary endings.

4. Motor Innervation

a) γ-Innervation. Although early morphological studies demonstrated the presence of an efferent innervation to intrafusal muscle fibres in mammalian spindles, the motor control of this sense organ received little attention until much later. B.H.C. MATTHEWS (1933) found that some spindle afferent fibres showed discharges during contraction when the motor nerve was stimulated at a strength near maximal for contraction. He considered that this was due to intrafusal fibre contraction. A more recent study by HARVEY and P.B.C. MATTHEWS (1961a) suggests this was so.

The elucidation of the motor control of mammalian spindles followed investigation of the function of the small diameter motor fibres to skeletal muscle. ECCLES and SHERRINGTON (1930) examined the diameter distribution of myelinated fibres in deafferented nerves to cat hind limb muscles and found it to be bimodal. The larger diameter or α group ranged from about 9–19 μ and the smaller or γ group from about 3–8 μ. (The fibre diameter spectrum differs among various muscles. Fast pale muscles have slightly larger fibres than slow red muscles. More distal limb muscles tend to have smaller α fibres.) The α and γ groups are usually quite separate. Eccles and Sherrington assumed that the small fibres, like the larger ones, innervated extrafusal muscle fibres and produced muscle tension. O'Leary, Heinbecher and Bishop (1934) determined the tension produced by muscle as a stimulus to the muscle nerve was progressively increased. Twitch tension became maximal when all the α fibres were stimulated; no increment in tension was produced by stimulating in addition the smaller γ fibres. They first made the suggestion that the latter might provide the motor innervation to muscle spindles.

Leksell (1945) was able, using pressure, to block conduction in most of the larger motor fibres without interruptung conduction in the smaller fibres. Under these conditions he found that stimulation of the smaller fibres produced no measureable increase in muscle tension but did cause a detectable increase in afferent discharge recorded from muscle nerve. This, he concluded, came from muscle spindles as the result of activation of intrafusal muscle fibres.

Kuffler, Hunt and Quilliam (1951) and Hunt and Kuffler (1951) subsequently investigated the effect of small diameter motor fibres by stimulating isolated, individual motor fibres of slow conduction velocity in ventral root filaments. Even on repetitive stimulation single γ fibres, alone or in combination, caused no detectable contraction and elicited no action potentials in extrafusal fibres. They introduced the method of recording from single afferent fibres in dorsal root filaments. Fibres from muscle spindles were identified by their responses to stretch and their "in parallel" behavior during muscle contraction. Stimulation of individual small diameter (γ) ventral root fibres was found to increase the afferent discharge from particular spindle afferent fibres. Up to six ventral root fibres could be isolated which influenced the discharge from a single spindle afferent fibre. In these early studies most of the spindle afferent fibres investigated were Ia fibres, the muscles studied being soleus, gastrocnemius medialis and tenuissimus. Among the prominent features noted in the spindle afferent response to small motor (γ) fibre stimulation were: facilitation of the response with repetitive stimulation, single stimuli being relatively ineffective; a progressive increase in response to successive stimuli until, in some cases, a 1:1 relationship between motor fiber stimulation and afferent discharge was established ("driving"); an effective frequency range of motor stimulation extending considerably above the fusion frequency of extrafusal fibres; a synergistic interaction between stretch and intrafusal fibre stimulation.

Matthews (1962) found two distinct types of motor axons to the spindle, the dynamic and static fusimotor fibres. On stimulation these were found to affect the response of the primary ending during stretch in quite distinct ways: The dynamic fusimotor fibre increased the discharge frequency from primary endings when the muscle was held at a constant length, but had a more striking effect on the dynamic, velocity-sensitive component of its response to stretch. The static fusimotor fibre, on the other hand, augmented the steady state frequency, at any muscle length, and did not increase or decrease the dynamic response of the primary ending.

These findings have been confirmed and extended by several investigators (Crowe and Matthews, 1964; Brown et al., 1965; Bessou et al., 1966). Appelberg, Bessou and Laporte (1966) found that secondary endings were generally not influenced by dynamic fusimotor fibres but their frequency during static stretch was enhanced by stimulation of static fusimotor fibres. BROWN, ENGBERG and MATTHEWS (1967) showed that the usually small dynamic response of secondary endings was not influenced by static fusimotor fibres.

The effect of stimulating a dynamic fusimotor fibre on primary endings may be seen in Fig. 8. On the left are shown responses from four spindles in the tenuissimus muscle. The instantaneous frequency curves show the characteristic, although slightly different, responses of primary endings during the application of stretch. On the right may be noted the responses to the same stretch during stimulation of a dynamic fusimotor fibre at a frequency of 100/sec. In all cases the dynamic response was enhanced by the fusimotor stimulation. Fig. 9, also taken from BESSOU, LAPORTE and PAGÈS (1966), shows the frequency changes in four primary endings resulting when dynamic fusimotor stimulation began prior to ramp-and-hold stretch. Dynamic fusimotor stimulation produced a relatively small increase in discharge frequency at the initial length. Following the end of stimulation there was a slow decline in impulse frequency at the new length. Although primary endings usually exhibit a rapid drop in frequency following the completion of a ramp stretch, during dynamic fusimotor stimulation the discharge often decayed more slowly from its peak as may be seen in Figs. 8(3) and 9(3 and 4). In a number of muscles examined the effect of stimulating a dynamic fusimotor fibre was found to be similar on all the spindles examined.

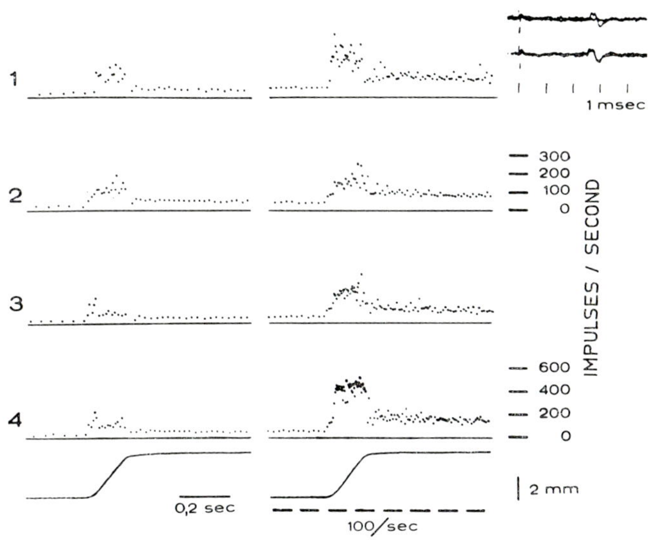

Fig. 8. Effect of stimulation of a single dynamic γ fibre on 4 different primary endings in tenuissimus muscle. Left: responses to ramp-and-hold stretch in the absence of stimulation. Right: responses to stretch during fusimotor stimulation at 100/sec. Insert shows action potential of γ fibre and conduction time from ventral root to muscle nerve. From BESSOU, LAPORTE and PAGÈS (1966)

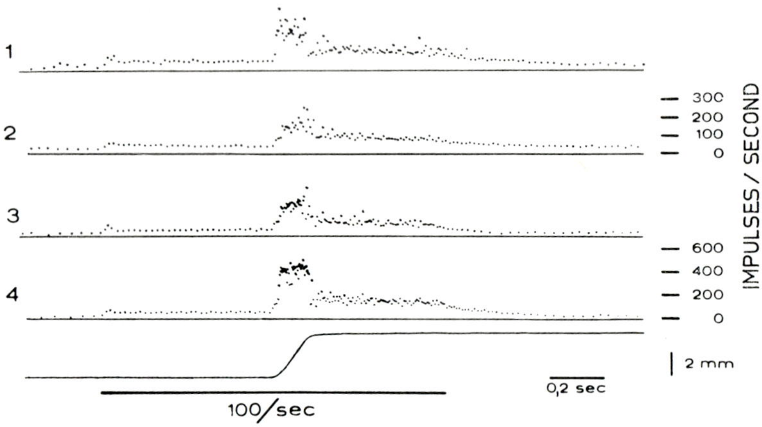

Fig. 9. Responses of 4 primary endings to stimulation of a single dynamic γ axon. From BESSOU, LAPORTE and PAGÈS (1966)

The relation between dynamic index of a primary ending and the velocity of stretch was altered by stimulation of a dynamic fusimotor fibre in the manner shown in Fig. 10. The curve was displaced by an amount which depended on the frequency of stimulation. In soleus spindles the slope was usually increased (CROWE and MATTHEWS, 1964), while in tibialis posterior units the curve was simply displaced. Stimulation of a dynamic fusimotor fibre altered, in a similar manner, the relationship between both peak frequency and mean frequency during extension and the velocity of stretch.

On stimulation of dynamic fusimotor fibres (ANDERSSON et al., 1968; LENNERSTRAND and THODEN, 1967a, b) the "velocity response" was found to increase with the frequency of stimulation, particularly the "slow" component (see above). At the highest levels of stretch the response often levelled off, apparently approaching a frequency ceiling. The slope of the steady state frequency-length curve for primary endings was found to be sometimes decreased during dynamic fusimotor stimulation, although the discharge rate at "zero" muscle length was slightly increased (see CROWE and MATTHEWS, 1964).

The principal effect of stimulating static fusimotor fibres is to enhance the discharge frequency of both primary and secondary endings during maintained extension (BESSOU et al., 1962; BESSOU and LAPORTE, 1966; BESSOU and PAGÈS, 1969; BROWN et al., 1969). During stretch the dynamic response of a primary ending may be unaltered or reduced. An example of the effects of stimulating a static fusimotor fibre on the response to stretch of four primary endings may be seen in Fig. 11. Here stimulation at 170/sec increased the discharge frequency at the initial length and at the new length but the dynamic effect of stretch was relatively reduced. The magnitude of these effects increases with increasing frequency of stimulation. The dynamic index is usually diminished and the relation between dynamic index and velocity of stretch is reduced by stimulation of a static fusimotor fibre (Fig. 10).

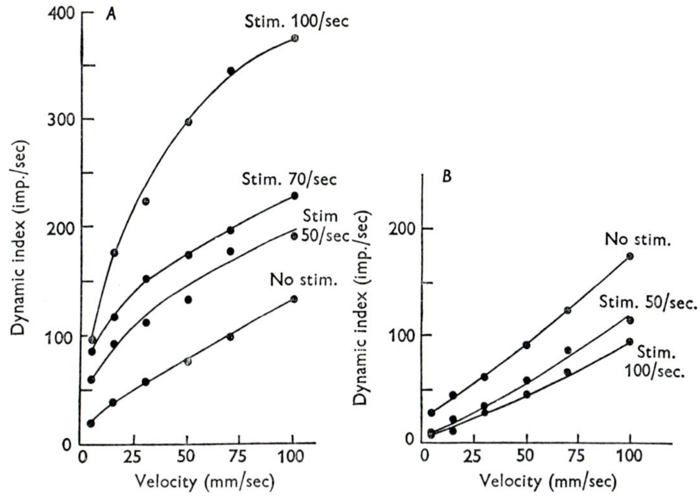

Fig. 10. Effect of stimulation of a fusimotor fibre on the relation between dynamic index and velocity of stretch of a primary ending. A Stimulation of dynamic fusimotor fibre. B Stimulation of static fusimotor fibre. From CROWE and MATTHEWS (1964)

Stimulation of static fusimotor fibres was found to have a marked effect on the discharge of secondary endings, causing a large increase in their discharge frequency (APPELBERG et al., 1966). An example of the effect of stimulating a single static fusimotor fibre on both a primary and secondary ending of the same spindle may be seen in Fig. 12. In the absence of fusimotor stimulation the primary and

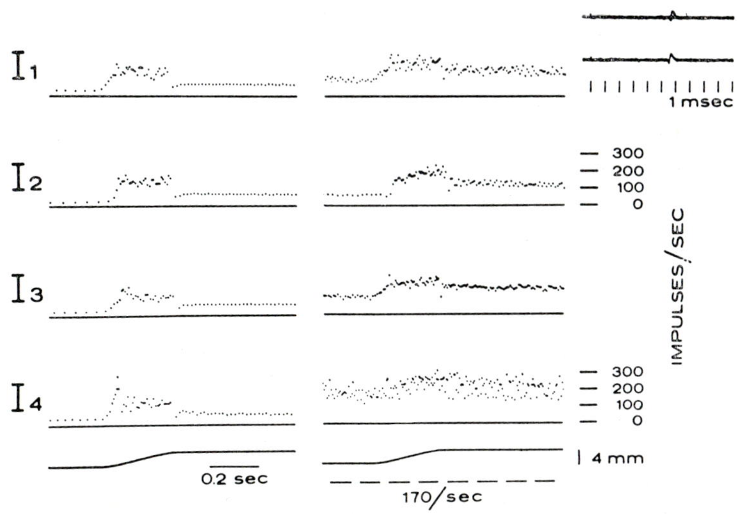

Fig. 11. Effect of stimulating a single static γ fusimotor fibre on 4 primary endings. Left: responses to ramp-and-hold stretch in the absence of stimulation. Right: responses during fusimotor stimulation at 170/sec. Insert shows conduction time of fusimotor fibre from root to muscle nerve. From BESSOU, LAPORTE and PAGÈS (1966)

secondary endings show their typical dynamic and static responses. On static fusimotor stimulation at the initial length, impulse frequency increases from both endings, more conspicuously from the primary. During stimulation, ramp stretch produces a proportionately smaller dynamic response in the primary ending while the secondary ending retains its static behavior.

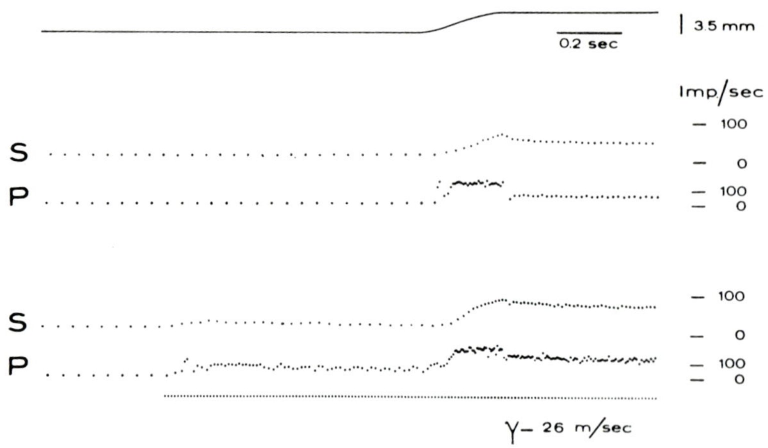

Fig. 12. Effect of stimulating a single static γ fibre on a primary and secondary ending of the same tenuissimus spindle. Upper pair of records shows responses of the primary P and secondary S endings to ramp-and-hold stretch. Lower pair of records shows responses with stimulation of the γ fibre. From BESSOU and LAPORTE (1966)

Stimulation of static, but not dynamic, fusimotor fibres often leads to "driving" of primary endings and secondary endings (BESSOU and LAPORTE, 1966). This is a phenomenon in which the afferent discharge frequency becomes identical with the fusimotor stimulus frequency, following an initial period of facilitation (HUNT and KUFFLER, 1951). Static fusimotor γ fibres have also been found in the rabbit (EMONET-DÉNAND et al., 1964).

Stimulation of static fusimotor fibres was found by LENNERSTRAND and THODEN (1968b) to increase discharge of primary endings at "zero" length and sometimes to increase the slope of the steady state frequency-length relationship (type II effect) and sometimes not (type I effect). Secondary endings from cat extensor muscles seldom showed type II effects; those from flexor muscles showed both.

Fusimotor stimulation enhances the sensitivity of primary endings to vibration (CROWE and MATTHEWS, 1964). This was found to result from stimulation of either static or dynamic fusimotor fibres. Although fusimotor stimulation also increases the response of secondary endings to vibration (BROWN et al., 1967) they remain relatively insensitive compared to the primary endings.

MATTHEWS and STEIN (1969) presented a careful analysis of the responses of primary and secondary endings to sinusoidal stretching in spindles receiving background fusimotor activity. The range of movement over which discharge frequency remained proportionately related to amplitude of stretch was smaller for primary than for secondary endings.

Low frequencies of sinusoidal stretch of muscle produce a modulation of discharge frequency of primary endings. Stimulation of dynamic fusimotor fibres enhances the periodic frequency variation while stimulation of a static fusimotor fibre increases the discharge from a primary ending but diminishes its periodic variation by the sinusoidal stretch (CROWE and MATTHEWS, 1964). This striking enhancement of dynamic sensitivity of dynamic fusimotor stimulation and diminution by static fusimotor stimulation may be seen in Fig. 13.

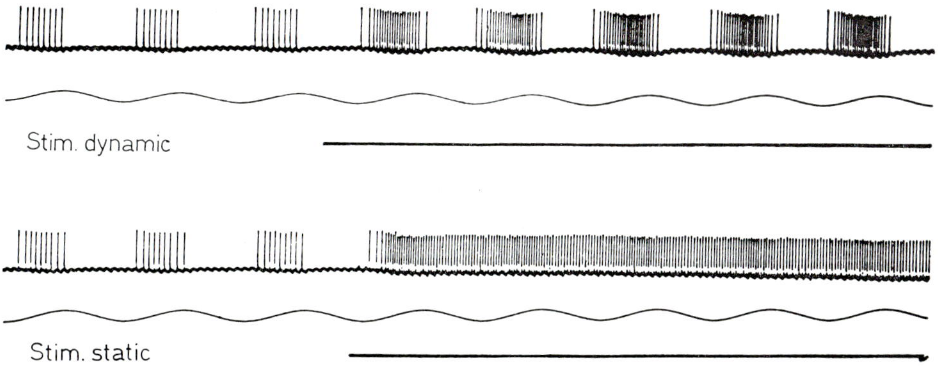

Fig. 13. Effect of stimulating fusimotor fibres on response of a primary ending to sinusoidal stretching at 3 c/s. Upper records: stimulation of dynamic γ axon at 100/sec. Lower records: stimulation of static fusimotor axon at 100/sec. From CROWE and MATTHEWS (1964)

On the basis of the modification of the responses to stretch by static and dynamic fusimotor fibres, and the morphological evidence that primary endings terminate on both nuclear-bag and chain fibers while secondary endings are principally on chain fibers, MATTHEWS (1962) suggested that dynamic fusimotor fibers innervated nuclear-bag fibers and static fusimotor fibers the nuclear-chain intrafusal fibers. This attractive hypothesis provided an explanation for many aspects of static and dynamic fusimotor effects. However, it now appears that many γ axons terminate on both bag and chain intrafusal fibres so that the situation is more complex (see below).

The mechanism by which stimulation of static fusimotor fibres decreases the dynamic response of the primary ending to applied stretch is not known but there are several possibilities: If static fusimotor stimulation caused shortening of nuclear-chain fibres, there might result an unloading of the nuclear-bag fibres. Since the primary terminations on the bag fibres may be responsible for the dynamic sensitivity to stretch this slackening could diminish the dynamic response. Alternatively, without actual slackening of the bag fibres, static fusimotor fibre stimulation could cause the chain fibres to become much stiffer than the bag fibres, redistributing the forces produced by applied stretch so that the distortion of the primary terminations on the bag fibre would be less. Another possibility is that there might be some change in the impulse generating site when the primary terminations on the chain fibres are highly depolarized by static fusimotor fibre stimulation. The evidence needed to decide among these several alternatives is not yet available.

EMONET-DÉNAND et al. (1972) have recently shown that some γ fusimotor fibres produce typical static fusimotor effects, with a diminution in dynamic response, when stimulated at frequencies of 60 to 150/sec while stimulation at a lower frequency (20–30/sec) produces a dynamic fusimotor effect, enhancing the dynamic response with little or no increase in static sensitivity. Such effects were found by recording changes in frequency of a primary ending during sinusoidal stretch (Fig. 14) and have been seen in spindles from tenuissimus, soleus and

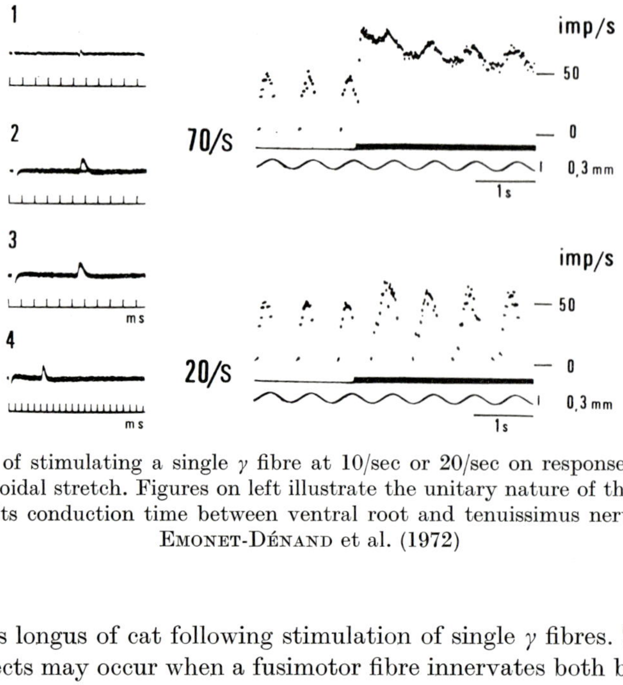

Fig. 14. Effect of stimulating a single γ fibre at 10/sec or 20/sec on responses of a primary ending to sinusoidal stretch. Figures on left illustrate the unitary nature of the γ fibre action potential and its conduction time between ventral root and tenuissimus nerve (1–4). From EMONET-DÉNAND et al. (1972)

flexor hallicus longus of cat following stimulation of single γ fibres. They suggest that such effects may occur when a fusimotor fibre innervates both bag and chain fibres. Low frequency stimulation would then cause more effective contraction in bag fibres while at higher frequencies contraction in the chain fibres would be greater.

Recently BARKER, EMONET-DÉNAND, LAPORTE, PROSKE and STACEY (1970, 1971, 1973) have shown clearly that static fusimotor fibres terminate in trail endings (see pp. 71). After sectioning all but one γ ventral root axon and allowing degeneration of the sectioned fibres, physiological and morphological studies showed that static γ axons were associated with trail endings in spindles of the tenuissimus muscle. These were distributed to both bag and chain fibres (see pp. 68).

Following repetitive stimulation of a fusimotor fibre, the static discharge frequency in a spindle afferent fibre may be appreciably higher than prior to stimulation (KUFFLER et al., 1951). This poststimulatory enhancement may last for minutes and is abolished by stretch. BROWN, GOODWIN and MATTHEWS (1969) found an enhancement in the initial burst in the discharge from primary endings to a ramp stretch when delivered after fusimotor stimulation. The poststimulation facilitation lasted for many minutes and was abolished by stretch of the spindle. The effect was found in both primary and secondary endings. If, following fusimotor stimulation, a muscle was slackened and then subjected to a ramp stretch, the burst occurred in the ramp at the length at which fusimotor stimulation was delivered. Both dynamic and static fusimotor fibres produced the effect independently, as suggested by the following observation: If one fusimotor fibre was stimulated at a particular muscle length and another fusimotor fibre stimulated at a shorter length, a subsequent ramp stretch revealed two "initial" bursts. While there is no direct evidence as to the mechanism underlying this effect, it appears due to the establishment of cross linkages between thick and thin filaments in the intrafusal fibres activated (MATTHEWS, 1972). If dynamic and static fusimotor stimulation produced facilitation in the same fibre, one would not expect independence of the effects as shown in the above experiment.

b) Frequencygrams. Single stimuli to static fusimotor fibres produce relatively little change in discharge frequency in a primary fibre. When a primary ending is discharging at fairly low frequency, a single static fusimotor stimulus usually changes, at most, the interval between only 2 or 3 impulses. BESSOU, LAPORTE and PAGÈS (1968) recorded the superimposed, instantaneous impulse frequency changes to a number of single stimuli delivered at random times with respect to the background afferent impulses. Summing these resulted in a cumulative instantaneous frequency plot termed a frequencygram.

The frequencygrams of different primary endings to single stimulation of static fusimotor fibres showed some variation. As shown in Fig. 15, there was: (1) no

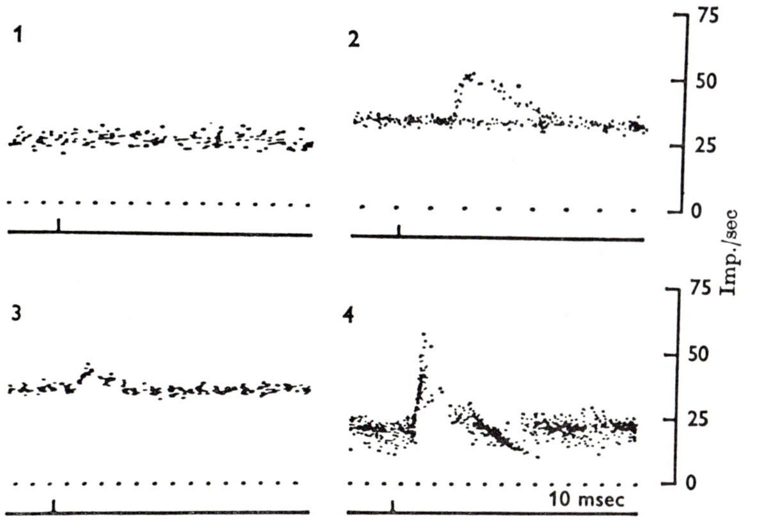

Fig. 15. Responses of 4 primary endings to stimulation of single static γ fusimotor fibres. About 30 responses to single stimuli superimposed in 1, 2 and 3 and about 80 in 4. From BESSOU, LAPORTE and PAGÈS (1968)

effect, (2) an all-or-none type of intermittent effect, or (3–4) a consistent rise in frequency of variable amplitude. The large responses (4) showed a rising phase as short as 5 msec and a falling phase lasting 30–45 msec. In the latter case, there was a pause following the peak frequency, and at the end of the falling phase, the frequency often dropped below the baseline level. Increase in static stretch augmented the amplitude and slightly reduced the duration of the response, particularly in responses of large amplitude.

Repetitive stimulation of a static fusimotor fibre produced frequencygrams of primary endings in which successive responses showed facilitation and, when the repetition rate was sufficiently high, summation. In those cases in which a primary ending gave no response or an intermittent response to stimulation of a static fusimotor fibre, the facilitation was particularly striking. Even at high frequencies of stimulation the plateau of the response showed periodic increments in frequency at the stimulus frequency.

Turning now to dynamic fusimotor fibres, the frequencygrams of primary endings following single stimuli to such fibres was found to differ in several respects from the responses to static fusimotor activation: The frequencygram showed no response or only a small rise in frequency following single stimuli to a dynamic fusimotor fibre (BESSOU et al., 1968). The response was slower in time course than that to static fusimotor fibre stimulation, the rising phase being about 15–25 msec and the falling phase 40–60 msec. Increase in steady tension diminished the response to single stimuli, in contrast to the opposite effect of stretch on static fusimotor responses. Following repetitive stimulation of a dynamic fusimotor fibre, single stimuli to the fibre produced a larger response than in the control period. The frequencygram obtained during repetitive stimulation showed summation. Increments with each stimulus occurred only at low rates as might be expected from the long duration of the response to single stimuli.

A similar study, was carried out by BESSOU and PAGÈS (1969) on the responses of secondary endings to stimulation of static fusimotor fibres. The frequencygrams to single stimuli were in many respects similar to those obtained from primary endings: some gave responses to single stimuli, others did not. Stretch augmented responses to single stimuli. However, there were also some differences: the time course of the response of the secondary ending was slightly slower than that of the primary ending. When double shocks were given the response to the second was conspicuously reduced at intervals up to several hundred milliseconds (Fig. 16). On repetitive stimulation the secondary ending showed summation and facilitation of the response but distinct increments were not seen during maintained stimulation at as high frequencies as in primary endings. The response to single stimuli was reduced in an all-or-none manner or with one intermediate step, indicating that static fusimotor activation is twitch-like rather than graded.

c) β-Innervation; Skeleto-Fusimotor Fibres. While it is clear that γ fibres provide the motor innervation to most intrafusal fibres, the occurrence of intrafusal fibre innervation by collaterals of axon of α motoneurons has been more difficult to evaluate.

It must be demonstrated that stimulation of a particular α motor axon causes an increase in discharge of a spindle afferent fibre which is due to intrafusal fibre

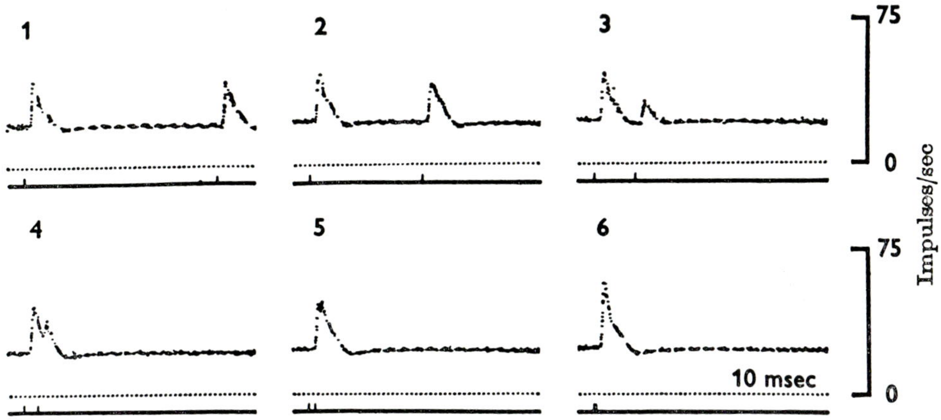

Fig. 16. Responses of a secondary ending to 2 stimuli at various intervals to a static γ fusimotor axon. From BESSOU and PAGÈS (1969)

contraction and not to contraction in extrafusal fibres. The problem would be simple if extrafusal fibres were always "in parallel" with muscle spindles and on contraction always unloaded the spindles.

There are instances in which discharge occurs in spindle afferent fibres, on stimulation of the motor nerve or ventral root, during the early part of the isometric tension rise. This "early discharge", consisting of a brief high frequency burst of impulses, was considered by HUNT and KUFFLER (1951) to result from extrafusal rather than intrafusal contraction for the following reasons: the early discharge often appeared when a number of α axons were stimulated together but not separately. This would not be expected if intrafusal contraction, produced by a specific α axon, caused the early discharge.

Subsequently GRANIT, POMPEIANO and WALTMAN (1959) interpreted the early discharge to be the result of activation of intrafusal contraction by α motor axons.

The first clear physiological evidence of innervation of intrafusal fibres by collaterals of α motor axons came from studies of BESSOU, EMONET-DÉNAND and LAPORTE (1963, 1965). Using the first lumbrical muscle of cat, which has only a few motor fibres, they found that some slowly conducting α (or β) fibres[1] (conduction velocity about 60 m/sec) which caused extrafusal contraction also increased spindle afferent discharge. The effects may be seen in Fig. 17. There was no relation between muscle tension and discharge in the Ia fibre: as stimulus frequency increased the latter continued to increase after tetanic tension produced by extrafusal fibres had become fused and maximal. Critical curarization blocked extrafusal contraction leaving the spindle afferent response intact.

Stimulation of these β fibres produced typical dynamic fusimotor effects, i.e. the dynamic index was increased. The same spindle could be innervated by static and dynamic γ fibres as well as by the β dynamic fibre.

1 The 2 groups of fibres innervating skeletal muscle have usually been termed α and γ. Since the motor fibres that are skeletofusimotor are usually among the smaller of the large fibre group, they have also been referred to as β. It should not be taken to indicate that all fibres of that velocity range are fusimotor as well as skeletomotor.

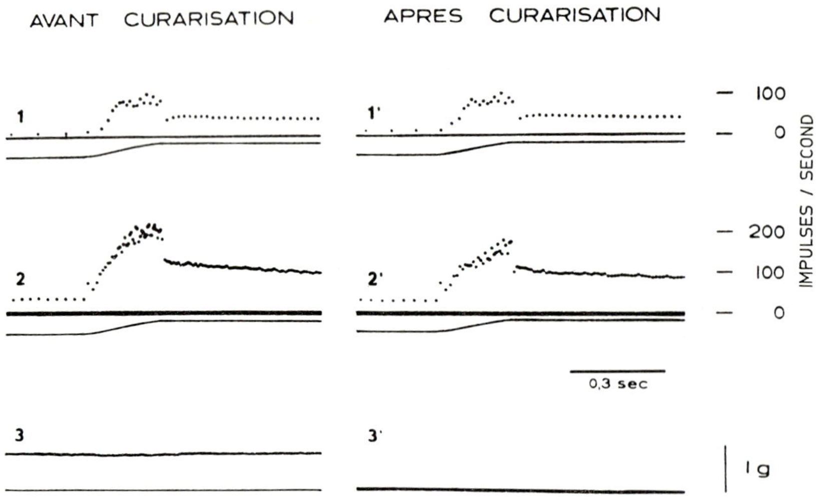

Fig. 17. Effect of stimulating a single B fibre on the response of a primary ending to ramp-and-hold stretch, before and after curarization. 1 and 1′ Responses in absence of stimulation, 2 and 2′ Responses during stimulation of B fibre at 190/sec. 3 and 3′ Tension developed by the motor unit. From Emonet-Dénand and Laporte (1966)

The β fibres producing intrafusal activation did not evoke an "early discharge". They also found some α fibres which increased spindle afferent discharge during, and apparently as a consequence of, extrafusal contraction. In this case both tension and afferent discharge showed the same dependence on stimulus frequency; both disappeared at the same time on curarization. Emonet-Dénand, Jankowska and Laporte (1969) also demonstrated β innervation in lumbrical muscles of rabbit.

On the basis of this and other physiological evidence (Kidd, 1964; Brown et al., 1965) and anatomical studies (see pp. 53) it is clear that β innervation does occur in mammalian spindles. Bessou, Emonet-Dènand and Laporte noted that its occurrence was infrequent and irregular in their lumbrical muscle preparation.

It has been more difficult to demonstrate β skeletofusimotor innervation of spindles in more proximal limb muscles. However, recent studies (Laporte, Emonet-Dénand, Jami and Laporte, 1973) on hind limb muscle in which all but a small branch of a muscle nerve has been cut have permitted studies similar to those carried out on the first lumbrical. That is, a restricted number of motor axons could all be examined for evidence of fusimotor innervation. These studies indicate that β innervation is more common than has been supposed. Whether all P1 plates are derived from β axons, as suggested by degeneration studies (see pp. 53), is not known from physiological experiments. If so β innervation may be quite common indeed.

Although the effective frequency of stimulation of β axons in producing fusimotor effects is much higher than that causing maximal fused tetanic contraction of extrafusal fibres, it seems unlikely that such β motoneurons fire reflexly at higher than the fusion frequency for extrafusal contraction (ca. 20/sec). This may limit the effectiveness of the β innervation. However, the coactivation of extrafusal and intrafusal fibres by impulses in β axons may occur commonly enough for this innervation to play an important role.

d) The Question of Exclusively Fusimotor α Fibres. KUFFLER, HUNT and QUILLIAM (1951) stimulated single α motor axons to soleus, medial gastrocnemius and tenuissimus muscles of cat. These always produced extrafusal contraction. Recently HAASE and coworkers (BARRIOS et al., 1967; BARRIOS et al., 1966; BOES et al., 1968; HAASE and SCHLEGEL, 1966; HAASE et al., 1966; RUTLEDGE and HAASE, 1961) have suggested that many spindles in cat pretibial muscles are innervated by α fibres that are exclusively fusimotor. An extensive study by ELLAWAY, EMONET-DÉNAND, JOFFROY and LAPORTE (1972) of 1813 α motor axons (conducting between 50 and 105 m/sec) to tibialis anterior, extensor digitorum longus and flexor hallucis longus showed that all but 20, probably injured, axons produced extrafusal activation. It seems clear that no significant number of α motor axons are exclusively fusimotor to these muscles.

e) Excitation and Contraction in Intrafusal Muscle Fibres. Potentials in intrafusal fibres resulting from fusimotor stimulation have been recorded both extracellularly and intracellularly. On stimulation of static fusimotor fibres extracellular recording indicated all-or-none potential responses (BESSOU and LAPORTE, 1965; see also EYZAGUIRRE, 1960). Intracellular recordings show that the responding intrafusal fibres have both junctional and all-or-none potentials (BESSOU and PAGÈS, 1971, 1972). In contrast, stimulation of dynamic fibres elicits external monophasic potentials near the equatorial region suggesting junctional potentials (BESSOU and LAPORTE, 1965) and intracellular recording confirms this (BESSOU and PAGÈS, 1972). These are illustrated in Figs. 18 and 19.

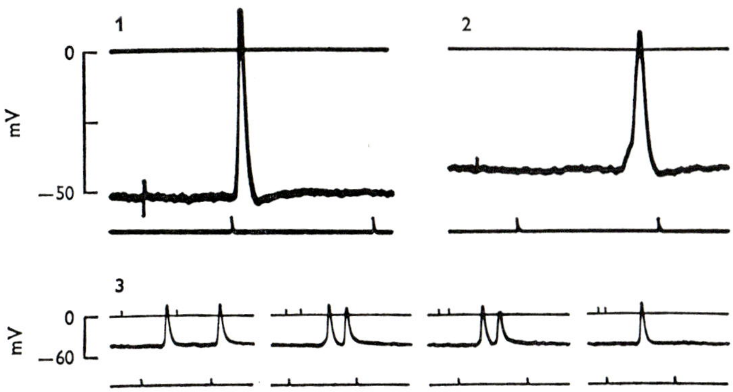

Fig. 18. Intracellular recordings from intrafusal muscle fibres on stimulation of single static fusimotor fibres. Time marks: 10 msec. 1 and 2 Single stimuli different spindles. 3 Double stimuli at varyiig intervals. From BESSOU and PAGÈS (1972)

An important step has been made recently in identifying the type of intrafusal fibre responding to static or dynamic fusimotor stimulation and characterizing its response. By applying Procion Yellow through an intracellular recording pipette,

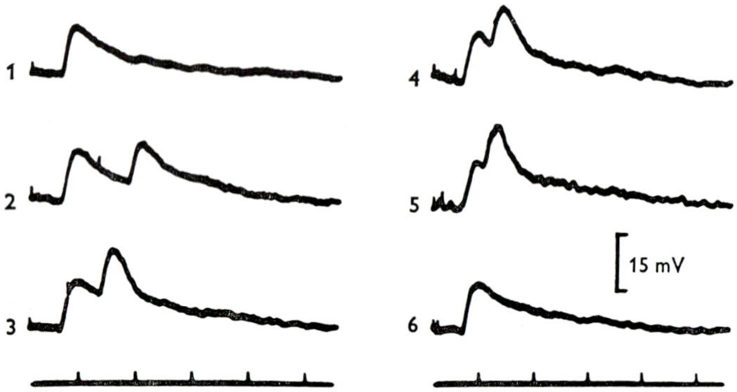

Fig. 19. Intracellular recordings of intrafusal muscle fibre potentials evoked by stimulation of a dynamic fusimotor fibre. 1 Single stimulus. 2–6 Double stimuli. Time marks: 10 msec. From Bessou and Pagès (1972)

nuclear-bag fibres have been shown to respond to dynamic fusimotor stimulation (Barker et al., unpublished observations). Bag fibres can also be activated by static fusimotor axons. Chain fibres are thought to give local or propagated potentials (Bessou and Pagès, 1972).

Boyd (1966) indicates that direct visualization of the spindle showed no contractile response of intrafusal fibres to single stimuli to γ axons. Repetitive stimulation caused contraction only to one side of the equatorial region. He also describes contractions as being "local" or "focal". It seems clear that better optical resolution and high speed photography will be needed to resolve the time course and spatial spread of contraction in intrafusal fibres.

Direct observation of intrafusal fibres shows contraction to be more rapid in chain fibres than in bag fibres (Smith, 1966; Boyd, 1966; Diete-Spiff, 1967). Both type fibres contract in response to single direct stimuli; contraction in both increases markedly with repetitive stimuli. In both fibre types, stimulated directly, contraction appears to occur toward one pole (Beacham and Hunt, unpublished).

Movement of intrafusal fibres appears to be only toward one pole on stimulation of single γ fibres (Boyd, 1970, 1971). Action potentials in chain fibres apparently do not propagate through the sensory region and activate the other pole, as can occur in frog and snake spindles.

Recent studies indicate that individual static γ axons may activate either chain fibres alone or bag and chain fibres. Boyd, Gladden and McWilliam (1973) reported that of 8 single static γ axons, 4 produced contraction in chain fibres alone and 4 in both bag and chain fibres. Bessou and Pagès (1973) also found some static γ axons which caused visable contraction in chain fibres or in bag and chain fibres; dynamic γ axons produced detectable contraction in bag fibres only.

Bessou and Pagès (1967) measured the displacement of the distal polar region of semi-isolated spindles in tenuissimus, using an optical method, in response to stimulation of a static fusimotor fibre. Movement of the spindle showed a latency of 15–30 msec, a rising phase of 35–100 msec and a total duration of 300–450 msec. The duration of these movements is much longer than the response seen in the frequencygrams to similar stimulation. This is not surprising, since the mechanical conditions are quite different and the time required for shortening may give much greater values for the time course of contraction with this optical method.

f) Static and Dynamic Fusimotor Effects; The Role of Plate and Trail Innervation. Following the categorization of fusimotor fibres as static or dynamic, the suggestion was made that trail endings restricted to chain fibres and plate endings confined to bag fibres produced these respective actions (Boyd, 1962). Subsequent morphological and physiological studies have made it clear that the situation is more complex. As has been described, trail endings are not confined to chain fibres nor are plate endings restricted to bag fibres. Further, it has recently been shown that many fusimotor fibres are not exclusively static or dynamic but the effects produced depend upon the frequency of stimulation (Emonet-Dénand et al., 1972).

The clearest evidence linking trail endings with static γ fusimotor fibres comes from experiments in which all but a single fusimotor fibre to tenuissimus were sectioned and allowed to degenerate. Following physiological identification of the type of fusimotor action produced by the fibre, histological study determined the type motor endings remaining (Barker et al., 1970, 1973). Trail endings only were found in all static fusimotor fibres remaining. These were frequently found on bag as well as chain fibres. Such a distribution might account for the dual, dynamic and static, effects evoked by stimulating some single γ fibres at different frequencies if at one frequency contraction in bag fibres predominated while at others contraction in chain fibres was greater.

Dynamic γ fusimotor fibres are considered to terminate in p_2 plates. While these end principally on nuclear-bag fibres, a small number also terminate on chain fibres (see pp. 56). However, stimulation of dynamic γ fusimotor fibres has been found to have no effect on secondary endings (Appelberg et al., 1966). Perhaps the effects produced by the plate endings on chain fibres are too small to cause changes in response of the secondary terminations on such fibres.

As has been noted, the proportion of fibres ending in p_1 plates and p_2 plates on spindles of peroneus longus (P.L.) and flexor hallucis longus (F.H.L.) of cat are quite different. Barker, Stacey and Adal (1970) found the ratio of p_1 to p_2 plates was about 5:1 in F.H.L. and 1:7 in P.L. p_2 plates have been considered to come from dynamic γ fibres and p_1 plates from β fibres (skeletofusimotor). Since static fusimotor γ fibres terminate in trail endings, the number of dynamic γ axons to P.L. should be much larger than to F.H.L. Ellaway, Emonet-Dénand, Jami and Joffroy (1972) found, however, that the proportion of static and dynamic γ fusimotor axons was the same to both muscles. These findings appear to be in conflict with the hypothesis that p_2 plates are the only terminations of dynamic γ fusimotor fibres.

The identification of all p_1 plates as the terminations of β fibres rests largely on the time of their degeneration following motor axon section (see pp. 53). The large number of p_1 plates in muscles such as soleus, flexor hallucis longus and peroneal muscles is in contrast to the amount of β innervation thus far demonstrated physiologically in these muscles. Perhaps skeletofusimotor fibres are more common than has been found. Alternatively, p_1 endings might not be derived exclusively from β fibres but also from the larger of the γ dynamic fusimotor fibres.

At the present time available evidence suggests that static and dynamic fusimotor effects result from activation of nuclear chain and nuclear bag fibres respectively, regardless of the type of motor ending that is involved. The situation is probably more complex for, although the classification of intrafusal fibres as bag or chain undoubtedly is valid, there appear to be significant numbers of intermediate type intrafusal fibres in cat and in rabbit (see pp. 18). In this context it is of interest that β fibres in rabbit have been shown to produce static fusimotor effects as well as dynamic effects typical of most such fibres in cat (Emonet-Dénand et al., 1970).

g) Summary. The motor innervation of the muscle spindle is far from well understood but a number of features are becoming clear. Two types of γ axon are present: One terminates in trail endings; in many cases an axon supplies trail endings to both bag and chain intrafusal fibres. The other terminates in plate (p2) endings; most are on bag fibres and some are on chain fibres. Whether a γ axon produces dynamic or static fusimotor effects probably depends on whether excitation is produced mainly in bag or chain fibres. Static fusimotor effects appear to result from stimulation of γ axons which terminate in trail endings; these presumably evoke contraction mainly in chain fibres. The type ending associated with γ axons producing dynamic fusimotor effects has not yet been experimentally identified; presumably plate endings, predominantly on bag fibres, elicit dynamic fusimotor effects. Certain γ axons may produce static fusimotor effects at one frequency of stimulation and dynamic fusimotor effects at another. It seems likely that the stimulus frequency causing dynamic effects produces more contractile activation in bag fibres and that causing static effects evokes more contraction in chain fibres. The type ending associated with such dual effects has not been determined. The demonstration that single γ axons can evoke both dynamic and static fusimotor effects suggests that both bag and chain fibres can be coactivated by the same motor axon.

Plate endings derived from β axons appear to innervate bag fibres predominantly. They generally produce dynamic fusimotor effects.

The nature of neuromuscular transmission at trail and plate endings and of the post-synaptic responses evoked in bag and chain fibres is only beginning to be understood. Dynamic fusimotor axons appear to cause junctional potentials while static fusimotor axons are associated with graded junctional potentials or all-or-none action potentials.

Two techniques have been used to identify the type intrafusal fibre responding to static or dynamic fusimotor stimulation. One is to record intracellularly the electrical response from an intrafusal fibre to stimulation of a particular γ axon. Procion Yellow injected via the intracellular pipette allows the fibre to be identified. The other technique is to use glycogen depletion as an index of contractile activation (Brown and Butler, 1973). These approaches have provided some important information: the association of junctional potentials in nuclear-bag fibres with dynamic fusimotor effects and of action potentials or graded potentials in nuclear-chain fibres with static fusimotor effects. A number of interesting problems remain: Are action potentials in the nuclear-chain fibres propagated and, if so, how far? Direct visualization suggests that contraction often occurs toward one pole on fusimotor stimulation. Glycogen depletion is non-uniform along the nuclear-chain fibre following static fusimotor stimulation, even one pole does not appear to contract uniformly. No information is available on the threshold for contractile activation or its time course. It is curious that nuclear-chain fibres which evidently have action potentials have such a widespread distributed motor innervation. In extrafusal fibres, that type of motor innervation is usually associated with local junctional potentials and tonic type contraction. Nuclear-bag fibres, on the other hand, have a preponderance of plate type innervation. They appear to have only local junctional potentials yet this type of innervation in extrafusal fibres is typically associated with propagated action potentials and twitch type contraction.

5. Sympathetic Effects on Muscle Spindles

The effects of stimulation of the sympathetic trunk on the afferent discharge from cat muscle spindles was studied by ELDRED, SCHNITZLEIN and BUCHWALD (1960) and by HUNT (1960). Both studies showed that at constant muscle length sympathetic stimulation caused an initial rise in discharge frequency followed by a fall, often to cessation of discharge. ELDRED et al. obtained similar effects on occluding the blood supply to the muscle (gastrocnemius) and considered the effects of sympathetic stimulation to arise mainly from vasoconstriction. On the other hand, HUNT found the sympathetic effect was similar in spindles of the tenuissimus muscle to which the circulation had been eliminated. He also noted these effects in both group Ia and II fibres.

The significance of these effects is not clear. Sympathetic fibres do enter the spindle in connection with blood vessels. Effects not secondary to decreased blood flow might occur as a result of released adrenergic transmitters diffusing within the capsular space and acting on the sensory terminals, but there is no direct evidence for this. It seems unlikely that this is an important mechanism for regulation of spindle activity.

B. The Golgi Tendon Organ

The Golgi tendon organ has received less attention than the muscle spindle. This receptor, located near the musculo-tendinous junction, is effectively in series with certain of the extrafusal muscle fibres. The threshold of the tendon organ to external stretch varies considerably but is generally much higher than that of muscle spindle receptors. External stretch of several hundred grams applied to a muscle tendon is often required to produce a maintained discharge in tendon organs of muscles such as gastrocnemius or soleus (cat) although occasional units may have appreciably lower thresholds.

The receptor characteristically discharges during contraction of the muscle (B.H.C. MATTHEWS, 1933; and others). The increased frequency of discharge is usually most prominent during the rising phase of the isometric twitch.

Responses of Golgi tendon organs in soleus and tibialis anterior of cat have been compared by ALNAES (1967). Anterior tibial units all responded to passive stretch within the limits of full physiological extension, in contrast to tendon organs in soleus some of which failed to discharge within the normal range of extension (JANSEN and RUDJORD, 1964). Most anterior tibial units had the same tension threshold to muscle contraction and passive stretch. Most soleus tendon organs had a lower threshold to contraction than to passive extension; in many units threshold to contraction was less than 50 gm. While the tendon organ discharge frequency during stretch of soleus was greater at the end of dynamic stretch than during its maintenance, during fused tetanic contraction the discharge showed little or no adaptation.

Dynamic responses have been studied in anterior tibial and soleus tendon organs by using sinusoidal as well as ramp and hold stretches (Alnaes, 1967). The periodic variation in impulse frequency to sinusoidal stretch leads the rise in muscle tension indicating some degree of dynamic sensitivity. The receptors in the two muscles did not appear to differ in this regard. Although the tendon organ has some dynamic sensitivity, this appears to be far less than that seen in primary endings of muscle spindles.

Houk and Henneman (1967) studied the responses of Golgi tendon organs during contraction of different motor units in soleus muscle. Subdivisions of ventral root were used to stimulate individual motor axons or aggregates thereof. Some tendon organs showed a pause during contraction when part of the ventral root outflow was stimulated, indicating that the sense organ lay effectively in parallel with certain muscle fibres. Tendon organs showed acceleration of discharge on stimulation of individual motor units. Under these circumstances the threshold for discharge was at a much lower tension than that seen when the whole motor outflow was stimulated. From 4 to 15 motor units were found to drive an individual tendon organ. These studies indicate that tendon organ discharge may be evoked on activation of only one or a few motor units and, hence, may come into play at much lower levels of contraction than previously thought. It appears that tendon organs are effectively in series not with all the contracting muscle elements but with a number of fascicles.

Houk and Henneman (1967) also investigated the relation between tendon organ discharge frequency and tension, when motor axons were tetanically stimulated. Discharge frequency rose to a peak, as compared to the force record, then declined toward a steady state value. The amount of adaptation seen was greater when the initial tension was less. These findings were interpreted as indicating that the tendon organ had a significant dynamic component in its response. However, the variation in the apparent dynamic response as a function of initial tension suggests that mechanical factors rather than inherent dynamic sensitivity may have accounted for the findings. Houk (1967) has suggested that tendon organs have more dynamic sensitivity than previously supposed. However, the low sensitivity of the tendon organ to vibration suggests that the receptor itself does not have a large dynamic response.

The afferent axons of tendon organs have conduction velocities in the 70–120 m/sec range (Hunt, 1954) indicating their diameters are in the group I range. The fibres from tendon organs are usually designated as Ib fibres.

There is no direct evidence about the mechanisms of impulse initiation in Golgi tendon organs. The afferent axon divides into a number of branches which end in terminal arborizations on the multiple small fascicles of the tendon organ. Stretch deformation of these terminals presumably produces a receptor or generator potential that initiates impulse activity.

C. Other Receptors

1. Receptors of Small Myelinated Fibres (Group III)

The trimodal distribution of myelinated afferent fibres from muscle has been noted. Muscle spindle afferent fibres account for groups Ia and most, if not all, of group II, while tendon organ fibres are principally in group Ia. Group III fibres, ranging in size from 1 to 4 μ appear to terminate mainly in high threshold receptors, although a few are from muscle spindles.

Two independent studies of group III fibres appeared at nearly the same time (PAINTAL, 1960; BESSOU and LAPORTE, 1960, 1961). PAINTAL recorded discharge in nerve filaments from lateral gastrocnemius-soleus and tibialis anterior, while BESSOU and LAPORTE recorded afferent discharge from lateral gastrocnemius and soleus in fine dorsal root filaments. In both studies conduction velocity was used to estimate fibre diameter. The majority of the fibres studied by PAINTAL responded to strong localized pressure on the muscle and very irregularly, if at all, to large stretches of the muscle. The responses adapted rapidly. Some fired with contraction of the muscle. Asphyxia was generally not an effective stimulus. Injection of hypertonic NaCl, locally, excited the receptors. PAINTAL noted a few "pressure-pain receptors" associated with group I and II fibres. The difficulty in interpreting such findings is that the stimuli used might excite impulses in nerve fibres rather than receptors.

BESSOU and LAPORTE also found that many group III fibres discharged in response to high threshold mechanical stimuli. In some cases the discharge was slowly adapting. Some receptors appeared to be located near the musculotendinous junction, some on the surface of the muscle or tendon. A few responded to stretch.

Among the group III fibres several were found which showed responses characteristic of muscle spindles (PAINTAL, 1960; BESSOU and LAPORTE, 1961). These probably come from secondary endings. Also, BESSOU and LAPORTE found 2 (of 105) fibres which exhibited responses typical of Golgi tendon organs.

It should be noted (see pp. 144) that some large sensory axons, even of group I size, end in free nerve terminals within the muscle. The lack of physiological evidence on such fibres is probably due to their being ignored. Rather than using the strong mechanical stimuli probably needed to excite such receptors one might assume they are damaged fibres from a more usual receptor.

2. Receptors of Sensory C Fibres

Using the collision technique of DOUGLAS and RITCHIE (1957), BESSOU and LAPORTE (1958) studied the responses of unmyelinated afferent fibres from soleus and lateral gastrocnemius. The reduction in size of a muscle nerve action potential of C fibres was reduced by activity produced in such fibres initiated by stimulation of their receptor terminations. The combination of asphyxia and contraction was found to be an effective stimulus to these receptors.

In addition to the well known sensation of muscle pain, there are afferent inputs from muscle, activated by contraction, which bring about respiratory and cardiovascular reflex adjustments. The latter might involve receptors sensitive to chemical alterations associated with contraction. There is no clear evidence on the receptors that might be involved. While the unmyelinated fibre group is probably concerned with muscle pain, particularly that associated with ischemic contraction, this fibre group may subserve other functions. A unitary analysis of the receptor characteristics of C fibres would be informative in this regard.

III. Reptilian Muscle Spindles

Muscle spindles of snakes and lizards have only a single intrafusal fibre and a single sensory ending. Hence they are considerably simpler than those of other vertebrates. Their structure has been studied for over a century but they have been examined physiologically only in the last several years. Anatomical studies have shown that two types of reptilian spindle can be recognized: (1) the long capsule spindle which has a long encapsulated sensory region around an intrafusal fibre which shows only a slight increase in diameter as compared to the polar regions and contains myofilaments throughout the sensory region. (2)The short capsule spindle which has a shorter encapsulated sensory region with an intrafusal fibre that has a marked increase in diameter; this region is filled with various cytoplasmic organelles but myofilaments do not continue into it (see pp. 105). The motor innervation is known to come from collaterals of motor axons innervating extrafusal motor endings, either from motor axons with en grappe or en plaque extrafusal motor endings. The description of the physiological characteristics of reptilian spindles which follows will consider first the sensory responses evoked by stretch and then the responses to motor stimulation.

A. Sensory Responses

Responses have been recorded from the afferent fibres of reptilian spindles in isolated preparations from snake (HUNT and WYLIE, 1970; FUKAMI, 1970a) and from *in vivo* as well as isolated muscles from the lizard (PROSKE, 1969). In the case of snake muscle spindles, studies have also been carried out on individual isolated muscle spindles.

In the snake, and in the lizard, individual spindle afferent fibres show two types of response to applied stretch. One type of spindle has a greater dynamic response than the other. This, in turn, is correlated with two distinct structural types. When external stretch of a ramp and hold configuration is applied to the muscle, the discharge in an afferent fibre from a short capsule spindle shows a large dynamic component in its response, that is, the frequency of discharge is much higher during the application of the stretch than during its maintenance. In contrast, the long capsule, tonic type spindle shows an afferent discharge exhibiting a smaller dynamic index: the steady state frequency is not much lower than the peak frequency attained during application of the stretch.

In the costocutaneous muscle of snake, individual long or short capsule spindles can be visualized and can be isolated. Short capsule spindles show a more prominent dynamic response than long capsule spindles. FUKAMI and HUNT (1970) and FUKAMI (1970a) referred to these as phasic and tonic type spindles. (For the sake of uniformity it seems preferable to adopt the term short and long capsule spindle.) Examples of the response to stretch from the two spindle types may be seen in Fig. 20. In the lizard PROSKE (1969) found that the scalenus muscle con-

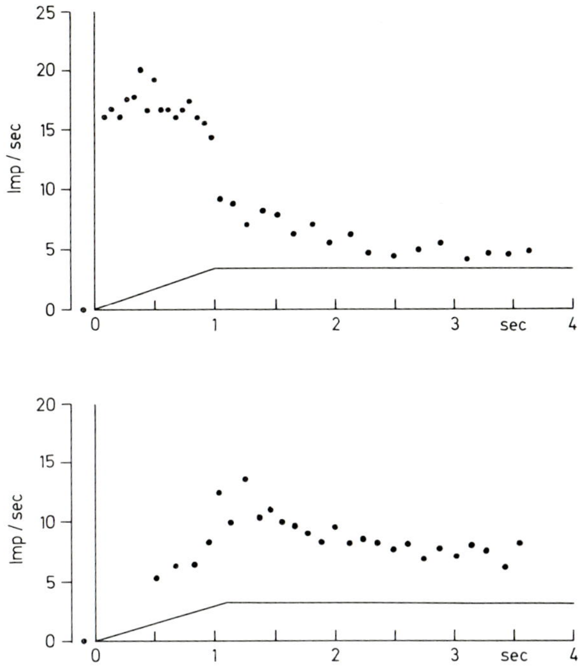

Fig. 20. Response of a phasic (short capsule) snake spindle (above) and a tonic (long capsule) spindle (below) to ramp-and-hold stretch. From FUKAMI (1970)

tained only short capsule spindles, whereas the semimembranosus contained only long capsule spindles. A comparison of the responses to stretch showed that spindles from the scalenus muscle behaved more phasically, that is, they had a considerably higher dynamic index, than those from semimembranosus. Short capsule spindles in the lizard are analogous to the "phasic" spindles in snake, while the long capsule spindles in lizard are similar to the "tonic" type snake spindle (FUKAMI, 1970a). However, it is not yet clear that structure of snake and lizard spindles are identical.

The dynamic behavior of the short capsule spindle is characterized by an increase in frequency of discharge as the rate of stretch becomes greater. This is reflected in an increase in dynamic index with increasing rate of stretch. In the snake, FUKAMI (1970) found that the dynamic index increased linearly with rate of stretch in phasic type (short capsule) units but in tonic type units (long capsule spindles) the dynamic index was not only smaller but showed little increase with rate of stretch until the latter exceeded 10 mm/sec. In contrast, PROSKE (1969) noted that the dynamic index of short capsule spindles increased rapidly as the rate of stretch was increased up to about 10 mm/sec and then more gradually, while the long capsule spindles showed a smaller but linear increase in dynamic index as stretch was increased. The basis for this difference in the results from snake and lizard spindles is not yet clear.

The structural differences between long and short capsule spindles in both snake and lizard are in keeping with the idea that the dynamic component of the sensory response might depend upon mechanical properties of the intrafusal fibre upon which the sensory terminations are applied. The presence of myofilaments and the small number of nuclei in the sensory region of the intrafusal fibre of the long capsule spindle is compatible with the notion that this portion of the spindle might have visco-elastic properties similar to the intrafusal fibre in the more polar regions. The structure might, therefore, be expected to behave with little time dependence in its deformation to stretch. On the other hand, the sensory region of the short capsule spindle where the intrafusal fibre is enlarged, lacks myofilaments and contains an accumulation of nuclei and other inclusions might differ in its physical properties from the polar regions. This suggests that a greater viscosity in the polar regions might render the deformation of the sensory terminals greater during the application of stretch than during its maintenance and hence account for its dynamic behavior.

This hypothesis has now been put to a more direct test. Using differential interference contrast microscopy of the Nomarski type, it has been possible to observe and photograph in some detail the structures within the sensory region of isolated phasic and tonic spindles (FUKAMI and HUNT, unpublished). Positive identification of structures giving rise to images in the interference microscope has been made by electron microscopic studies of the same spindle. One can, with assurance, identify nuclei, sarcomeres, bands of connective tissue, sensory terminations and other structures. High speed photographs of both phasic and tonic spindles before and during the application of a ramp and hold stretch have been taken simultaneous with the discharge from the afferent nerve fibre. Plots of the displacement of elements in the sensory region of phasic and tonic spindles in response to external stretch failed to show a component of the displacement which could account for the dynamic behavior of the phasic spindle.

Another possible mechanism for the dynamic component of the phasic spindle discharge might lie in the properties of the impulse generating site. Evidence now available suggests this may not be the case. FUKAMI (1970) has shown that external current passed through the impulse generating site influences the discharge frequency in the same way and with the same time course in both tonic and phasic spindles. Thus, depolarizing current increases discharge frequency and hyperpolarizing current diminishes discharge frequency with the same brief time constant in both types of spindle. No accommodative component of the phasic spindle impulse generating site has yet been demonstrated that would account for the higher discharge frequency during application of stretch than during its maintenance.

This leaves the receptor potential and its underlying conductance changes as the apparent determinant of the phasic or tonic behavior of the reptilian spindle. To date it has not been possible to compare the receptor potential from long and short capsule reptilian spindles. On the basis of information presently available one might expect the short capsule spindle to show a receptor potential similar to that of the frog spindle or the primary ending of the mammalian spindle, i. e. with a prominent dynamic component while in the long capsule spindle such a component might be expected to be small or absent.

FUKAMI and RIDGE (1971) examined the response of snake spindles after denervation of the muscle. Responses to stretch persisted long after failure of neuromuscular transmission and, prior to disappearance of the sensory response the spindle exhibited only phasic responses to stretch. FUKAMI (1972) has studied the reinnervation of spindles in these muscles. About 20 days after crushing the nerve to the muscle responses to stretch reappear. At first these are only phasic in both long and short capsule spindles, later static responses reappear (after about 60 days). Passage of current, at a time when only phasic responses to stretch are seen, reveals a rapid accommodation at the impulse initiating site in contrast to the normal spindle. This appears to explain the phasic behavior of spindles early in reinnervation although it is possible that there may be as well changes in the receptor potential. FUKAMI found the spindles late in reinnervation to have the same appearance of sensory terminals under electron microscopy as normal spindles.

B. Motor Responses

In both snake and lizard spindles the motor innervation of the intrafusal fibre is provided by collaterals of axons which innervate extrafusal muscle fibres. The effect of motor fibre stimulation on the sensory discharge from spindles has been studied in both lizard and snake. In the latter, studies have also been carried out on the activity produced by motor stimulation in the intrafusal fibres of isolated single muscle spindles. Since the results obtained in lizard differ in some respects from those obtained in snake they will be presented separately.

When recording from spindles in the iliofibularis muscle of the lizard, PROSKE (1969) found that stimulation of one of the motor axons supplying that muscle caused an acceleration of the spindle discharge during the rising phase of contraction; all other motor axons produced on stimulation a cessation or decrease in discharge during the period of muscle shortening. The discharge produced by the motor axon consisted of a burst of impulses with evident facilitation on repetitive stimulation.

Some lizard spindles exhibited evidence of multiple motor innervation, acceleration of the discharge occurring on stimulation of several motor axons. Fig. 21 shows a record published by PROSKE (1969) from a spindle in the semitendinosus muscle. The burst of sensory impulses was of longer duration but lower in frequency than that elicited in spindles innervated by a single motor axon.

The motor responses of snake muscle spindles have been studied by HUNT and WYLIE (1970) and FUKAMI (1970). Stimulation of specific motor axons was found to evoke a burst of impulses during the rising phase of muscular contraction. Repetitive stimulation revealed a conspicuous subnormality following a single motor response. The discharge in response to motor stimulation in snake spindles appeared to occur at lower frequency than in lizard spindles. However, only a few muscles have thus far been studied in the snake. In these studies it is not always clear whether the responses were from long or short capsule spindles.

CLIFF and RIDGE (1973) have examined the innervation of reptilian spindles by motoneurons which innervate twitch or tonic (slow) extrafusal muscle fibres. A single motor axon was stimulated by a microelectrode placed near one of its intramuscular branches. The tension developed by the motor unit was recorded. Following critical curarization, intrafusal contraction was observed in long capsule

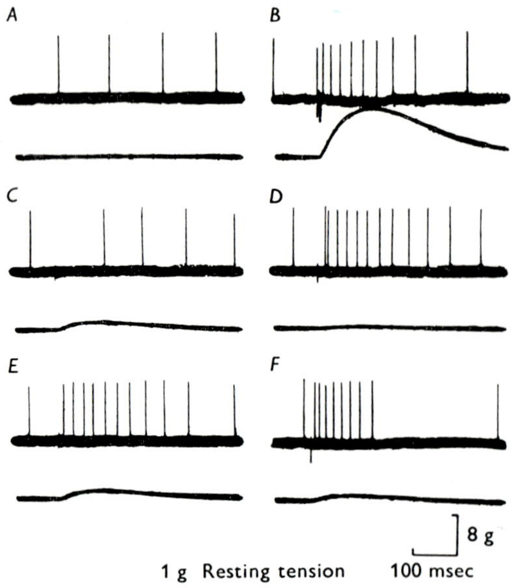

Fig. 21. Responses of lizard muscle spindles in semitendinosus muscle. A Resting discharge. B Maximal stimulus to muscle nerve. C–F Stimulation of different nerve filaments. Upper traces: afferent responses. Lower traces: tension. From Proske (1969)

spindles and the effect on spindle afferent discharge recorded. Both twitch and tonic type motor units were found to innervate long capsule spindles; their effect on the afferent discharge was similar. Individual spindles were sometimes innervated by more than one motor unit, often of different types.

In isolated snake spindles the intrafusal muscle fibre has been shown to exhibit a propagated action potential following motor axon stimulation (Fukami, 1970a). Extracellular recording showed the speed of conduction to be about 0.2 to 0.7 m/sec at 25° C. The impulse is conducted through the sensory region appeared to accelerate where the fibre diameter increased. Double shock stimulation shows the intrafusal fibre action potential to have a longer refractory period and subnormality than extrafusal fibres. Whether these responses were from long or short capsule spindles is not certain.

IV. Frog and Toad Muscle Receptors

A. Muscle Spindles

The frog muscle spindle has been the subject of many neurophysiological investigations since the early studies of Adrian and Zotterman (1926) and B.H.C. Matthews (1931a, b). These revealed many features of this slowly adapting mechanoreceptor: the higher frequency of discharge during application of stretch than during its maintenance and an adaptation occurring during maintained stretch to a steady level of discharge. As is suggested by morphological and functional evidence, there appears to be only one type of sensory ending in frog spindles. The myelinated sensory axon divides and redivides, ending in multiple slender unmyelineated axons with bulb-like expansions applied to the intrafusal fibres (see pp. 96).

Stretch produces a depolarization of the sensory terminals, the receptor or generator potential, which, in turn, initiates impulses in the afferent fibre. OTTOSON and SHEPHERD's article in volume I of this Handbook deals with these processes in detail. Hence, only a few of the more recent findings will be considered here. More attention will be devoted to studies on the motor innervation of the frog spindle.

1. Factors Determining the Response to Stretch

The application of stretch of ramp and hold configuration produces a response with both dynamic and static components. This is evident in both the discharge frequency and in the receptor potential recorded from frog spindles. The most generally accepted explanation of the dynamic sensitivity has been that the greater viscosity of the polar regions of the intrafusal muscle fibres cause the applied stretch to produce a velocity sensitive deformation of the sensory terminals. A similar explanation has been offered for the dynamic sensitivity of primary endings in mammalian spindles (see MATTHEWS, 1972). Recent evidence makes it seem unlikely that this is the basis for the dynamic sensitivity of frog spindles. Stretch to a constant length shows some dynamic overshoot of tension but when the spindle is stretched to a constant tension rather than length, the dynamic component of the receptor potential is only 15—20% smaller, for stretches giving less than the maximal dynamic peak frequency (HUSMARK and OTTOSON, 1971a). More direct evidence comes from stroboscopic microphotographs of the sensory region of frog spindles during applied stretch (OTTOSON and SHEPHERD, 1970). No dynamic component is seen in the resultant displacement of the sensory region. These experiments have been criticized on the basis that the spindles in the preparations used were truncated, i. e. the intrafusal bundles tied to the pulling device close to the sensory region (MATTHEWS, 1972). However, such preparations do exhibit essentially the same features in their sensory response as normal full-length frog spindles and it seems unwarranted to dismiss the findings for this reason. Further, a similar lack of a dynamic component has been found in high speed photographs of the displacement of the sensory regions of phasic type reptilian spindles and primary endings of mammalian spindles (see above). The overall mechanical properties of intrafusal fibres such as would result from a higher viscosity in the polar regions do not appear to explain dynamic sensitivity in any of these spindles although mechanical factors within the sensory region might be responsible. The possible role of ionic factors has been considered by HUSMARK and OTTOSON (1971b).

2. Motor Responses

The motor innervation to frog intrafusal fibres appears to come exclusively from collaterals of motor axons which innervate extrafusal muscle fibres. The two types of frog extrafusal muscle fibres, twitch and slow, are innervated by different groups of motor axons, the motor axons to slow fibres being smaller in diameter. The intrafusal fibres are also of two morphological types (see pp. 96). One is similar to twitch extrafusal fibres, except for a less well organized sarcoplasmic reticulum and fewer triads. The other has features intermediate between extrafusal twitch and slow fibres.

Stimulation of large motor axons results in an acceleration of the sensory discharge. After block of extrafusal contraction by critical curarization the sensory response consists of a train of impulses in response to a single motor stimulus (KATZ, 1949). Repetitive stimulation of a large motor axon causes a large increase in afferent discharge frequency; on cessation of stimulation the frequency falls abruptly and transiently below the baseline level. Stimulation of small motor axons cause a slower increase in afferent discharge frequency and, on cessation of stimulation, discharge frequency only gradually falls to the control level (EYZAGUIRRE, 1957, 1958).

Propagated action potentials are known to occur in some frog intrafusal fibres (KATZ, 1949). The conduction velocity of these intrafusal muscle action potentials is about 0.3 to 0.6 m/sec. They can propagate through the region of sensory innervation, i. e. from one pole to another, although at high frequencies impulses may fail to propagate through the sensory region (SMITH, 1964). Stimulation of large motor axons can produce propagated intrafusal action potentials, clearly seen in intracellular records (EYZAGUIRRE, 1957; EYZAGUIRRE and VIAL, 1956; KOKETSU and NISHI, 1957). KOKETSU and NISHI (1957a, b) recorded both multiple junctional potentials and impulse activity from intrafusal fibres following nerve stimulation but did not determine whether these resulted from activation of large or small motor axons.

SMITH (1964a, b) observed both the electrical and mechanical activity of intrafusal fibres in *Xenopus* spindles. Recording electrical activity by a microelectrode inserted in the capsule, he found evidence of propagated impulses in both large and small diameter muscle fibres and for separate motor innervation to the two sized fibres. Conduction through the sensory region failed in the larger fibres at stimulation frequencies of greater than 30/sec and in the small fibres above 5/sec. Cinephotography showed that contraction in the larger fibre was more rapid than in the small fibre. The large intrafusal fibres were activated by stimulation of large diameter motor axons and the small intrafusal fibres by small motor axons.

The presence of two types of intrafusal fibre and two types of motor ending is probably correlated with large and small motor axon innervation. Thus, twitch type fibres having plate-like motor endings appear to be innervated by large motor axons and intermediate type fibres with "en grappe"-like motor endings appear to be innervated by small motor axons. The former contract more rapidly than the latter. The intrafusal fibres innervated by small motor fibres appear to differ from the slow extrafusal fibres innervated by these axons in having some ability to give propagated action potentials.

In the toad it has been shown that individual motor axons can activate a number of spindle afferent fibres. Conversely, individual sensory units can be activated by a number of motor axons (EYZAGUIRRE, 1958).

B. Other Frog Muscle Receptors

In addition to muscle spindles, other type receptors occur in tendon or in relation to extrafusal fibres (Huber and Dewitt, 1900; Ito et al., 1964) (see pp. 104). Tendon receptors have been studied by recording from nerve branches to the pelvic or distal end of the frog sartorius, the main muscle nerve being used for stimulation (Ito, 1967, 1968). These receptors discharge in response to stretch and muscle contraction, most showing large dynamic responses but no static discharge. Other units, considered to come from leaf-like receptors lying in relation to a number of extrafusal fibres (Ito et al., 1964), gave responses resembling those from tendon receptors (Ito, 1968).

V. Avian Muscle Receptors

Apart from the study of Dorward (1970) there appear to be no published reports on the physiology of avian muscle receptors. Dorward examined the responses of afferent fibres from two muscles in the duck, gastrocnemius lateralis (GL) and flexor perforans et perforatus digiti III (FPP3). Afferent units were isolated in filaments of sciatic nerve and stimuli were applied to muscle nerves. Individual motor units were not isolated. Stretch receptors showing response patterns typical of muscle spindles or of tendon organs were found. Muscle spindle units showed a pause during contraction especially at low initial tensions. Many showed an "early discharge". Increased discharge during contraction was brought about by stimulating high threshold, presumably γ, fibres. Also, some apparently fusimotor effects were associated with "α" fibre excitation. Repetitive stimulation and increased initial tension enhanced these fusimotor effects. The maximal effective frequency of fusimotor stimulation was between 140 and 195/sec.

The distribution of conduction velocities (and calculated fibre diameters) of afferent fibres showed no evidence of a bimodal distribution. However, the number of fibres was not large enough to indicate with certainty that the distribution is unimodal. Morphological evidence is not clear as to whether or not there are 2 types of sensory ending (see pp. 119).

In some avian muscles, spindles appear to have only one type of intrafusal fibre. Also, each intrafusal fibre becomes separately encapsulated in the sensory region (see pp. 120). These features make the avian spindle an interesting subject for further physiological studies. It seems likely that avian spindles may receive a substantial α (or β) innervation as well as a motor innervation from smaller fibres.

Dorward (1970) found typical tendon organ behavior in a large number of afferent fibres. The threshold for maintained discharge in response to steady stretch was high for GL units, but often quite low for FPP3 units. The "in series" behavior was comparable to that seen in mammalian tendon organs. The fibre diameter (calculated) spectrum showed a unimodal distribution. No other types of muscle receptor were described.

Acknowledgement

Supported by NIH grant NS07907.

References

Adrian, E.D., Zotterman, Y.: The impulses produced by sensory nerve endings. Pt. 2. The response of a single end-organ. J. Physiol. (Lond.) **61**, 151–171 (1926).

Alnaes, E.: Static and dynamic properties of Golgi tendon organs in the anterior tibial and soleus muscles of the cat. Acta physiol. scand. **70**, 176–187 (1967).

Andersson, B.F., Lennerstrand, G., Thoden, U.: Fusimotor effects on position and velocity of spindle endings in the external intercostal muscle of the cat. Acta physiol. scand. **74**, 285–300 (1968a).

Andersson, B.F., Lennerstrand, G., Thoden, U.: Response characteristics of muscle spindle endings at constant length to variations in fusimotor activation. Acta physiol. scand. **74**, 301–318 (1968b).

Appelberg, B., Bessou, P., Laporte, Y.: Action of static and dynamic fusimotor fibres on secondary endings of cat's spindles. J. Physiol. (Lond.) **185**, 160–171 (1966).

Barker, D., Emonet-Dénand, F., Laporte, Y., Proske, U., Stacey, M.: Identification des terminaisons motrices des fibres fusimotrices statiques chez le chat. C.R. Acad. Sci. (Paris) **271**, 1203–1206 (1970).

Barker, D., Emonet-Dénand, F., Laporte, Y., Proske, U., Stacey, M.: Morphological identification and intrafusal distribution of the endings of static fusimotor axons in the cat. J. Physiol. (Lond.) **230**, 405–427 (1973).

Barrios, P., Haase, J., Heinrich, W.: Fusimotorische Alpha-Reflexe an prätibialen Flexorenspindeln der Katze. Pflügers Arch. ges. Physiol. **296**, 49–69 (1967).

Barrios, P., Haase, J., Heinrich, W., Schlegel, H.J.: Die Entladungen deafferentierter Katzen-Flexor-Muskelspindeln mit fusimotorischer-Innervation bei repetierender Reizung von Ventralwurzelfilamenten. Pflügers Arch. ges. Physiol. **200**, 101–113 (1966).

Beacham, W.S., Fukami, Y., Hunt, C.C., Ottoson, D.: Displacement of elements within mammalian muscle spindles. Proc. I.U.P.S. IX. Abstr. XXV. Int. Congr. Munich (1971).

Bessou, P., Emonet-Dénand, F., Laporte, Y.: Effects of the stimulation of slow gamma-fusimotor fibres on the secondary endings of neuromuscular spindles. C. R. Acad. Sci. (Paris) **156**, 1154–1158 (1962).

Bessou, P., Emonet-Dénand, F., Laporte, Y.: Occurence of intrafusal muscle fibre innervation by branches of slow α-motor fibres in the cat. Nature (Lond.) **198**, 594–595 (1963).

Bessou, P., Emonet-Dénand, F., Laporte, Y.: Motor fibres innervating extrafusal and intrafusal muscle fibres in the cat. J. Physiol. (Lond.) **180**, 649–672 (1965).

Bessou, P., Laporte, Y.: Activation des fibres afférentes amyllinguées d'origine musculaire. C. R. Acad. Sci. (Paris) **152**, 1587–1590 (1958).

Bessou, P., Laporte, Y.: Etude des récepteurs musculaires innervés par les fibres afférentes du groupe III (fibres myélinisées fines) chez le chat. Arch. ital. Biol. **99**, 292–321 (1961).

Bessou, P., Laporte, Y.: Responses from primary and secondary endings of the same neuromuscular spindle of the tenuissimus muscle of the cat. In: Symposium on Muscle Receptors. Barker, D., Ed., pp. 105–119. Hong Kong: Hong Kong University Press. 1962.

Bessou, P., Laporte, Y.: Potentials fusoriaux provoqués par la stimulation de fibres fusimotrices chez le chat. C. R. Acad. Sci. (Paris) **260**, 4827–4830 (1965).

Bessou, P., Laporte, Y.: Observations on static fusimotor fibres. In: Muscular Afferents and Motor Control. Granit, R., Ed., pp. 81–89. Stockholm: Almqvist and Wiksell. 1966

Bessou, P., Laporte, Y., Pagès, B.: Similitude des effets (statiques ou dynamiques) exercés par les fibres fusimotrices uniques sur les terminaisons primaires de plusieurs fuseaux chez le chat. J. Physiol. (Paris) **58**, 31–39 (1966).

Bessou, P., Laporte, Y., Pagès, B.: Frequencygrams of spindle primary endings elicited by stimulation of static and dynamic fusimotor fibres. J. Physiol. (Lond.) **196**, 47–63 (1968).

Bessou, P., Pagès, B.: Enregistrement de mouvements de fuseaux neuromusculaires partiellement disséqués consécutifs à la stimulation de fibres fusimotrices statiques chez le chat. C.R. Acad. Sci. (Paris) **265**, 351–353 (1967).

Bessou, P., Pagès, B.: Spindle secondary ending responses elicited by stimulation of static fusimotor axons. J. Physiol. (Lond.) **202**, 569–584 (1969a).

Bessou, P., Pagès, B.: Potentiels endocellulaires de fibres musculaires intrafusales provoqués par la stimulation d'axones fusimoteurs statiques et dynamiques chez le chat. J. Physiol. (Paris) **63**, 115 (1971).

Bessou, P., Pagès, B.: Intracellular potentials from intrafusal muscle fibres evoked by stimulation of static and dynamic fusimotor axons in the cat. J. Physiol. (Lond.) **227**, 709–727 (1972).

Bessou, P., Pagès, B.: Tableau général de distribution des axones fusimoteurs aux fibres musculaires intrafusales. C. R. Acad. Sci. (Paris) (in press).

Bianconi, R., Meulen, J.P., van der: The responses to vibration of the end organs of mammalian muscle spindles. J. Neurophysiol. **26**, 177–190 (1963).

Boes, A., Haase, J., Muhlberg, B., Ropte, H., Sontag, K.H.: Dynamische und statische Aktivität prätibialer Katzen-Muskelspindeln und ihre fusimotorischen α-Reflexe vor und nach intercolliculärer Dezerebrierung. Pflügers Arch. **302**, 193–205 (1968).

Boyd, I.A.: The structure and innervation of the nuclear bag muscle fibre system and the nuclear chain muscle fibre system in mammalian muscle spindles. Phil. Trans. B **245**, 81–136 (1962).

Boyd, I.A.: The behaviour of isolated mammalian muscle spindles with intact innervation. J. Physiol. (Lond.) **186**, 109–110 (1966).

Boyd, I.A.: The muscle spindle (film), distributed by J. Wiley & Sons, Chichester, England (1970).

Boyd, I.A.: The mammalian muscle spindle — an advanced study (film). J. Physiol. (Lond.) **214**, 1 (1971).

Boyd, I.A., Gladden, M.H., McWilliam, P.N., Ward, J.: Static and dynamic fusimotor action in isolated cat muscle spindles with intact nerve and blood supply. J. Physiol. (Lond.) **230**, 29 (1973).

Boyd, I.A., Ward, J.: The response of isolated cat muscle spindles to passive stretch. J. Physiol. (Lond.) **200**, 104–105 (1969).

Brown, M.C., Butler, R.G.: Depletion of intrafusal muscle fibre glycogen by stimulation of fusimotor fibres. J. Physiol. (Lond.) **229**, 25–26 (1973).

Brown, M.C., Crowe, A., Matthews, P.B.C.: Observations on the fusimotor fibres of the tibialis posterior muscle of the cat. J. Physiol. (Lond.) **177**, 140–159 (1965).

Brown, M.C., Engberg, I., Matthews, P.B.C.: Fusimotor stimulation and the dynamic sensitivity of the secondary ending of the muscle spindle. J. Physiol. (Lond.) **189**, 545–550 (1967).

Brown, M.C., Engberg, I., Matthews, P.B.C.: The relative sensitivity to vibration of muscle receptors of the cat. J. Physiol. (Lond.) **192**, 773–800 (1967).

Brown, M.C., Goodwin, G.M., Matthews, P.B.C.: After-effects of fusimotor stimulation on the response of muscle spindle primary afferent endings. J. Physiol. (Lond.) **205**, 677–694 (1969).

Brown, M.C., Laurence, D.G., Matthews, P.B.C.: Static fusimotor fibres and the position sensitivity of muscle spindle receptors. Brain Res. **14**, 143–184 (1969).

Cliff, G.S., Ridge, R.M.A.P.: Innervation of extrafusal and intrafusal fibres in snake muscle. J. Physiol. (Lond.) (in press).

Cooper, S.: The responses of the primary and secondary endings of muscle spindles with intact motor innervation during applied stretch. J. exp. Physiol. **46**, 389–398 (1961).

Crowe, A., Matthews, P.B.C.: The effects of stimulation of static and dynamic fusimotor fibres on the response to stretching of the primary endings of muscle spindles. J. Physiol. (Lond.) **174**, 109–131 (1964a).

Crowe, A., Matthews, P.B.C.: Further studies of static and dynamic fusimotor fibres. J. Physiol. (Lond.) **174**, 132–151 (1964b).

Diete-Spiff, K.: Tension development by isolated muscle spindles of the cat. J. Physiol. (Lond.) **193**, 31–43 (1967a).

Diete-Spiff, K.: Time course of mammalian intrafusal muscle contraction as revealed by cinephotography of isolated lumbrical muscle spindles of the dog. Arch. ital. Biol. **104**, 387–405 (1967b).

Doward, P.K.: Response characteristics of muscle afferents in the domestic duck. J. Physiol. (Lond.) **211**, 1–17 (1970).

Douglas, W.W., Ritchie, J.M.: A technique for recording functional activity in specific groups of medullated and non-medullated fibres in whole nerve trunks. J. Physiol. (Lond.) **138**, 19–30 (1957).

Eccles, J.C., Sherrington, C.S.: Numbers and contration-values of individual motor units examined in some muscles of the limb. Proc. roy. Soc. B. **106**, 326–357 (1930).

Eldred, E., Granit, R., Merton, P.A.: Supraspinal control of the muscle spindle and its significance. J. Physiol. (Lond.) **122**, 498–523 (1953).

Eldred, E., Schnitzlein, H.N., Buchwald, J.: Response of muscle spindles to stimulation of the sympathetic trunk. Exp. Neurol. **2**, 13–25 (1960).

Ellaway, P., Emonet-Dénand, F., Jami, L., Joffroy, M.: Proportion des fibres fusimotrices statiques et dynamiques dans les muscles peroneus longus et flexor hallucis longus du chat. C. R. Acad. Sci. (Paris) **274**, 3591–3600 (1972).

Ellaway, P., Emonet-Dénand, F., Joffroy, M., Laporte, Y.: Lack of exclusively fusimotor α-axons in flexor and extensor leg muscles of the cat. J. Neurophysiol. **35**, 149–153 (1972).

Emonet-Dénand, F., Houk, J.: Some effects of polarizing current on discharges from muscle spindle receptors. Amer. J. Physiol. **216**, 404–406 (1969).

Emonet-Dénand, F., Jami, L., Laporte, Y.: Mise en évidence d'axones squeletto-fusimoteurs dans le muscle flexor hallucis longus chez le chat. C. R. Acad. Sci. (Paris) (in press).

Emonet-Dénand, F., Jankowska, E., Laporte, Y.: Mise en évidence de fibres squeletto-fusimotrices (fibres β) dans les muscles lombricaux du lapin. J. Physiol. (Paris) **61**, Suppl. 2, 285 (1969).

Emonet-Dénand, F., Jankowska, E., Laporte, Y.: Skeleto-fusimotor fibres in the rabbit. J. Physiol. (Lond.) **210**, 669–680 (1970).

Emonet-Dénand, F., Joffroy, M., Laporte, Y.: Fibres fusimotrices dont l'action sur la sensibilité phasique des terminaisons primaires dépend de leur fréquence de stimulation. C. R. Acad. Sci. (Paris) **275**, 89–92 (1972).

Emonet-Dénand, F., Laporte, Y., Pagès, B.: Mise en évidence de fibres fusimotrices statiques chez le lapin. C. R. Acad. Sci. (Paris) **259**, 2690–2693 (1964).

Eyzaguirre, C.: Functional organization of neuromuscular spindle in toad. J. Neurophysiol. **20**, 523–542 (1957).

Eyzaguirre, C.: Modulation of sensory discharges by efferent spindle excitation. J. Neurophysiol. **21**, 465–480 (1958).

Eyzaguirre, C.: The electrical activity of mammalian intrafusal fibres. J. Physiol. (Lond.) **150**, 169–185 (1960).

Eyzaguirre, C., Vial, J.D.: Electrical activity of the intrafusal muscle fibres. Nature (Lond.) **178**, 317–318 (1956).

Fukami, Y.: Tonic and phasic muscle spindles in the snake. J. Neurophysiol. **33**, 28–35 (1970).

Fukami, Y.: Electrical and mechanical factors in the adaptation of reinnervated muscle spindles in the snake. In: Research in Muscle Development and the Muscle Spindle. Excerpta Medica Int. Congr. Series **240**, 379–399 (1972).

Fukami, Y.: Accomodation in afferent nerve terminals of snake muscle spindle. J. Neurophysiol. **33**, 475–489 (1970).

Fukami, Y., Hunt, C.C.: Structure of snake muscle spindles. J. Neurophysiol. **33**, 9–27 (1970).

Fukami, Y., Ridge, R.M.A.P.: The effect of chronic denervation on spindle discharge patterns and morphology in isolated costocutaneous muscles of garter snakes. J. comp. Neurol. **143**, 137 (1971).

Granit, R., Henatsch, H.D.: Gamma control of dynamic properties of muscle spindles. J. Neurophysiol. **19**, 356–366 (1956).

Granit, R., Pompeiano, O., Waltman, B.: The early discharge of mammalian muscle spindles at onset of contraction. J. Physiol. (Lond.) **147**, 399–418 (1959).

Haase, J., Meuser, P., Tan, U.: Die Konvergenz fusimotorischer α-Impulse auf de-efferentierte Flexorspindeln der Katze. Pflügers Arch. ges. Physiol. **289**, 50–58 (1966).

Haase, J., Schlegel, H.J.: Einige funktionelle Merkmale von alpha-innervierten Extensor- und Flexorspindeln der Katze. Pflügers Arch. ges. Physiol. **287**, 163–175 (1966).

Harvey, R., Matthews, P.B.C.: Some effects of stimulation of the muscle nerve on afferent endings of muscle spindles, and the classification of their responses into types A1 and A2. J. Physiol. (Lond.) **156**, 470–497 (1961).

HARVEY, R., MATTHEWS, P.B.C.: The response of de-efferented muscle spindle endings in the cat's soleus to slow extension of the muscle. J. Physiol. (Lond.) **157**, 370–392 (1961).

HENATSCH, H.D., SCHÄFER, S.S.: Fusimotor sensor and fusimotor trigger functions: a re-interpretation of the dual control of mammalian muscle spindles. Brain Res. **6**, 385–387 (1967).

HOUK, J.: A viscoelastic interaction which produces one component of adaptation in responses of Golgi tendon organs. J. Neurophysiol. **30**, 1482–1493 (1967).

HOUK, J., HENNEMAN, E.: Responses of Golgi tendon organs to active contractions of the soleus muscle of the cat. J. Neurophysiol. **30**, 466–481 (1967).

HUBER, G.C., DE WITT, L.M.A.: A contribution on the nerve terminations in neurotendinous end organs. J. comp. Neurol. **10**, 159–208 (1900).

HUNT, C.C.: Relation of function to diameter in afferent fibres of muscle nerves. J. gen. Physiol. **38**, 111–131 (1954).

HUNT, C.C.: The effect of sympathetic stimulation on mammalian muscle spindles. J. Physiol. (Lond.) **151**, 332–341 (1960).

HUNT, C.C., KUFFLER, S.W.: Further study of efferent small-nerve fibres to mammalian muscle spindles; multiple spindle innervation and activity during contraction. J. Physiol. (Lond.) **113**, 283–297 (1951a).

HUNT, C.C., KUFFLER, S.W.: Stretch receptor discharge during muscular contraction. J. Physiol. (Lond.) **113**, 298–315 (1951b).

HUNT, C.C., OTTOSON, D.: Receptor potentials and impulse activity in isolated mammalian spindles. J. Physiol. (Lond.) **230**, 49 (1973).

HUNT, C.C., WYLIE, R.M.: Responses of snake muscle spindles to stretch and intrafusal muscle fibre contraction. J. Neurophysiol. **33**, 1–8 (1970).

HUSMARK, I., OTTOSON, D.: The contribution of mechanical factors to the early adaptation of the spindle response. J. Physiol. (Lond.) **212**, 577–592 (1971).

HUSMARK, I., OTTOSON, D.: Is the adaptation of the muscle spindle of ionic origin? Acta physiol. scand. **81**, 138–140 (1971).

ITO, F.: Differentiation between spindle, leaf-like and tendon receptors in the frog by their recovery curves of excitability after antidromic discharge. J. Physiol. Soc. Jap. **29**, 598–599 (1967).

ITO, F.: Functional properties of tendon receptors in the frog. Jap. J. Physiol. **18**, 576–589 (1968).

ITO, F., TOYAMA, K., ITO, R.: A comparative study on structure and function between the extrafusal receptor and the spindle receptor in the frog. Jap. J. Physiol. **14**, 12–33 (1964).

JANSEN, J.K., RUDJORD, T.: On the silent period and Golgi tendon organs of the soleus muscle of the cat. Acta physiol. scand. **62**, 364–379 (1964).

KATZ, B.: The efferent regulation of the muscle spindle in the frog. J. exp. Biol. **26**, 201–217 (1949).

KIDD, G.L.: Excitation of primary muscle spindle endings by beta-axon stimulation. Nature (Lond.) **203**, 1248–1251 (1964).

KOKETSU, K., NISHI, S.: Action potentials of single intrafusal muscle fibres of frogs. J. Physiol. (Lond.) **137**, 193–209 (1951a).

KOKETSU, K., NISHI, S.: An analysis of junctional potentials of intrafusal muscle fibres in frogs. J. Physiol. (Lond.). **139**, 15–26 (1951b).

KUFFLER, S.W., HUNT, C.C., QUILLIAM, J.P.: Function of medullated small-nerve fibres in mammalian ventral roots: Efferent muscle spindle innervation. J. Neurophysiol. **14**, 29–54 (1951).

LEKSELL, L.: The action potential and excitatory effects of the small ventral root fibres to skeletal muscle. Acta physiol. scand. **10**, Suppl. **21**, 1–84 (1945).

LENNERSTRAND, G.: Position and velocity sensitivity of muscle spindles in the cat. I. Primary and secondary endings deprived of fusimotor activation. Acta physiol. scand. **73**, 281–299 (1968).

LENNERSTRAND, G., THODEN, U.: Fusimotor effects on position and velocity sensitivity of muscle spindles. Experientia (Basel) **23**, 205–206 (1967a).

LENNERSTRAND, G., THODEN, U.: Responses of primary muscle spindle endings at constant or changing muscle length to variations in fusimotor activation. Experientia (Basel) **23**, 396–398 (1967b)

Lennerstrand, G., Thoden, U.: Dynamic analysis of muscle spindle endings in the cat using length changes of different length-time relations. Acta physiol. scand. **73**, 234–250 (1968).

Lippold, C.C.J., Nicholls, J.W.T., Redfearn, I.W.T.: Electrical and mechanical factors in the adaptation of a mammalian muscle spindle. J. Physiol. (Lond.) **153**, 209–217 (1960).

Lloyd, D.P.C., Chang, H.T.: Afferent fibres in muscle nerves. J. Neurophysiol. **2**, 199–208 (1948).

Matthews, B.H.C.: The response of a single end-organ. J. Physiol. (Lond.) **71**, 64–110 (1931).

Matthews, B.H.C.: The response of a muscle spindle during active contraction of a muscle. J. Physiol. (Lond.) **72**, 153–174 (1931).

Matthews, B.H.C.: Nerve endings in mammalian muscles. J. Physiol. (Lond.) **18**, 1–53 (1933).

Matthews, P.B.C.: The differentiation of two types of fusimotor fibres by their effects on the dynamic response of muscle spindle primary endings. Quart. J. exp. Physiol. **47**, 324–333 (1962).

Matthews, P.B.C.: The response of de-efferented muscle spindle receptors to stretching at different velocities. J. Physiol. (Lond.) **168**, 660–678 (1963).

Matthews, P.B.C.: Mammalian muscle receptors and their central actions. Baltimore: Williams & Wilkins 1972.

Matthews, P.B.C., Stein, R.B.: The sensitivity of muscle spindle afferents to small sinusoidal changes of length. J. Physiol. (Lond.) **200**, 723–743 (1969).

Merton, P.A.: Slowly conducting muscle spindle afferents. Acta physiol. scand. **29**, 87–88 (1953).

O'Leary, J., Heinbecker, P., Bishop, G.H.: Analysis of function of a nerve to muscle. Amer. J. Physiol. **110**, 636–658 (1934).

Ottoson, D., Shepherd, G.M.: Length changes within isolated frog muscle spindle during and after stretching. J. Physiol. (Lond.) **207**, 747–759 (1970).

Paintal, A.S.: Functional analysis of group III afferent fibres of mammalian muscles. J. Physiol. (Lond.) **152**, 250–270 (1960).

Proske, U.: An electrophysiological analysis of responses from lizard muscle spindles. J. Physiol. (Lond.) **205**, 289–307 (1969).

Rutledge, L.T., Haase, J.: Flexor muscle spindles and reflex firing of early discharging units. J. Neurophysiol. **24**, 182–192 (1961).

Schäfer, S.S.: The acceleration response of a primary muscle spindle ending to ramp stretch of the extrafusal muscle. Experientia (Basel) **23**, 1026–1027 (1967).

Smith, R.S.: Activity of intrafusal muscle fibres in muscle spindles of Xenopus laevis. Acta physiol. scand. **60**, 223–239 (1964a).

Smith, R.S.: Contraction in intrafusal muscle fibres of Xenopus laevis following stimulation of their motor nerves. Acta physiol. scand. **62**, 195–208 (1964b).

Smith, R.S.: Properties of intrafusal muscle fibres. In: Muscular Afferents and Motor Control. Granit, R. Ed., pp. 69–80. Stockholm: Almqvist & Wiksell 1966.

Chapter 3

Central Actions of Impulses in Muscle Afferent Fibres

By

A. K. McIntyre, Monash University (Australia)

With 8 Figures

Contents

I. Introduction

Skeletal muscle and its tendinous attachments constitute a major fraction — probably approaching half — of the total body mass. The rich inflow of afferent signals from the extensive array of receptors within and in relation to these tissues must be expected to have widespread and far-reaching effects at many levels of central nervous activity, of particular relevance to the control and guidance of the muscles themselves in their maintenance of posture and execution of movement. Indeed, long before the incontrovertible demonstration of the existence of this sensory equipment (Sherrington, 1894), the general notion of muscle as a source of afferent signals had been entertained. According to Liddell (1960), Charles Bell first formulated the idea of such sensory feedback, writing in 1826: "Between the brain and the muscle there is a circle of nerves; one nerve conveys the influence of the brain to the muscle, another gives the sense of the condition of the muscle to the brain". Several generations later this notion was rediscovered and made respectable by mathematics and the coining of a new word, cybernetics.

There is now a bewildering mass of data about the central actions of afferent input from muscle and tendon, a large part of which relates to spinal and hind-brain mechanisms, presumably involved in unconscious, moment-to-moment reflex adjustments of motoneurone output. However, it is interesting to recall that, despite Sherrington's own early recognition of the reflex significance of muscle and tendon receptors, for many years attention was largely directed towards the presumed proprioceptive role of this innervation in the conscious awareness of position and movement, to which the widespread use of the term "muscular sense" bears witness (Sherrington, in Schäfer, 1900). After the classical investigations of Liddell and Sherrington (1924, 1925) on the stretch reflex in decerebrate preparations, and the subsequent development of electro-physiological techniques, the major thrust of experimental effort was for many years largely concerned with the detailed analysis of central reflex mechanisms set in train by muscle afferent impulses, especially those in the spinal cord, with relatively little attention being paid to their possible role in conscious proprio-ception. Indeed, until fairly recently, available evidence indicated that impulses from spindle primary or tendon organ receptors were unlikely to have immediate access by simple pathways to the cerebral cortex, and several other lines of investigation suggested that sensory feedback contributing to such awareness of position and movement arises in tissues other than muscle and tendon, notably in those of joints (Mountcastle et al., 1952; McIntyre, 1962), and that central monitoring of motor effort (the "felt will" of Helmholtz) is also important (McIntyre, 1963; Merton, 1964). However, in the light of further studies, including some still in progress, a reassessment has become necessary of the possible contribution of muscle and tendon afferent input to the higher level cortical systems controlling muscles. Thus current opinion is swinging back towards the view that information from muscle receptors may play an important role at the cerebral cortical level in addition to its undoubted participation in spinal and cerebellar reflex machinery.

In this chapter, no attempt can be made to deal comprehensively with all aspects of the central actions of muscle afferent impulses; for example, the recent rapid accumulation of new experimental data on monosynaptic excitatory mechanisms and the associated development of sophisticated models to explain such action on motoneurones would alone justify a major review. Much the same could be said about other particular aspects of muscle afferent central actions, such as the ascending pathways and synaptic mechanisms through which information from muscles may reach the cerebellum. The aim, therefore, is to avoid detailed discussion of synaptic machinery and to present an overall view of the central projections and actions of impulses in the various categories of afferent fibres present in muscle nerves insofar as present understanding permits. In the interests of perspective and clarification of current terminology, brief mention will be included of some of the stages through which this understanding has progressed. Because of the sparseness of information about central actions of muscle afferent impulses in lower vertebrates, experimental findings in mammals, especially the cat, provide the bulk of the material discussed in the pages that follow. It has not been considered appropriate to discuss analogous invertebrate mechanisms in this context, despite the similarities presented by peripheral receptor mechanisms in some invertebrate species.

II. Muscle Nerve Afferent Input to the Central Nervous System

The task of discussing central actions of muscle afferent impulses would be greatly simplified if the different classes of receptors were served by separate and distinguishable groups of afferent axons, and if there were completely reliable means of selectively stimulating each receptor-type or its own particular group of afferent fibres. However, this unfortunately is not the case, except to a very limited extent, there being much overlap in diameter, and consequently in electrical threshold and conduction velocity, of afferent fibres from different receptors. Furthermore, only limited selectivity can be achieved by natural stimulation. It is therefore necessary to preface any discussion of the central actions of signals from the different classes of receptors described by Barker and by Hunt (this Vol.) by discussion of what is known about the afferent fibre-content of nerves supplying muscles, and of the various methods which aim to achieve a pure or relatively pure input of impulses from a selected receptor-type.

A. Afferent Fibre Groups and Function

It has been known for a long time that from one third to one half of the myelinated fibres in mammalian muscle nerves are afferent (Sherrington, 1894). In this and subsequent work, the wide size-range of these afferent fibres (1.5–20 μm or more) was disclosed, and it was noted that some exceeded in diameter the largest of the motor fibres present in a corresponding deafferented nerve. It was also observed that the largest muscle afferent fibres substantially exceeded in diameter that of the largest cutaneous fibres. Histograms of their size-distribution showed two prominent peaks, one between 3 and 5 μm and another at about 14–15 μm; a smaller and less well defined intermediate group of afferent fibres was also apparent between 6 and 12 μm (Eccles and Sherrington, 1930). Ample confirmation of the presence of a group of uniquely large afferent fibres in muscle nerves as compared with cutaneous nerves has since been obtained by many other workers (e.g., O'Leary et al., 1935; Lloyd and Chang, 1948; Rexed and Therman, 1948; Hagbarth and Wohlfart, 1952; Barker et al., 1962; Boyd and Davey, 1968). All the different fibres referred to above are myelinated and therefore belong to the "A" category of Erlanger and Gasser (1937). However, nerves supplying muscle also contain, as do other somatic nerves, an impressive complement of unmyelinated afferent C fibres of dorsal root origin, so that the total afferent fibre content of muscle nerves far exceeds the motor fibre component (see Barker, pp. 2, 168, this Vol.).

As a result of difficulties and ambiguities in use of the early, empirical classification-scheme for nerve fibres, which did not distinguish between those of motor and sensory function, a new scheme of nomenclature for afferent fibres was proposed in 1943 with the aim of relating afferent fibre-groups of different size-range more closely to the central actions of their impulses. In this scheme, as initially proposed, the group of largest myelinated afferent fibres of diameters ranging from 12–20 or more μm were designated as Group I, those between 6–12 μm as Group II, and those less than 6 μm comprised Group III (Lloyd, 1943b). Lloyd made the point that the Group I fibres, with a distribution peak around 15–16 μm, are characteristic of muscle nerves. He also stated that Group II fibres, with a mode at 8–9 μm, occur in cutaneous as well as muscle nerves, but are much more prominent in the former; and that the Group III fibres, peaking at about 3–4 μm, are plentiful in both skin and muscle nerves, and correspond to the A delta pile in Gasser's terminology. The further designation of Group IV for the unmyelinated C fibres was also added to this nomenclature scheme.

Lloyd's nomenclature has since been widely used for muscle afferent fibres, though less consistently for those of cutaneous nerves; according to Boyd and Davey (1968), it was recommended for general adoption in 1965 during the First Nobel Symposium on "muscle afferents and motor control". Difficulty has arisen, however, because the original definition appeared to be based strictly on fibre-size. Thus literal application of Lloyd's classification to nerves other than those of muscle — for example, cutaneous nerves, in which occur many fibres with diameters extending well into the Group I size-range (e.g., Gasser and Grundfest, 1939; Hunt and McIntyre, 1960b), has led some authors to refer to these fibres as belonging to Group I. However, it is the author's view that Lloyd's nomenclature implies some categorization related to functional differences between fibre-groups, and is not based solely on differences in fibre-diameter. Indeed, the observation of a unique central action of the largest muscle afferent fibres, namely, monosynaptic excitation of homonymous and synergic motoneurones (Lloyd, 1943a, b), was an important reason for proposing the new classification. Thus the label "Group I" should be attached only to the uniquely large fibres serving stretch receptors of muscle and tendon. Such fibres may sometimes be found in other than strictly muscle nerves, for example, in the nerve to the interosseous membrane or in articular nerves, but these can properly be called Group I fibres because of their origin in typical receptors of muscle or tendon (Hunt and McIntyre, 1960a; Barker, 1962; Boyd and Davey, 1968).

So far, this discussion of afferent fibres has been largely restricted to direct histological measurements and the diameter-distribution of fibre-populations. It is clear even from these studies (e.g., LLOYD and CHANG, 1948) that the boundaries between the three groups of myelinated fibres are to a considerable extent arbitrary; for example, Lloyd and Chang's histograms show the Group III fibres peaking at 2–3 μm, with a range of 1–4 μm, and not as originally defined. Furthermore, there is considerable variation in the maximum diameters of Group I fibres and in the proportions of fibres in the three groups, in different muscle-nerves, and from animal to animal. Despite the problems raised by the arbitrary nature of the sub-divisions between groups and the measure of ambiguity arising from their partly functional, partly structural basis, Lloyd's scheme has proved to be a useful verbal device when used with awareness of its limitations, one of which is that it is derived from measurements in the cat. Interpreted broadly, however, by taking into account the peaks of fibre-size distribution as well as the span of diameters in each pile, and where possible relating fibres to receptors of origin, the system can be applied usefully to different muscle nerves of the cat, and even to those of other mammals, despite substantial differences in the range of afferent fibre diameters (BOYD and DAVEY, 1968).

As already described in Hunt's contribution (p. 193) Matthews' classical work disclosing the functional characteristics of single mammalian muscle receptors, and his classification of the majority into "A" (in parallel) and "B" (in series) types interpreted as responses of muscle spindles and tendon organs respectively (MATTHEWS, 1933), paved the way for more detailed studies in which functionally single afferent fibres were isolated, their receptor of origin determined and the diameter of each fibre calculated from its measured conduction-velocity (HURSH, 1939). In this way, the functional inhomogeneity of the Group I band demonstrated by its duality of reflex action (LAPORTE and LLOYD, 1952) could be related to specific receptors, namely spindle primary (annulospiral) endings and tendon organs, both with afferent fibres in the Group I range (HUNT and KUFFLER, 1951; HUNT, 1954; LAPORTE and BESSOU, 1957; HUNT and MCINTYRE, 1960a; COPPIN et al., 1969; MACLENNAN, 1971, 1972).

Amongst these studies the careful analysis by HUNT (1954) of afferent fibres in the nerves to *soleus* and *gastrocnemius medialis* is especially important. In this work, particular effort was devoted to the representative sampling of fibres throughout as wide a range of diameters as possible and the additional criterion of response to fusimotor stimulation was included in the identification of spindle afferents. The synthetic fibre populations so derived for these two muscle nerves agreed closely with actual histological afferent spectra, the Group I band being fully accounted for by fibres from spindles (Ia fibres) and from tendon organs (Ib fibres)[1]. Similarly the Group II band (4–12 μm) appeared to consist almost wholly of spindle afferents presumably from secondary endings; some spindle fibres also belonged to Group III (1–4 μm).

The extent to which fibres from spindle secondary endings account for the population of Group II fibres varies in different muscle nerves, but so far only a

1 It is the author's view that the sub-designations "Ia" and "Ib" should be used strictly to define Group I fibres serving spindle or tendon organ receptors, as in HUNT (1954), and not as labels for deflexions in a recorded afferent volley.

limited number have been studied adequately. In some muscle nerves the afferent fibre composition seems to be more complex than in the nerves of *triceps surae*. For example, although the nerves to *flexor digitorum longus* and *flexor hallucis longus* show distributions of spindle and tendon organ fibres similar to those of *soleus* and *gastrocnemius*, there are in addition other fibres in both the Group I and the Group II range, some of which appear to come from outside the muscle (Hunt and McIntyre, 1960a). They include fibres from Pacinian corpuscles and two other types of receptor ("tap" and "tension") of which the latter are probably unusually sited tendon organs (Barker, 1962). Furthermore, a purely afferent branch of the *flexor digitorum longus* nerve contains many fibres of Group I and Group II calibre from the interosseous cluster of Pacinian corpuscles and from tap and tension receptors (Hunt, 1961), and this branch is likely to be included in anatomical or physiological studies of this nerve.

Even with the seemingly simpler situation in *triceps surae*, it is difficult to exclude the presence of some fibres in the Group II range derived from receptors other than muscle spindles, such as paciniform lamellated receptors or free endings; for example, in de-efferented *soleus* preparations, comparison of total counts of histologically identified receptors in the muscle with the afferent fibres in its nerve trunk suggests the presence of an appreciable number of Group II fibres not attributable to spindles (Barker et al., 1962; but see Hunt, 1954). Occasional fibres from Pacinian corpuscles, of conduction velocity usually in the Group II range, may be encountered in many muscle nerves, and in some, afferents from joints may occur (Skoglund, 1956).

Because a precise knowledge of the afferent fibre composition of a wide range of muscle nerves is essential for the satisfactory interpretation of central actions of afferent volleys therein (Hunt and McIntyre, 1960a; McIntyre, 1965), the method of systematic analysis pioneered by Hunt (1954) has been applied to other muscle nerves in recent years, and extended in some cases to include fibre-threshold as well as conduction velocity measurements. Thus detailed information is now available, at least for afferent fibres in the Group I and Group II range, for *semitendinosus*, *peroneus longus* and *tibialis anterior*, as well as for *triceps surae*, *flexor digitorum longus* and *flexor hallucis longus* (Hunt and McIntyre, 1960a; Sumner, 1961; Coppin et al., 1969; MacLennan, 1971, 1972; see also Matthews, 1972). In general, the findings for *triceps surae* afferent fibres are applicable in that spindle (A type) afferent fibres occur throughout Groups I and II, and those from tendon organs (B type) are largely restricted to Group I. However, substantial differences are apparent in the relative positions of the Ia and Ib fibres in the afferent spectra of these muscle nerves. Furthermore, variable numbers of fibres from receptors other than spindles or tendon organs contribute to Group II and to some extent the Group I band. This is especially marked in the case of *tibialis anterior*, in which many such fibres of Group I and Group II velocity are present, some from unidentified endings in the muscle, others from extramuscular receptors, including Pacinian corpuscles, possibly in the region of the interosseous membrane, in periosteum of the leg bones or in the ankle joint (MacLennan, 1971, 1972). The implications of these findings for the study of central actions will be discussed subsequently in the context of electrical stimulation.

Receptors served by the smallest (Group III) myelinated fibres in muscle nerves are less well documented, but their responses have been studied by BESSOU and LAPORTE (1960, 1961) and by PAINTAL (1960, 1962); HUNT (1954) and HUNT and MCINTYRE (1960a) have also observed responses of a small sample of muscle receptors with Group III afferent fibres. Histological investigations (e.g., BARKER, 1962, 1967; BARKER et al., 1962; STACEY, 1969) indicate that Group III fibres supply a variable but usually small number of paciniform lamellated end organs, and an extensive array of freely branching, unencapsulated endings, found throughout the muscle but especially in regions of tendons, in aponeuroses and fascial sheaths, and in the adventitia of blood vessels; a full account is given in the section by BARKER (Chapter 1). It should be noted that some paciniform lamellated organs and some free endings may have larger afferent fibres of Group II calibre (BARKER et al., 1962); also, as already mentioned, some of the smallest fibres from spindle secondary endings may be found in the Group III range.

Physiologically, as described in Hunts' contribution (p. 221), most of the receptors with Group III afferent fibres appear to be nociceptive in function, or at any rate of high mechanical threshold, responding to strong pressure or squeezing or to other unusual stimulation such as the injection of hypertonic solutions. Exceptions are the few paciniform corpuscles, which are highly phasic and respond to weak mechanical stimulation, and the small number of spindle secondary endings with Group III fibres. The extensive unmyelinated Group IV component of the afferent fibres from muscles appears to end as freely branching endings, and are probably also nociceptive in function (BESSOU and LAPORTE, 1958; IGGO, 1961). Their endings appear to be slowly adapting, are fired by local application of fairly strong mechanical pressure or of hypertonic solutions, and by local warming or cooling; they also fire impulses if the muscle is made to work under ischaemic conditions (BESSOU and LAPORTE, 1958).

B. Approaches to Selective Activation of Functionally Different Muscle Afferents

The above outline of the afferent fibre population of muscle nerves and their receptive endings, though to some extent repetitive of material presented more fully in other sections, is a necessary background to discussion of experimental attempts to achieve selective activation of functionally distinct groups for study of their central actions, especially by the most widely used approach, namely electrical stimulation of the nerve trunks. Apart from its simplicity, ease of control and convenience, electrical stimulation has the major advantage of readily eliciting simultaneous action in many afferent fibres, so that relatively massive volleys, with the impulses at least initially in synchrony, can be despatched into the central nervous system. Despite the abnormal nature of such volleys, their central actions are powerful and less likely to escape observation than those of subtler and more physiological inputs; furthermore, their synchronous nature permits precise timing of events such as the volley's arrival and spread within the central nervous system and the onset of its post-synaptic actions. By far the greater part of our basic knowledge of the central actions of muscle afferent input is derived from experiments employing electrical stimulation of muscle nerves.

However, such stimulation even when carefully controlled and graded can only achieve discrimination by way of the far from clear-cut differences in threshold between the different functional groups of afferent fibres. Furthermore, because of the well known relationship between electrical threshold and size of fibre (Erlanger and Gasser, 1937), recruitment of fibres with increasing stimulus-strength can only proceed in one direction, from those of largest to those of smallest diameter, unless some additional procedure capable of selectively blocking the larger fibres can be adopted. With the usual electrical stimulation procedure, therefore, the central actions set up by the smaller fibres can only be seen as different or additional effects superimposed on those already elicited by the larger fibres inevitably also engaged by the stimulus, and leading the volley by virtue of their more rapid conduction. This would be so even if the afferent fibre constituents of different muscle nerves were identical and relatively simple as in those of the *triceps surae* muscle group. However, the considerable individual differences in the various afferent fibre groups which have already been briefly discussed, and the presence in some of afferent fibres of extramuscular origin travelling as passengers in the nerve, greatly complicates the situation.

These considerations apply especially to the larger fibres of lower threshold in muscle nerves. Even though there is no clear-cut difference in thresholds of the smaller Group II and larger Group III fibres, it is not difficult to elicit a muscle afferent volley engaging a substantial number of the larger Group II fibres but none of those in Group III (Brock et al., 1951; Eccles and Lundberg, 1959a). Similarly, volleys can readily be set up which engage most of the myelinated fibres, but exclude those of Group IV. Stimuli of strength adequate to recruit all Group II fibres will inevitably excite some of those in Group III, but because histological studies indicate that many fibres from both groups take origin from the same kind of terminal, this may not be a serious barrier to the analysis of central actions, although the possibility of functional differences amongst these endings must be borne in mind. The real difficulties lie in the extent to which electrical stimulation can discriminate between Group I fibres of spindle (Ia) and those of tendon organ (Ib) origin, and between the smaller Group I and larger Group II fibres.

1. Controlled Electrical Stimulation

Laporte and Lloyd (1952) were the first to demonstrate duality of central reflex response to Group I volleys, through the use of carefully graded electrical stimulation not much above threshold. They observed that thresholds of the two functional types of Group I fibres showed little separation, as would be expected from the results of Hunt and Kuffler (1951), Hunt (1954) and Hunt and McIntyre (1960a), at least for certain muscle nerves of the hind limb. However, a much larger threshold separation between Group I fibres of spindle and tendon organ origin was proposed by Bradley and Eccles (1953), and by Eccles et al. (1957a). Bradley and Eccles observed that the low threshold afferent volley from muscle nerves of the thigh, especially of *quadriceps* and flexor hamstrings, is commonly split into fast and slow components as recorded at the dorsal root entry zone. By ingenious use of double volleys and refractoriness, they were able to define stimulus intensities which only recruited fibres of the fast component, and others which engaged varying numbers of the slow wave fibres. The central actions of such graded volleys led them to conclude that the fast wave comprised fibres from spindle primary endings, while impulses in fibres of tendon organ origin made up the slow component. Perhaps unfortunately, the fast and slow waves were labelled "Ia" and "Ib" respectively. Certainly this terminology has been widely interpreted as literally meaning input volleys from spindle primary and tendon organ endings respectively. LAPORTE and BESSOU (1957) studied a population of Group I muscle afferent units from thigh muscles in cats selected to show Group I volleys of clearly double contour, and their results partly confirmed Bradley and Eccles' interpretation. In their population, the fast wave was largely composed of spindle impulses with a small number of tendon organ responses, whereas the slow component included many spindle impulses as well as the bulk of those from tendon organs. Eccles and his colleagues have also described splitting of Group I volleys from muscle nerves of the leg such as those of *triceps surae*, at least in some preparations (ECCLES et al., 1957a), and the interpretation of a number of investigations involving stimulation of these nerves has depended on the reliability of such double contoured afferent volleys as indicators of spindle primary and tendon organ input.

However, a number of studies in which monosynaptic reflex output was related to size of graded Group I input volley (e.g., HUNT, 1955) have shown that reflexly effective input fibres are represented throughout the Group I range, so that, despite some contrary reports (RALL, 1955; ECCLES et al., 1957a), the degree of discrimination suggested by Eccles and his colleagues seemed unlikely, especially in the more distal muscle nerves. Furthermore, it had been reported by BROCK et al. (1951) that Group II fibres begin to be recruited with increasing stimulus strength at intensities as low as 1.2–1.3 times threshold and yielding Group I responses not more than about half maximal.

These uncertainties and apparent contradictions emphasize the need for as precise a knowledge as possible of the kinds of afferent fibres and their electrical thresholds in the whole range of muscle nerves commonly used as input-channels for reflex and other neurophysiological studies. As already noted, detailed information has recently become available for a number of muscle nerves in the cat's hind limb. SUMNER (1961) measured the threshold and conduction velocity of 216 Group I fibres from *gastrocnemius medialis* nerve, 128 of the units being spindle receptors and the remainder tendon organs. His conduction velocity measurements closely matched those previously described by HUNT (1954) for this nerve. Since threshold of each unit was determined in relation to threshold and maximum stimulation-intensities for the whole nerve in each preparation, the pooled results could be plotted to show the proportion of spindle primary and tendon organ afferent fibres which would be engaged by each intensity of stimulus between the nerve threshold and Group I maximum. A similar analysis of 110 Group I fibres in the same nerve by Coppin (personal communication) has given closely similar results; together, the experiments show a systematic difference in the threshold of Ia and Ib fibres in the nerve to *gastrocnemius medialis*. This difference is small for the most excitable members of each group, and Group I volleys exceeding 8–10% of maximum would include impulses in some tendon organ fibres; however, with increase in stimulus strength, recruitment of Ia fibres keeps ahead of Ib engagement until the Group I volley nears its maximum. For example, a stimulus exciting half the Ia population will engage only about 10–20% of the Ib fibres in the nerve. Somewhat similar results have been obtained for Group I fibres in the nerves of *soleus* and *tibialis anterior* (MacLennan, 1971, 1972), although the threshold differentiation is slightly greater for *soleus*, slightly less for *tibialis anterior*.

Strikingly different findings have been reported for *semitendinosus* (Coppin et al., 1969; Coppin, personal communication), and for *peroneus longus* (Jack and MacLennan, 1971; MacLennan, 1971). In the distal nerve branch supplying *semitendinosus*, study of 154 Group I fibres has shown a much bigger threshold difference between fibres of spindle and of tendon organ origin. The results indicate that in this nerve, it may be possible to recruit more than 50% of the Ia fibre population before engaging any fibres of tendon organ origin; in practice, a Group I volley not exceeding 30% of maximum would in most preparations engage fibres solely from spindle primary endings (Coppin et al., 1969). Thus the conclusions of Bradley and Eccles (1953) have been confirmed in relation to the existence of a substantial threshold difference between Group I fibres serving spindles and tendon organs in the nerve to *semitendinosus*. However, the results of Laporte and Bessou (1957) have also been confirmed, in that fibres from primary spindle receptors in this muscle are distributed throughout the Group I range and therefore contribute substantially to the slower component of the Group I volley. Although corresponding detailed analyses including threshold measurements have not yet been carried out on other muscle nerves of the thigh (e.g., *quadriceps*, *biceps posterior*, and *semimembranosus*), it seems likely that similar considerations apply in view of the common occurrence of split Group I volleys in these nerves.

On the other hand, results for *peroneus longus* demonstrate very little difference in threshold between Group I fibres from spindles and from tendon organs (Jack and MacLennan, 1971; MacLennan, 1971). Fig. 1 presents a comparison between the relative thresholds and manner of recruitment with increasing stimulus-strength of Ia and Ib fibres in the nerves of *peroneus longus*, *gastrocnemius medialis* and *semitendinosus*. Each point on the three curves shows, for that muscle nerve, the proportion of the spindle Ia fibre-population recruited, plotted against the proportion of the tendon-organ population engaged (both expressed as percent of maximum) by a given intensity of stimulation. Were the Ia and Ib thresholds similarly distributed, the curve shown by the broken line would result. It can be seen that a stimulus engaging about 50% of Ia fibres in the *peroneus longus* nerve would also fire nearly 40% of those from tendon-organs; by contrast, a stimulus strong enough to recruit between 60–70% of *semitendinosus* Ia fibres would be needed to engage any of tendon-organ origin.

2. *Differential Block*

Various methods have been devised to achieve more or less complete blockade of conduction in certain groups of fibres in whole nerves, but so far there have been few attempts to achieve selective block of muscle afferent fibres. Methods

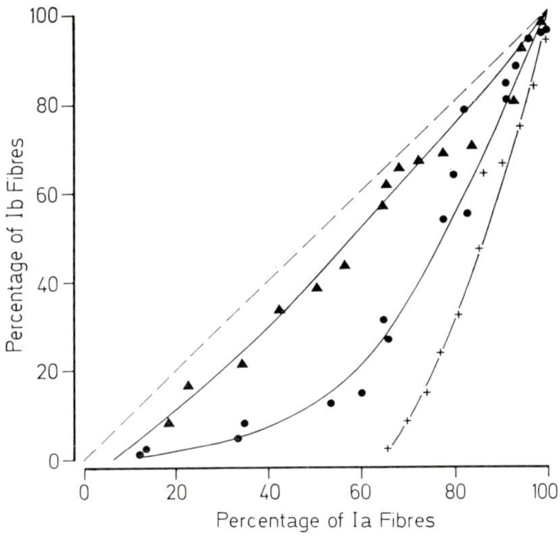

Fig. 1. Plots of the proportion of spindle and tendon organ Group I fibres recruited by different strengths of stimulation for three different muscle nerves. Abscissa, proportion of Ia fibres recruited at different stimulus intensities as percentage of maximum Ia recruitment; ordinate, proportion of Ib fibres recruited, also expressed as percentage of maximum Ib engagement. Triangles, for *peroneus longus* nerve. Circles, *medial gastrocnemius* (pooled results of COPPIN and SUMNER). Crosses, distal branch of *semitendinosus*. Redrawn and modified from Fig. 21 in MACLENNAN (1971). ▲ Peroneus Longus, ● Medial Gastrocnemius, + Semitendinosus

such as controlled mechanical compression, local anaesthesia, cooling, or passage of weak D.C. currents have been used with varying success in mixed, purely sensory or purely motor nerves. The procedures are difficult in practice and tend to be unreliable, the degree of selectivity being limited and the level of blockade difficult to maintain at a constant level. The best results seem to have been achieved with motor fibres, in which carefully controlled procaine application can block fusimotor fibres without interrupting conduction in a large proportion of alpha fibres (MATTHEWS and RUSHWORTH, 1957); also in amphibian ventral roots, it is possible to prevent conduction in large motor fibres by anodal blockade, whilst sparing conduction in the small tonic motor fibres (KUFFLER and VAUGHAN WILLIAMS, 1953). However, LAPORTE and BESSOU (1959) were able to block conduction apparently in all Group I fibres in some mammalian muscle nerves by repeated application of brief bursts of low frequency tetanic stimulation. For periods of half an hour or more, it proved possible by electrical or natural receptor stimulation below the block to send volleys centrally in afferent fibres smaller than those of Group I, without contamination by impulses in the latter. Quite recently, it has been shown (COPPIN et al., 1970; MACLENNAN, 1971) that selective blockade of Group I fibres from spindle primary endings can be achieved by episodes of appropriate low amplitude longitudinal vibration of certain muscles in the cat's hind limb; during the blockade, electrical stimulation at strengths below Group II threshold would evoke volleys restricted to fibres of tendon-organ origin.

3. Dissection of Filaments in Continuity

The surest way of limiting activity to afferent fibres from known receptors is to restrict by dissection a nerve or nerve root so as to leave one or at most a few intact fibres from an identifiable receptor-type. The technical difficulties of this approach are formidable indeed, and furthermore the central effects of such restricted, if pure, input are of course minimal. The technical feat has nevertheless now been performed by a number of workers for single Ia afferent fibres, either by dissecting a muscle nerve in continuity (Kuno, 1964), or by selecting dorsal root filaments and either stimulating chosen sub-divisions of muscle nerves or stretching an appropriate portion of a muscle (Jack et al., 1967; Mendell and Henneman, 1968, 1971; Jack et al., 1970, 1971). As a result of this achievement, the effect of minimal, but pure, input from spindle primary endings on motoneurones has now been extensively studied, and the technique will undoubtedly be further exploited.

4. Selective Stimulation of Receptors

Many attempts have been made to excite selectively one or other class of muscle stretch receptor, without restricting functional afferent channels by dissection or by differential blockade. Successful achievement of such selective activation would make it possible to bombard the central nervous system with a relatively massive input from a given receptor-type, with the likelihood of readily detectable central actions. On the other hand, such volleys being necessarily repetitive and asynchronous, precise timing of central effects would be difficult if not impossible.

Leaving aside crude attempts to separate spindle and tendon-organ actions by removing gross tendons, or directly stimulating different regions of a muscle presumed to contain predominantly one or another kind of receptor, some fairly successful methods have been devised. One of these involves the specific activation of spindle receptors by stimulation of a large number of isolated fusimotor fibres to a given muscle (Hunt, 1953). However, though tendon-organs and other endings are not affected, the procedure increases the firing of both primary and secondary spindle endings. The most successful attempts to stimulate spindle and tendon-organ receptors selectively, without fibre-blockade or fibre-dissection, are those of Lundberg and Winsbury (1960a) and Stuart et al. (1970). With the devices used by these workers, it is possible to excite Ia receptors by brief stretch of the muscle, and to engage Ib endings fairly selectively by applying a brief pull during contraction of the muscle.

The careful studies of Matthews and his colleagues (e.g., Brown et al., 1967; Matthews, 1972) of the relative sensitivity of different muscle stretch receptors to longitudinal vibration of the muscle, as described in Hunts' article (p. 198) have made it possible to activate with a high degree of selectivity spindle primary endings, at least in the soleus muscle, by use of vibration of carefully chosen amplitude and frequency. Exploitation of this method of selectively evoking Group Ia discharge is now proceeding in many laboratories.

III. Central Terminations of Muscle Afferent Fibres

Classical anatomical studies can not reveal the extent to which a particular component of the total dorsal root afferent inflow penetrates into the central nervous system; however, the reach of primary afferent fibres from muscle nerves obviously lies within the range determined by degeneration studies for whole dorsal roots, that is, within the segment of entry and a few adjacent segments, and in the appropriate dorsal column nucleus by way of long ascending processes. Electrophysiological methods can provide more precise information by tracing the central spread of impulse-volleys set up in muscle nerves, together with the more sensitive procedure of mapping the central boundaries of primary afferent fibre-spread through stimulation at different sites within the central nervous system and recording antidromic volleys in peripheral muscle nerves (LLOYD and McINTYRE, 1950). The refinement of these electrophysiological techniques, stemming directly from the pioneer work of Adrian, Gasser, Matthews, Zotterman and their colleagues, and skilfully exploited especially by their pupils and those of the Sherrington school, has been responsible for most of our present knowledge of the central projection of muscle and other primary afferent fibres, so far surpassing in resolution the contribution of neuroanatomy. It may prove, however, that advances in neurohistological techniques (for example, by the labelling of single, functionally identified fibres with tracer substances) will lead to a new level of precision in fine structural studies.

A. Segmental Distribution

Afferent fibres of muscle nerves contribute substantially to the dorsal columns near their level of entry, and many of them send collaterals to motoneurones and interneurones, including some giving origin to long ascending tract systems. Most appear to terminate ipsilaterally, but at least in some situations, collaterals cross to the opposite side (SPRAGUE, 1958). However, most of the detailed information about termination of muscle afferent fibres relates to Group I, and little is known of the central distribution of the smaller fibres (Groups II, III and IV).

That some large dorsal root fibres give off direct reflexo-motor collaterals to motoneurones, with branches to interneurones, was established long ago by CAJAL (1909), but their unique peripheral origin from spindle primary endings was not determined until comparatively recent times. Functional confirmation and analysis of this uniquely monosynaptic pathway, and the demonstration that the receptors concerned are restricted to skeletal muscle, were amongst the earliest achievements of the then new electrophysiological techniques (ECCLES and PRITCHARD, 1937; RENSHAW, 1940; LLOYD, 1943a). More recent direct anatomical study of these Group Ia collaterals and their branches, by following their degeneration after dorsal root section, has shown them to end on both the soma and dendrites of motoneurones and on interneurones (SZENTÁGOTHAI, 1958; SPRAGUE, 1958; STERLING and KUYPERS, 1967; CONRADI, 1970). The distribution of these endings on motoneurones appears to be sparse, perhaps amounting to less than one percent of the total synaptic scale. However, this may prove to be an underestimate because of the vagaries of staining techniques, and ILLIS (1967) describes a considerably denser distribution of monosynaptic boutons on the motoneurone surface. Functional analysis of the monosynaptic pathway will be discussed subsequently.

Less is known of the detailed distribution of muscle afferent fibre branches to interneurones, most being closely intermingled with those from other receptors. However, some from the Group Ia collaterals can be seen to end on interneurones in Rexed's lamina VII (the intermediate region of Cajal) and lamina VIII, the "non-motor" parts of the ventral horn (Szentágothai, 1967). More extensive connections with interneurones, especially in the intermediate region, are affected by Group Ib and higher threshold fibres of muscle and other origin, but the details of these linkages are obscure. Most of the evidence is electrophysiological, and some of the conclusions about Ia and Ib central connections depend on the questionable criteria already discussed (Eccles et al., 1954; Eccles et al., 1956; Eccles et al., 1960; Eccles, 1962; Lundberg, 1966). A notable exception is the recent combined electrophysiological and histological achievement of staining individual functionally identified interneurones, for example, those of the Ia inhibitory system (Jankowska and Lindström, 1972).

B. Extrasegmental Distribution

The limitations of neuroanatomical techniques in defining segmental distribution apply at least as much to the problem of determining spread of muscle afferent fibre ramifications beyond the boundaries of their segments of entry. For example, Cajal (1909) observed the entry into Clarke's nucleus of fibres from primary afferent collaterals of fairly large diameter ascending in the dorsal column, but their peripheral origin remained undetermined until very much later. Electrophysiological methods then established that afferent fibres of muscle origin, including those of Group I, are prominent amongst those ending on Clarke's cells (e.g., Grundfest and Campbell, 1942; Lloyd and McIntyre, 1950). The same methods were necessary to demonstrate the nature of the afferent input to cells of origin of the ventral spinocerebellar tract (Oscarsson, 1957).

The fact that by no means all primary fibres in the dorsal root inflow send ascending collaterals to the dorsal column nuclei (only 25% do so to the *nucleus gracilis*) was established histologically (Glees and Soler, 1951), but by this means alone no functional significance could be ascribed to this exclusion. However, electrophysiological analysis demonstrates that at least in the cat, very few muscle afferent fibres from the hind limb send dorsal column collaterals above the lower thoracic segments (Lloyd and McIntyre, 1950; Wall, 1961; McIntyre, 1962). By contrast, it is now well established that Group I afferent fibres from the forelimb do send collaterals in the dorsal column to cells of the cuneate nucleus (Oscarsson and Rosén, 1963).

There is little definite information about descending collaterals of muscle primary afferent fibres, except that those of Ia origin appear to extend no further than the most caudal extremities of their homonymous, synergic or antagonist motoneurone pools. Some of the smaller muscle afferent fibre collaterals may extend somewhat further rostrally or caudally to engage interneurone pools such as those mediating flexor or crossed extensor reflex patterns, long propriospinal links, or other long ascending pathways.

In summary, then, a combination of neurohistological and electrophysiological methods have defined many sites of termination of the larger muscle afferent fibres within the spinal cord. Prominent amongst these synaptic targets are motoneurones of the same or adjoining segments, to which Group Ia collaterals form the sole primary afferent fibre-projection; cells of origin of the dorsal spinocerebellar tract (Clarke's cells) and of the ventral spinocerebellar tract, with which many Group I fibres make close contact; and many interneurones subserving a variety of functions in the segments of entry and adjacent segments. However, there remains much uncertainty about the precise central distribution of many muscle afferent fibres, especially those of small diameter.

IV. Spinal Reflex Actions of Impulses in Muscle Afferent Fibres

In this section, a survey is presented of present understanding of the reflex actions within the spinal cord of input engendered in the various components of the afferent equipment of skeletal muscles. Much of the experimental material is derived from experiments on the lumbosacral cord of the cat, but there is no doubt that essentially similar reflex systems operate at other spinal segmental levels, and in other mammals including man. This account aims at filling the gap of 12 years since the appearance of the review by HUNT and PERL (1960) on spinal reflex mechanisms, and will necessarily be incomplete; but other authoritative articles on various aspects of this broad field have appeared in the intervening period, for example, ECCLES (1964), LUNDBERG (1966), GRANIT (1970), and very recently the admirable monograph by MATTHEWS (1972). The central reflex actions will be discussed in relation to the different functional subdivisions of muscle afferent inflow described in preceding pages.

A. Group Ia Impulses

The central effects of impulses in the Ia fibres from spindle primary endings are better known than those of most other primary afferent fibres, the segmental reflex actions including the monosynaptic myotatic reflex, reciprocal inhibition of antagonist motoneurones, and other less well defined effects by way of segmental interneurones.

1. Monosynaptic Excitation of Motoneurones

The functional significance of Cajal's direct monosynaptic links between primary afferent fibres and motoneurones was disclosed by Lloyd in a series of meticulous experiments which provided a firm foundation for the many greatly expanded and refined experimental analyses of spinal reflex function carried out since then in many laboratories. The essential facts demonstrated by Lloyd's experiments are that monosynaptic connections are made only by the largest dorsal root fibres; that these come solely from muscle receptors; that the peripheral distribution of monosynaptic discharge exactly matches the distribution of the stretch-reflex response described by LIDDELL and SHERRINGTON (1924), that is, it engages motoneurones only of the muscle containing the executant receptors; and finally, that brief, controlled stretch of muscles evokes motoneurone discharge in monosynaptic tempo (Lloyd, 1943a, b, c). Since this pioneer work, it has been generally accepted that the long familiar, phasic tendon-jerk response is mediated by this simplest of neural circuits. Its very short latency had been noted long before, but was considered too brief for the response to represent a true reflex (Sherrington, 1900); however, on the basic of accurate latency measurements Jolly (1911) suggested that the knee-jerk might be a monosynaptic response, and later work in man supported this conclusion (Hoffmann, 1922). The development of thought on this topic, including the curious reluctance of the earlier physiologists to recognise the reflex nature of the tendon-jerk, is well outlined by Matthews (1972) and will not be pursued further here.

Agreement was also slow on the question of which muscle receptor is responsible for the myotatic reflex; it is not long since some standard textbooks cited the tendon organ as the structure in question. Lloyd carefully avoided speculation about the specific myotatic reflex receptor, since accurate information was lacking in the nineteen forty's about conduction velocity in relation to receptors of origin (e.g., Lloyd, 1952); however, the duality of Group I fibre function was soon established, as already noted (Hunt and Kuffler, 1951). These findings in conjunction with Lloyd's earlier work simplified the problem to a clear choice between tendon organs and primary spindle endings as the specific receptor for the phasic stretch reflex; but because of the low mechanical threshold of both the reflex and the spindle endings, assignment of this role to spindle primary receptors seemed reasonable even at that time. Other less direct evidence supported this conclusion; for instance, the finding that selective increase of spindle discharge, evoked by isolated fusimotor fibre stimulation, facilitates monosynaptic test responses of homonymous and synergic motoneurones (Hunt, 1952). However, in these experiments, increase in secondary ending discharge is an unavoidable concomitant of fusimotor stimulation. Another argument is based on the work of Szentágothai (1948), who showed that unipolar cells in the mesencephalic trigeminal nucleus are the primary afferent neurones for spindle receptors in the masticatory muscles, and also supply direct collaterals to trigeminal motoneurones. As described by Cajal (1909), the peripheral processes of these unipolar mesencephalic nucleus cells reach the mandibular division by way of the motor root of the trigeminal, and hence would remain intact after interruption of the sensory root in which tendon organ afferent fibres presumably run, along with many others. Szentágothai's findings together with the persistence of a stretch reflex in jaw-closing muscles after section of the trigeminal sensory root (McIntyre, 1951; McIntyre and Robinson, 1959) may be regarded as additional evidence implicating spindle receptors in the myotatic reflex, but cannot distinguish between primary and secondary endings.

Once intracellular recording techniques for motoneurones became available, the way was open to more direct experimentation, and the elicitation of monosynaptic depolarizing synaptic potentials (EPSP's) in motoneurones by weak stimulation of appropriate muscle nerves strongly supported the view that spindle primary endings are the receptors for the myotatic reflex (Brock et al., 1952). However, because of the previously discussed uncertainties about relative thresholds of Ia and Ib fibres, conclusive direct evidence could only be provided by selective adequate stimulation of spindle primary endings. This was first achieved by LUNDBERG and WINSBURY (1960a), to which reference has already been made. They used brief, weak muscle-stretch, and proved by direct testing on adequate samples of single stretch-receptor units that their stimulation procedure elicited discharge only from spindle primary endings. Fig. 2, reproduced from their paper, shows that pure Ia input elicited in this way sets up in impaled homonymous motoneurones monosynaptic excitatory synaptic potentials (EPSP's) almost identical with those elicited by weak electrical stimulation of the muscle nerve. Recently, similar results have been obtained by STUART et al. (1971). Thus there can be no doubt that the specific stretch receptors which form the starting point of the phasic monosynaptic stretch reflex arc are the primary endings of muscle spindles. Further evidence implicating spindle primary terminals (but not monosynapticity of path) in the stretch-evoked response is the reflex contraction of a muscle to longitudinal vibration of frequency and amplitude appropriate for selective activation of primary endings (MATTHEWS, 1967; BROWN et al., 1967).

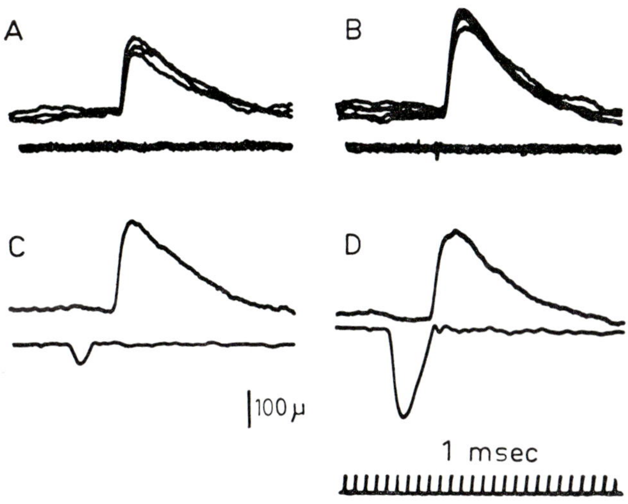

Fig. 2. Monosynaptic excitatory synaptic potentials (upper traces) recorded intracellularly from a *soleus* motoneurone. A and B evoked by stimulation of the intact *soleus* nerve at two intensities (lower traces show afferent volleys recorded from the dorsal root-cord junction). Five sweeps superimposed in each. C and D EPSP's elicited by weak, brief stretch of *soleus*, shown on lower traces: C stretch of about 50 μ, D 250 μ. Reproduced with permission from Fig. 6 in LUNDBERG and WINSBURY, Acta physiol. scand. **49**, 155–164 (1960)

On the other hand, none of the experimental evidence so far cited excludes the participation in stretch reflex responses of more complex parallel reflex pathways involving interneurones; indeed, evidence has been cited by some authors supporting the view that Ia impulses may also excite motoneurones polysynaptically (ECCLES et al., 1960; GRANIT, 1970). As already noted, reflexo-motor collaterals of Ia fibres send branches to many interneurones, and though functions for some of these have been defined, others could contribute to the postulated polysynaptic autogenetic excitatory pathways which might play an important role in sustained myotatic responses. It may be that the infrequency with which polysynaptic components in Ia-evoked EPSP's have been detected relates to the common use of single volleys of impulses whereas the opening of an interneuronally-mediated excitatory path might require temporal summation through repetitive afferent input of the kind which must obtain during normal activity. That secondary endings may also contribute through such pathways to tonic stretch reflex responses as manifest in the decerebrate preparation has recently been proposed (Matthews, 1969); this will be discussed in relation to the central actions of muscle Group II impulses.

A number of searching investigations have now been made on the central actions of minimal Ia afferent input on individual motoneurones, which have amply confirmed their unique monosynaptic excitatory action. The difficult feat of restricting afferent input to a single impulse in a spindle primary afferent fibre, as mentioned earlier, has been achieved by a number of workers, of whom Kuno (1964) was the first to record the effect of such minimal input on individual impaled motoneurones. Kuno's experiments, and since then those of a number of other authors, demonstrate quite clearly that single Ia impulses bring about detectable monosynaptic EPSP's in homonymous and synergic motoneurones. Fig. 3, from the paper by Mendell and Henneman (1971), is a striking example of the monosynaptic EPSP elicited by a single Ia impulse in an impaled motoneurone revealed by computer summation technique.

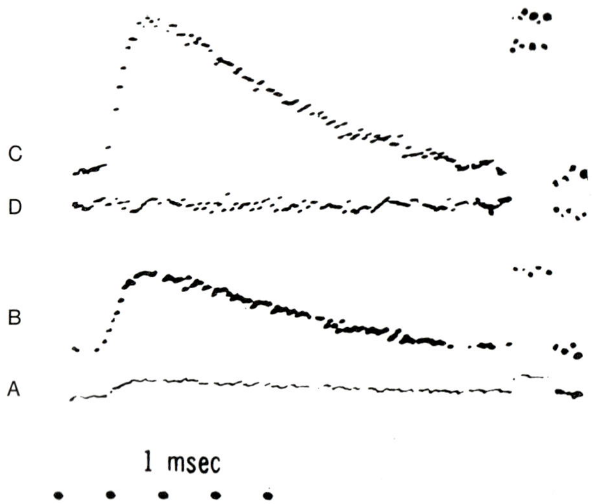

Fig. 3. Intracellularly recorded monosynaptic EPSP evoked by a single Ia impulse in a motoneurone of *gastrocnemius*, revealed by computer summation. A, B and C after repeating 100, 500 and 1,000 sweeps respectively. D control, showing the sum of 1,000 sweeps after removing microelectrode tip to a position just outside the cell. Calibration pulses 200 μV. Reproduced with permission from Fig. 3 in Mendell and Henneman, J. Neurophysiol. **34**, 171–187 (1971)

The thrust of these and related studies has been increasingly directed towards a detailed analysis at the cellular level of synaptic processes such as quantal aspects of transmitter action, the spatial distribution of presynaptic endings, and the biophysical properties of the motoneurones themselves in relation to their integrative function. As such, it would be inappropriate to discuss them in detail here, and various publications describing these new findings and their implications are already available, for example, TERZUOLO and LLINÁS (1966); JACK et al. (1967, 1971); KUNO (1971). However, brief mention should be made of certain conclusions from these studies about the distribution and density of Ia monosynaptic terminals, because this is relevant to the overall operation of monosynaptic reflex responses. In agreement with histological findings, electrophysiological analysis confirms that functional monosynaptic connections are made both with dendrites and cell bodies of motoneurones, although their density appears to be greater on cell bodies and more proximal dendrites than on finer peripheral dendritic branches. Individual Ia monosynaptic collaterals appear to undergo remarkably profuse branching, to spread widely through their target motoneurone pools. Mendell and Henneman's results indicate that a single Ia fibre from medial gastrocnemius would send terminals to nearly all its homonymous motoneurones, and that a single medial gastrocnemius motoneurone might receive monosynaptic terminals from nearly all the Ia fibres belonging to this muscle. Fewer monosynaptic terminals are supplied to synergic motoneurones, but there seems to be no significant difference in their spatial location on the motoneurone surface as compared with their homonymous counterparts (JACK et al., 1970).

The significance to normal motor function of the monosynaptic stretch reflex is still a matter of debate, and must remain at least partly speculative until much more is known about matters such as the role of fusimotor spindle innervation under normal conditions. One source of difficulty is the confusion which appears in many discussions between the stretch reflex in its phasic and tonic manifestations, and though there is no disagreement that the former is mediated monosynaptically, the extent of this pathway's role in the latter is still uncertain. However, even if the tonic stretch reflex is partly sustained by action in polysynaptic paths, as seems probable, it is reasonable to assign a special significance to the direct monosynaptic route, not only because of its rapidity, but because uniquely it allows the signals from spindle primary endings to influence motoneurones directly, by-passing paths involving interneurones, necessarily subject to many other converging influences. Because mechanisms are now known which can affect the transmitting effectiveness of presynaptic terminals, the independence of Ia monosynaptic access to motoneurones cannot be regarded as absolute; none the less, the pathway's directness and rapidity, its receptors' "in parallel" arrangement with extrafusal muscle fibres and their length-monitoring ability combine to make a compelling case for Merton's "length-servo" or some closely related hypothesis for at least one of its functions (MERTON, 1953). The great dynamic sensitivity of spindle primary endings adds weight to this contention, for very small, rapid deviations from a muscle length appropriate to the fusimotor-set condition of the spindle contractile systems will provide a powerful corrective, negative feedback input — either increase or decrease of Ia firing — via the monosynaptic myotatic reflex path. It is possible that the operation of certain muscles with major postural function, such as the axial musculature of the vertebral column, might ordinarily be largely controlled via the so-called "gamma loop", i.e., the primary motor output is the fusimotor setting of length, with the myotatic reflex operating in strictly "length servo" fashion.

But this is certainly not the case for many muscles, such as those operating the digits; for example, the threshold for cortical excitation of such motoneurones is unaffected by dorsal root section, indicating direct rather than "gamma loop" control (Lewis and Porter, 1971). The fact that the many spindles in small, distal limb muscles commonly receive so-called "beta" motor innervation, that is, an intrafusal motor supply from collaterals of the large axons innervating extrafusal fibres, also reduces the plausibility of a simple length-servo or follow up servomechanism in these muscles. The original suggestion of Kuffler and Hunt (1952) seems more likely, namely, that the intra- and extra-fusal motoneurones are co-activated in the generation of movement so that spindle receptors maintain their sensitivity, despite changes in muscle length; a more recent version of this describes the spindle-myotatic-reflex system as an assisted servomechanism. There is now much direct evidence demonstrating co-activation of small and large motoneurones in a variety of motor responses (see Granit, 1970), including voluntary movement in man (Vallbo, 1971). Perhaps the clearest example of such "alpha-gamma" linkage in normal movement is that seen in the intercostal muscles of respiration (Critchlow and v. Euler, 1953; Sears, 1954).

2. *Inhibition of Antagonistic Motoneurones*

Liddell and Sherrington (1925) showed that passive extension of hamstring knee flexors strongly inhibits a sustained myotatic reflex of *quadriceps*. Monosynaptic reflex testing (Lloyd, 1946a) and intracellular recording (Brock et al., 1952; R.M. Eccles and Lundberg, 1958a) revealed the existence, between antagonistic motoneurones, of a general pattern of Group I-dependent reciprocal inhibition into which this early example fits; however, presynaptic mechanisms probably also participate, because Liddell and Sherrington (1925) observed partial strychnine-resistance of their *quadriceps* inhibition from knee flexors. Lloyd (1946b) coined the term "myotatic unit" for the group of muscles acting at a given joint and linked through Group I reflex connections: thus impulses from stretch-reflex afferents from a given muscle excite their own motoneurones, facilitate those of synergists and inhibit antagonistic motoneurones. Although exceptions to this general rule have been described (Eccles et al., 1957b; R.M. Eccles and Lundberg, 1958a), the myotatic unit concept has proved generally useful and holds true in close approximation, especially for muscle groups acting on simple hinge-type joints such as the knee and elbow.

Since the input-output experiments of Lloyd (1943a) and of Hunt (1955), there has been general agreement that the Group I fibres responsible for inhibition within the myotatic unit are identical with those yielding monosynaptic excitation, the implication being a central branching of Group Ia fibres to bring about these different actions. However, the central pathway through which Ia impulses bring about reciprocal inhibition of antagonist motoneurones has been the subject of controversy. Lloyd's term "direct inhibition" was coined to distinguish it as a specific inhibitory process rather than mere depression through subnormality of elements common to conditioning and testing pathways. Because of its extremely short latency, it also came to be regarded as being exerted by monosynaptic action. This was entirely reasonable, especially since careful timing, using monosynaptic test reflexes, in some instances showed inhibition to be well established at "zero" time, that is, with conditioning and testing volleys arriving simultaneously in the segment containing the test motoneurone pool. This seemed to exclude the possibility of an interneuronal link (LLOYD and WILSON, 1959). However, intracellular records from motoneurones reveal that the latency of Ia inhibitory hyperpolarization exceeds that of excitatory synaptic depolarization by amounts ranging from about 0.5 to 0.8 msec (ECCLES et al., 1956; CURTIS et al., 1958; FRANK and SPRAGUE, 1959), a finding most readily explained by the presence of an interneurone in the pathway. On the other hand, the matching of input-output curves for monosynaptic facilitation and direct inhibition found by LLOYD (1943a) and by HUNT (1955) suggested identity of pathway for both actions; if the inhibitory path included an interneurone, a different configuration of the curve might be expected reflecting, for example, a need for summation.

Further experiments have, however, resolved the question in favour of the interneurone hypothesis, and the original reservation of RENSHAW (1942) about the true minimal latency for "direct" inhibition is now widely accepted as explaining the discrepancy between latency measurements made with reflex testing and with intracellular recording. The crucial point is that the timing of inhibitory latency must be made in relation to discharge of those motoneurones which fire latest, yet contribute to the measured peak of the test reflex, and not to the onset of test reflex excitation. JACK (1957) carried out experiments relating size of Ia volley to amount of direct inhibition of a test monosynaptic reflex in unanaesthetized acutely spinal preparations. He demonstrated substantial latency-shift with change in volley-size, and input-output curves of form differing from those for Ia monosynaptic facilitation; together, these pointed to the presence of a synaptic interruption of the Ia inhibitory path. Further evidence for the disynaptic nature of this path was obtained by R.M. ECCLES and LUNDBERG (1958b), who demonstrated spatial facilitation therein, in anaesthetized preparations, by subdividing the inhibitory input and examining the interaction between its components. ARAKI et al. (1960) also showed that the timing of inhibitory suppression of reflex discharge is appropriate for a disynaptic pathway.

Since then, extensive studies have been made, especially by Lundberg and his colleagues, of individual interneurones with properties matching those predicted for this intercalated element in the "direct" inhibitory path (LUNDBERG, 1969a, 1970). The Group Ia synaptic linkage with these interneurones is powerful and of "afferent" type: the need for summation is minimal, and single volleys may evoke repetitive discharge. In recent elegant experiments, microelectrodes have been used to locate single interneurones fitting this role on the basis of response to Group I volleys and to antidromic activation from within the "directly" inhibited motoneurone pool. Direct iontophoretic stimulation of these interneurones by glutamate has also been shown to produce inhibitory potentials in motoneurones of the target motor nucleus (JANKOWSKA and ROBERTS, 1972). Although the selectivity of iontophoretic stimulation might be questioned, especially as many other spinal interneurones with inhibitory actions on motoneurones must be nearby, the case is nevertheless compelling that these belong to the Ia reciprocal inhibitory path, because of their response to the same Group I volleys which inhibit the antagonist motoneurones, and because of their responses to other inputs known to interact with the Ia inhibitory system. For example, the Ia interneurones are themselves inhibited by the recurrent axon collateral system, and excited by impulses in the cortico-spinal and other descending pathways. The range and complexity of this converging input on the Ia inhibitory interneurones has only recently been disclosed (Lundberg, 1969a, 1970; Hultborn et al., 1971; Hultborn, 1972), and only brief reference to it can be made in this context. It is especially interesting that the patterns which are emerging demonstrate an important integrative role for these interneurones. For example, those inhibiting flexor motoneurones are also facilitated by vestibulo-spinal impulses, and by crossed extensor reflex input from the opposite limb, thus contributing to the relaxation of flexors in both these extensor response-patterns. Inhibition of these interneurones through the Renshaw system, itself subject to descending and segmental controls, makes possible variations in the effectiveness of Ia reciprocal inhibitory linkage in accordance with differences in requirements for different patterns of motor response, even its complete suppression in situations demanding co-contraction of antagonistic muscles. Certainly, the Ia inhibitory interneurones can no longer be viewed as mere transmitter-switching appendages to Group Ia or other afferent collaterals, but play a significant integrative role in spinal cord function (Lundberg, 1969a, 1970; Hultborn, 1972).

Recently, a sample of these interneurones, identified by the same functional criteria, has been directly visualised by Procion yellow injection (Jankowska and Lindstrom, 1972). The cells were in the ventral horn in a location corresponding to the more ventral of the two interneurone groups stated by Szentágothai (1967) to receive Ia collateral branches. The functions of the more dorsal group of Ia innervated interneurones (in Rexed's lamina VII) have not yet been determined, but a role in the Ia reciprocal inhibitory path seems to be excluded (cf. Eccles et al., 1956). It has already been mentioned that some might belong to a postulated polysynaptic component of the stretch reflex. However, interneurones with a variety of connections are likely to be intermingled in this region, for example, some of those subserving reflex actions of tendon organ origin, and others concerned with responses to cutaneous and deep impulses in fibres of Groups II and III (Eccles et al., 1956; Coombs et al., 1956; Eccles et al., 1960).

3. Other Segmental Actions of Ia Impulses

There is little that can be said with certainty about actions of Ia input at segmental level other than those just discussed. Most of the stretch reflex and associated inhibitory phenomena, at least in muscles of the limbs, are unilateral; however, the reflex organization of tail muscles shows crossed reciprocal Ia inhibitory as well as some crossed monosynaptic excitatory component (Wilson and Lloyd, 1956); similar patterns would be expected in other axial muscles of the vertebral column with antagonistic or synergic function, as for those of the tail.

Ia impulses may also affect motoneurones of the contralateral limb. The experiments of Perl (1958, 1959) showed crossed reflex actions of Group I volleys, and of muscle stretch, on motoneurones of the muscle's contralateral fellow, including short latency inhibition interpreted as of Ia origin; however HOLMQVIST (1961) was unable to confirm this, nor could she assign to a particular component of Group I input the contralateral effects observed by her in response to brief trains of volleys. It is thus not possible to relate these findings with any certainty to the contralateral responses to stretch of *quadriceps* observed by SHERRINGTON (1909) in decerebrate preparations; however, the crossed inhibition seen by him to accompany the "shortening reaction" could well be a consequence of the spindle input responsible for the latter.

The possible reflex action of Ia input on fusimotor neurones is a question of particular interest, but so far information is scanty on this as on other aspects of fusimotor reflex organization. It would be surprising if there were an autogenetic excitatory link like that for "alpha" motoneurones, because of the positive feedback implications of such a situation. HUNT (1951) observed autogenetic fusimotor inhibition from muscle stretch, but this could not be confirmed in later experiments with spinal preparations (HUNT and PAINTAL, 1958). In these experiments, no distinction could be made between actions of different stretch receptors. However, BROWN et al. (1968) showed that selective spindle Ia excitation by vibration inhibited some spontaneously firing fusimotor neurones in ventral root filaments. Unfortunately the significance of this finding cannot be gauged, since the destination of the inhibited fusimotor discharge was unknown; it is also possible that the inhibition might have been effected through the recurrent axon collateral system as a result of "alpha" motoneurone firing evoked by the Ia input. Quite recently, PROSKE and LEWIS (1972) have shown that Ia firing selectively induced by vibration of one fraction of *triceps surae* can reduce background fusimotor discharge to a synergist, a finding consistent with the early observation of HUNT (1951); however, the effect is not powerful. It seems likely, therefore, that Ia input can bring about weak autogenetic inhibition of fusimotor neurones.

Impulses in Group Ia fibres have also been implicated in the generation of presynaptic inhibitory action and accompanying depolarization of primary afferent presynaptic terminals. Discussion of these phenomena and their possible roles in modulating spinal reflex activity-patterns will be deferred until after consideration of the spinal segmental actions of impulses in other muscle afferent fibres.

B. Group Ib Impulses

As would be expected from their "in series" arrangement and behaviour, tendon organ discharge evokes reflex responses quite different from those of spindle primary endings. However, much less is known about the full extent of their central actions than in the case of Ia impulses, and controversy continues about some of these. One reason for this is the difficulty of determining the extent of Ia and Ib contribution to electrically evoked Group I volleys, to which reference has been made (pp. 242–244). But despite the fact that satisfactory control of Group I afferent volley composition cannot regularly be achieved on the basis of electrical threshold differences alone, enough is now known about these to allow some conclusions to be drawn from such experiments in conjunction with the results of others using natural stimulation.

1. Autogenetic Inhibition

The existence has long been recognised of a reflex mechanism whereby impulses from stretch receptors can powerfully inhibit the motoneurones of the muscle in which they lie, for example, the "clasp-knife" relaxation (lengthening reaction) evoked by strongly stretching a tonically contracting extensor muscle in decerebrate preparations (Sherrington, 1909). There can now be little doubt that such autogenetic inhibitory action can be attributed, though perhaps not exclusively, to impulses arising in tendon organs. Granit (1950) used monosynaptic test reflexes to gauge the central actions of muscle afferent impulses set up by stretch or active contraction of a muscle upon motoneurones of the muscle itself or other muscles. Early facilitation, followed by inhibition, of the muscle's own motoneurones and those of its synergists was observed; inhibition could also be evoked in motoneurones supplying other muscles. Although these experiments could not establish with certainty the end organs responsible for these opposing effects, it was considered that afferent fibres of rapid conduction velocity were involved in both actions. The fibres responsible for facilitation were more susceptible to blockade by cold or pressure than those yielding inhibition, and tendon organs seemed the most likely candidates for the inhibitory receptors. Hunt (1952) demonstrated, also by monosynaptic testing, that *triceps surae* motoneurones are powerfully inhibited during the plateau of a fused tetanus of this muscle, a circumstance known to yield massive tendon organ discharge. Fig. 4 gives an

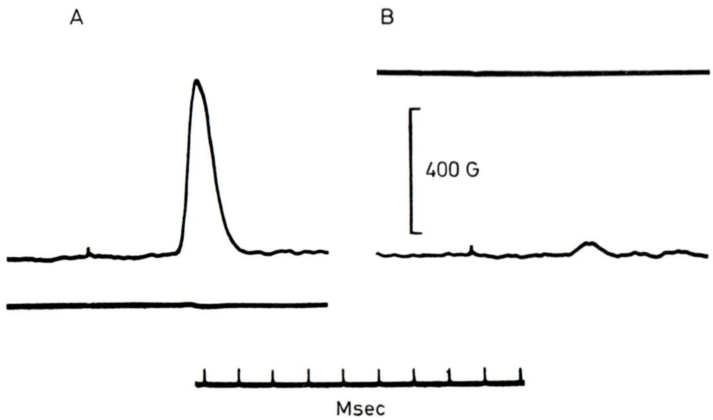

Fig. 4. Autogenetic inhibition of *triceps surae* monosynaptic reflex, recorded from the central end of the severed first sacral ventral root, during tetanic contraction of the *triceps surae* muscle group. A control reflex from Group I strength stimulation of (intact) *triceps surae* nerve. Lower trace, strain gauge record of initial tension (40 g). B identical stimulation of nerve during tetanic contraction of the muscle elicited by stimulation of the peripheral end of the first sacral ventral root at 200/sec, the tension developed (upper trace) being 770 g. (McIntyre, unpublished, 1951)

example of such an experiment showing almost complete autogenetic inhibition of a monosynaptic reflex of *triceps surae* motoneurones. Hunt also demonstrated concomitant inhibition of synergic and facilitation of antagonistic motoneurones. The case for these central actions being caused by tendon organ input thus became very strong; however, autogenetic inhibition as well as facilitation of motoneurones by spindle afferent discharge evoked selectively by fusimotor stimulation was also observed in these experiments; thus a possible role of spindle afferents in autogenetic inhibition and related actions could not be excluded.

Accurate information about central pathways for these actions can not be provided by experiments in which stretch or contraction of muscle provides the reflex input. LAPORTE and LLOYD (1952) filled this gap by experiments in which the central actions of electrically evoked muscle afferent volleys were measured by using monosynaptic test reflexes. By precise timing of the subliminal actions set up by afferent volleys of carefully graded amplitude, they demonstrated a strong inhibitory action of impulse in fibres of the Group I band upon synergic motoneurones, by way of a pathway including a minimum of one interneurone. Fig. 5 reproduced from their paper shows the effects, for two different synergic pairs, of three different amplitudes of Group I conditioning volleys on synergic motoneurones, the curves on the left being for an extensor, those on the right for a flexor synergic pair. In the latter, the weakest conditioning volley (A) gives the expected small monosynaptic facilitation of simple time-course; the larger volleys for B and C show stronger early monosynaptic facilitation which is inter-

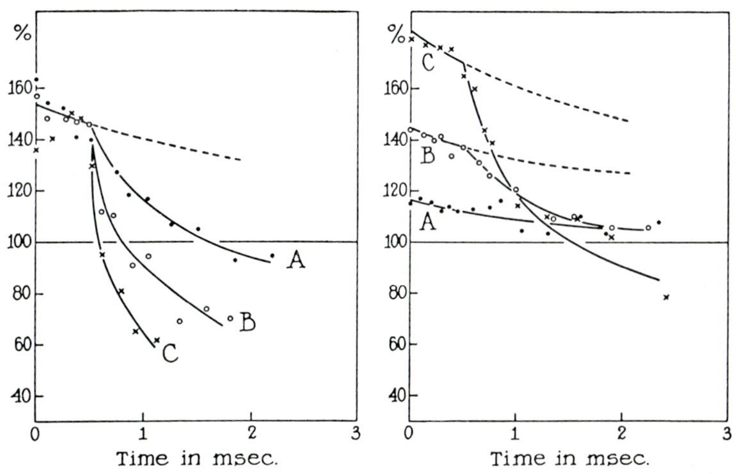

Fig. 5. Left hand curves, conditioning of monosynaptic reflex of the extensor *plantaris* by volleys in the nerve of a synergist, *flexor longus digitorum*, set up by three different intensities of stimulation. Action of the smallest volley is plotted by curve *A* (dots), that of the two successively larger volleys by curve *B* (circles) and curve *C* (crosses), in that order. Abscissa, interval in msec between arrival of the conditioning and testing volleys at the test reflex cord segment. Ordinate, size of test reflex as percentage of control amplitude. Right hand curves, similar plots of conditioning of monosynaptic reflex of the physiological flexor, *extensor longus digitorum*, by three different sizes of afferent volley in the nerve of its synergist, *tibialis anterior*. Symbols and labelling of the three curves, and ordinate and abscissa, as for the left hand curves. Reproduced with permission from Figs. 5 and 6 in LAPORTE and LLOYD, Amer. J. Physiol. **169**, 609–621 (1952)

rupted by deviation in the inhibitory direction after about 0.5 msec. Inhibition of similar latency is seen with the extensor pair, but the weakest volley (A) in this case produces some inhibitory deviation of the curve, and gives maximal early monosynaptic facilitation. Strictly speaking, this Group I inhibition can not be described as "autogenetic", the motoneurones tested being synergic rather than homonymous; however, since then many experiments have been carried out which leave no doubt that the inhibitory action of this Group I input, identifiable as Ib in the proper sense, is the same on homonymous and synergic motoneurones.

A wealth of data has been provided by intracellular recording, amongst the earliest of which were the results of Eccles et al. (1957c) and R.M. Eccles and Lundberg (1959a, b). Fig. 6 gives an example of intracellular potentials in a

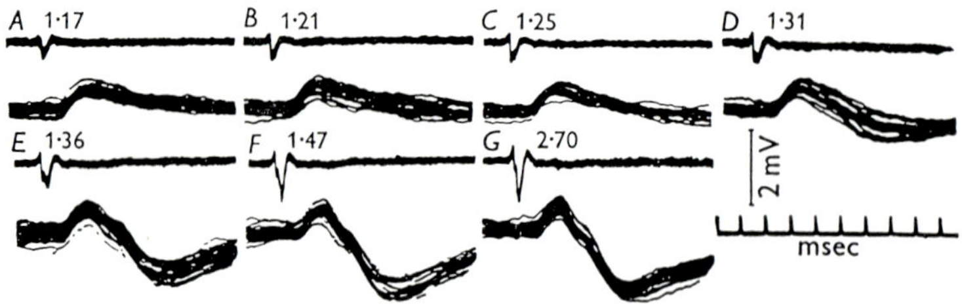

Fig. 6. Responses of a *plantaris* motoneurone to volleys set up in the nerve of *flexor digitorum longus* by stimuli of different intensity. Upper traces, afferent volleys simultaneously recorded at the dorsal root-cord junction; lower traces, intracellular recording of motoneurone response. Records obtained by superimposition of about 25 faint sweeps. A–G responses to increasing strength of stimulus, the intensity of each being indicated as a multiple of nerve threshold above each dorsal root record. Reproduced with permission from part of Fig. 12 in Eccles, Eccles and Lundberg, J. Physiol. (Lond.) **136**, 527–546 (1957a)

plantaris extensor motoneurone, in response to graded volleys in the nerve of its synergist, *flexor digitorum longus*; with volleys exceeding 1.25 times Group I threshold, the monosynaptic EPSP is distorted by a hyperpolarizing inhibitory potential beginning soon after its summit: this depolarization-hyperpolarization sequence should be compared with the left hand curves in Fig. 5. With further increase in size of the conditioning volley, this early hyperpolarization becomes more pronounced, and a second, larger inhibitory potential is added to it. There is no reason to doubt that impulses in Ib fibres account for the earliest phase of inhibitory hyperpolarization, but the impulses responsible for the second phase appearing at shock strengths 1.36 times threshold and above could include some in Group II fibres of low threshold.

Laporte and Lloyd also defined interneuronally mediated actions of tendon organ impulses other than inhibition of synergic motoneurones, namely, excitation of antagonists, and inhibition of motoneurones outside the myotatic unit — for example, of *triceps surae* by *quadriceps* volleys. The latter observation agrees with an early finding of Denny-Brown (1928) of a silent period in other extensors, as well as in the responding muscle, during a tendon jerk. Also well documented from myographic and electrophysiological studies is the appearance of excitation in antagonistic motoneurones during the course of autogenetic inhibition (e.g., CREED et al., 1932; GRANIT, 1952; HUNT, 1952). Intracellular studies agree in general with the earlier observations, but fail to support Laporte and Lloyd's conclusion that the excitation of antagonists is reciprocal, in that interneuronally mediated excitation of antagonist motoneurones could only be seen in those of flexors, and not in the reverse direction (ECCLES et al., 1957c). In fact the reflex pattern attributed to Ib action in these experiments closely match that of the classical flexor reflex, with widespread excitation of flexor and inhibition of extensor motoneurones; however there is evidence suggesting that a different set of interneurones is involved (R.M. ECCLES and LUNDBERG, 1959b).

Thus there is still uncertainty about the full extent and nature of Ib-evoked segmental reflex response-patterns, other than autogenetic inhibition. One possible reason for discrepancy in results is the type of preparation used in different experiments, unanaesthetised acute high spinal animals in Lloyd's work, as opposed to barbiturate anaesthesia, with or without spinal transection, in many intracellular studies. Furthermore, transections of the neuraxis at different levels may profoundly alter the responsiveness of different systems of spinal interneurones, as can activation of descending pathways from supraspinal levels. Thus it has been shown that, although Ib impulses from flexor muscles have little or no effect on motoneurones in anaesthetised preparations, during stimulation of the red nucleus patterns of Ib reflex action become apparent which more nearly match Laporte and Lloyd's findings (HONGO et al., 1969). Further experiments with better selectivity and control of tendon organ input in different types of preparation should help to clarify the position.

2. *Significance of Ib Segmental Reflex Actions*

Since the early experiments of SHERRINGTON (1909), a protective role has been postulated for the autogenetic inhibitory "clasp-knife" phenomenon, that is, the lengthening reaction, brought about by the development of relatively high tension in a passively stretched, contracting muscle. Tendon organs, while proved to be the executant receptors, have relatively high mechanical thresholds under these conditions, and the di- or tri-synaptic pathway to homonymous motoneurones is rapid and relatively powerful. However, more recent work on tendon organ thresholds to fractionated active muscle contraction, as would normally be the case in most natural movements, shows that many tendon organs are surprisingly sensitive to quite small tension developed by one or a few motor units sited favourably to pull on their particular tendon fascicles, and that such partial contractions may unload tendon organs in other parts of the muscle (HOUK and HENNEMAN, 1967). It is therefore very likely that tendon organ firing plays a much more subtle role than a simple protective brake, and must be involved in finely graded, delicate autogenetic inhibitory action in conjunction with the stretch reflex autogenetic excitatory mechanisms.

Functional roles for tendon organ reflex responses other than autogenetic inhibition are more difficult to discern, especially because of the uncertainty about their precise patterns. The strong inhibitory links shown between extensor muscles acting at different hind limb joints in anaesthetised, low spinal preparations, and the generalized excitation of flexors also ascribed to extensor Ib action, have been assigned plausible roles in the execution of stepping, such as inhibition of extensors and activation of flexors by tendon organ discharge from extensors firing at the peak of the extension phase of the step (Eccles and Lundberg, 1959a). The excitatory action of *quadriceps* Ib input on its contralateral fellow (Phillippson's reflex) may also be assigned a role in the mechanisms ensuring alternation of stepping movements in the two hind limbs (Sherrington, 1909, 1910).

However, the possibility cannot yet be dismissed that some actions attributed to Ib input may reflect a mixture of Group II impulses in the executant volley, a circumstances which would, of course, tend to evoke a general flexor reflex pattern as observed. The recognition of supraspinal mechanisms which may profoundly change the accessibility of certain reflex paths by "biassing" one or another set of interneurones, or by action on pre-synaptic terminals, so that a given input may be switched between alternative, though still stereotyped, response patterns greatly complicates the picture (Job, 1953; Eccles and Lundberg, 1959b; Kuno and Perl, 1960; Lundberg, 1967). Thus input from tendon organs, as from other receptors, may contribute to the co-ordination of movement in different ways under differing circumstances, and attempts to define their overall functional role cannot yet be made. For pertinent and imaginative discussion of tendon organ function, with particular reference to locomotion, the reader is referred to the papers by Lundberg (1969b) and by Stuart et al. (1972).

C. Impulses in Fibres of Groups II, III, and IV

1. Group II

It has been generally accepted for many years that the predominant reflex action of impulses in Group II muscle afferent fibres, like that of input in cutaneous fibres of similar size, is one of excitation of flexors with inhibition of extensors. This is in accord with the early observation of Sherrington (1910) that the receptive field for the flexor withdrawal response embraces deep as well as superficial structures, including muscle and tendon. The flexor pattern was first related to size-range of the muscle afferent fibres responsible by Lloyd (1943b) in his classic work on reflex responses to electrically evoked volleys in the acute spinal animal. Stimulation of muscle nerves at Group II strength evokes weak polysynaptic excitation of flexor and concomitant inhibition of extensor motoneurones (Lloyd, 1943b, 1946b, 1952), observations since confirmed by many others. As in the flexor response set up by cutaneous afferent input, the reflex pathway includes a minimum of one interneurone. However, the excitation of flexor motoneurones by Group II muscle afferent volleys is much weaker, and usually manifest only as facilitation of monosynaptic test reflexes or the accompanying synaptic depolarization recorded by intracellular microelectrodes. Group II inhibition of extensor motoneurones, also mediated polysynaptically, can be demonstrated by the same methods (Lloyd, 1946b; Eccles, 1962; R.M. Eccles and LUNDBERG, 1959a). In preparations under barbiturate anaesthesia, similar but still weaker actions are exerted by Group II input (BROCK et al., 1951). The same flexor pattern of response appears regardless of the flexor or extensor origin of the Group II volley.

However, the notion of all Group II reflex action being unvaryingly and inexorably flexor in pattern can no longer be sustained. Instances have been reported in which the usual action is replaced by flexor inhibition and even excitation of extensors (e.g., R.M. ECCLES and LUNDBERG, 1959a). WILSON and KATO (1965) have also shown an appreciable component of extensor excitation attributable to Group II impulses in nerves of *triceps surae* and flexor hamstrings; JACK and MCINTYRE (1957) have likewise observed mixed excitation and inhibition of *triceps surae* motoneurones from stimulating flexor hamstring nerves at Group II strength. Such observations are most simply explained by Group II functional heterogeneity, with the different components connecting with separate polysynaptic pathways of differing action. Alternatively, dichotomy has been postulated of an essentially homogeneous input into different, reciprocally connected interneuronal circuits of opposite action, with the bias of one or the other determined by supraspinal mechanisms (R.M. ECCLES and LUNDBERG, 1959a; HOLMQVIST and LUNDBERG, 1961). These notions are by no means mutually exclusive.

The extensive receptive field for the flexor response, the dominant spinal reflex pattern under many circumstances, led HOLMQVIST and LUNDBERG (1961) to introduce the designation "flexion reflex afferents" (FRA) to embrace fibres from all structures, skin, joint and other deep tissues, as well as those of Groups II and III of muscle, with flexor reflex action. However descriptively useful this verbal device may be, it carries an implication of homogeneity which may be misleading, and creates difficulties such as those now evident in the problem of muscle Group II central reflex actions. As discussed elsewhere, there is now direct evidence to show that fibres in the Group II range, though largely of spindle secondary ending origin, may include a number from other receptors, including some outside muscle, although the degree of this admixture probably varies greatly between different muscle nerves.

Many spindle secondary endings must ordinarily be engaged in providing information during a variety of motor responses. If their primary reflex action in the intact animal is indeed to provoke generalised limb flexion, no sensible role for these sensitive length-monitors can readily be discerned. However, in favour of a flexor action it has been reported that in spinal preparations the central actions of spindle discharge, selectively activated by fusimotor stimulation, includes facilitation of flexor motoneurones not attributable to Group Ia action (HUNT, 1953); and LAPORTE and BESSOU (1959) observed flexor-type responses to muscle-stretch appropriate for engagement of secondary endings during temporary blockade of Group I conduction in the muscle nerve. It is possible, however, that in these experiments some of the smaller Group I fibres had escaped blockade.

On balance, it seems most likely that the long-standing concept of the generalized flexor reflex response as a protective device, set in train by activation of nociceptors in virtually any tissue, may well prove to be correct for some fraction of the fibres lumped together as "flexion reflex afferents". Even a small number of fibres from such receptors in the Group II range (e.g., Paintal, 1960) could account for the flexor response set up by electrically evoked volleys in muscle nerves.

A factor which must also be taken into consideration is the existence of supraspinal control mechanisms which may have profound actions on interneuronally-mediated responses. For example, flexor reflex segmental mechanisms are strongly inhibited by supraspinal action in the decerebrate state, and released by spinal transection (Job, 1953; R.M. Eccles and Lundberg, 1959b; Kuno and Perl, 1960; Holmqvist and Lundberg, 1961), whereas cortical stimulation usually facilitates them by way of the pyramidal tract (Lundberg and Voorhoeve, 1962), but may lead to reversal of action by Group II input — that is, inhibition rather than excitation of flexors. This is also seen after low pontine section of the brainstem (Holmqvist and Lundberg, 1961), and other complex effects from different supraspinal regions on segmental reflex actions continue to be disclosed (e.g., Lundberg, 1967, 1969a; Hongo et al., 1969, 1972).

Matthews (1969) has recently presented evidence suggesting a different and more comprehensible role for impulses from spindle secondary endings, namely, autogenetic excitation via polysynaptic paths, in addition to that attributable to spindle primary impulses. From experiments in which reflex tension due to stretch and stretch plus vibration at different frequencies were measured, and on the assumption that all spindle primaries were driven by the vibration with little or no engagement of other receptors, Matthews calculated that during the tonic stretch-reflex more tension is developed than would be predicted from Ia action alone. The receptors most likely to account for this extra reflex tension are the spindle secondary endings. The hypothesis has been criticised by Grillner (1970) on the grounds that the lack of occlusion between the response to tonic stretch and to vibration might be explained by the length-tension relationship, i.e., the tension to be expected for a given amount of reflex motoneurone discharge would be greater at successively increasing muscle lengths. However, support for the hypothesis is afforded by recent work of Westbury (1972), who observed summation of the intracellularly recorded depolarization produced in motoneurones by stretch and by vibration of the appropriate muscle; and McGrath and Matthews (1970) have demonstrated early reduction of the response to combined stretch and vibration by procaine blockade, which blocks fusimotor fibres, hence reducing spindle receptor sensitivity, but should not interfere with vibration-induced spindle Ia discharge. On the other hand, Cangiano and Lutzemberger (1972) using constant current Group I block claim that spindle Group II impulses hyperpolarize *triceps surae* motoneurones in the decerebrate animal. A convincing defence of his hypothesis is made by Matthews (1972), but further work with identified spindle Group II input is needed before the question can be finally settled. Matthews' proposal that spindle Group II impulses participate in the tonic stretch reflex seems functionally more meaningful than a stereotyped flexor pattern of response, but given its firm establishment, it would still remain to be seen whether the usually discrepant actions of electrically evoked volleys can be explained by the presence of Group II afferents of true flexor reflex function, by the operation of supraspinal "switching" mechanisms which can change the central actions of spindle Group II impulses, or by both of these factors. It is also conceivable that spindle secondary input might increase tonic stretch reflex tension by suppressing Ib autogenetic inhibition rather than by additional interneuronally mediated excitation (Matthews, 1970).

The question of feedback from muscle stretch-receptors to homonymous or other motoneurones, either "alpha" or "gamma", is one of great interest because of the implications for the complex of control systems associated with muscle length and tension monitors. As already mentioned, Group Ia impulses may inhibit fusimotor neurones by unknown paths (Brown et al., 1968; Proske and Lewis, 1972); the short latency inhibitory synaptic potentials recorded by Grillner (1969) in a few fusimotor neurones in response to Group I volleys may be regarded as further evidence for this, but could have been of Ib origin. Stimulation of muscle nerves at Group II strength excited fusimotor neurones of flexors and either excited or depressed fusimotor neurones of the extensor gastrocnemius (Grillner, 1969), a pattern approximately in parallel with that seen in alpha motoneurones, and in general agreeing with the earlier results of Hunt (1951) on reflex fusimotor discharge. Whether the Group II fibres responsible are of spindle or other origin is quite uncertain.

Actions on contralateral segmental reflexes of electrically evoked Group II volleys have been observed (Perl, 1958; Holmqvist, 1961), but the effects vary widely with the nerves chosen and with state of the preparation, and no clear-cut pattern can yet be discerned.

2. Groups III and IV

Engagement of Group III fibres by electrical stimulation requires a shock-strength exceeding 8 to 10 times Group I threshold, and one of 30 times threshold may be necessary to recruit the least excitable members of the group (R.M. Eccles and Lundberg, 1959a). There is probably considerable overlap of threshold between the smaller Group II and larger Group III fibres, the distinction between which is blurred and arbitrary as previously noted; however, stimuli of strength below about 6 times threshold would probably not engage fibres of Group III. There is no satisfactory method of evoking by natural stimulation selective firing of Group III fibres, which in any case serve more than one kind of ending; however, Paintal (1961, 1962) has demonstrated flexor excitation as a result of natural stimulation (squeezing) known to elicit discharge from "pressure-pain" receptors the fibres of which belong predominantly to Group III.

In the acute spinal preparation, stimulation of any muscle nerve at Group III strength produces a flexor reflex response like that evoked by Group II input, but more powerful and of longer latency; reflex firing as well as facilitation of flexor motoneurones is usually elicited by a single Group III volley, and may continue as substantial after-discharge (Lloyd, 1943b). Ample confirmation of the general flexor action of Group III input, with accompanying inhibition of extensors, has been provided by many groups of investigators, in anaesthetised preparations with intact neuraxis as well as in the acute spinal state, and the interneurone system responsible appears to be the same for Group II and Group III action (e.g., R.M. Eccles and Lundberg, 1959a; Holmqvist and Lundberg, 1961). It is reasonable to regard the response as part of the general withdrawal reflex to noxious stimulation of any part of a limb, superficial or deep (Sherrington, 1910; Paintal, 1961). A closely similar response-pattern is elicited by activation of high threshold afferents in other deep nerves such as those of joints and interosseous membranes (R.M. Eccles and Lundberg, 1959a; McIntyre, 1961), and in cutaneous nerve fibres of comparable threshold. The designation of 'flexion reflex afferents' for these widely distributed small myelinated fibres may have more justification than its original broader coverage. But Group III fibres are not functionally homogeneous, even in muscle nerves, and may include many unrelated to the signalling of nociceptive events, so that caution in the use of the FRA abbreviation is necessary even in this limited context.

Supraspinal control of Group III-evoked segmental reflex action is also extensively documented. Just as for Group II action, the flexor excitation and extensor inhibition seen in spinal or anaesthetised preparations are largely suppressed in the decerebrate state. Conversely, excitation of the corticospinal and rubrospinal systems facilitates interneurones mediating segmental flexor responses, on which converge impulses in fibres both of Group II and Group III (R.M. Eccles and Lundberg, 1959b; Kuno and Perl, 1960; Lundberg, 1969a; Hongo et al., 1972).

Contralateral reflex actions of impulses in Group III fibres have been demonstrated by Perl (1957, 1958) and by Holmqvist (1961). Perl's experiments showed that the fibres responsible for the classical crossed extensor reflex (Sherrington, 1910) are amongst the myelinated fibres of smallest size-range, both from skin and muscle. In acute and chronic spinal preparations, especially in the latter, stimulation of muscle nerves at Group III strength usually yields prolonged facilitation of contralateral knee and ankle extensors and inhibition of flexors. However, crossed flexor responses can occur, but less commonly than when input is restricted to fibres of Group II. With the spinal cord intact, supraspinal influences can substantially affect the responses (Holmqvist, 1961).

Very little is known about segmental reflex action of impulses in unmyelinated (Group IV) fibres of muscle. On the basis of evidence largely provided by experiments on cutaneous afferent input, it seems likely that Group IV impulses from muscle, many of which serve nociceptive endings, would provoke a generalised flexor pattern of response, like those in many muscle Group III fibres. However, direct evidence is lacking, especially concerning the actions of muscle Group IV alone, for example, in the presence of a differential block of all myelinated afferents of the nerve under study. Stimulation of cutaneous nerves at intensities engaging unmyelinated fibres evokes in the spinal animal a strong additional flexor response of which the Group IV-dependent component is characterized by long latency and prolonged after-discharge (Laporte and Boer, 1955). Laporte and Bessou (1958) have also reported that Group IV volleys alone, set up below a region of total A fibre block in a cutaneous nerve, also elicit a flexor reflex pattern, hence do not require central priming by preceding input in myelinated afferent fibres for their action to be revealed. However, facilitation of extensor motoneurones was seen in some of these experiments and actual contraction of extensor muscles could occur in the decerebrate state. Such apparent reversal of flexor action is doubtless a manifestation of the difference in general flexor responsiveness between decerebrate and spinal preparations.

D. Control of Presynaptic Terminals

A phenomenon to which passing reference has been made must now be considered further, namely, the mechanism or mechanisms which can alter the excitatory or inhibitory effectiveness of impulses in presynaptic terminals, thus controlling on the pre-junctional side the extent of postsynaptic action exerted by them. The process being depressive is designated 'presynaptic inhibition'; however, when tonically active, enhancement or reduction of this background leads respectively to inhibition or facilitation of postsynaptic action by the controlled terminals. The cellular basis of this presynaptic action has been considered in a number of reviews (e.g., ECCLES, 1964; LUNDBERG, 1966; SCHMIDT, 1971) and will not be pursued here. It is generally believed that the action is exerted by axo-axonic endings made by interneurones on or near the terminals of the controlled presynaptic axons, and that depolarization of the latter reduces their transmitter output, as shown directly to occur at an invertebrate synapse (TAKEUCHI and TAKEUCHI, 1962). However, there are still differences of opinion as to the precise mechanism causing the reduction of transmitting effectiveness, and even whether all instances of inhibition unaccompanied by measurable postsynaptic inhibitory potentials or conductance change are presynaptic in origin (e.g., KELLERTH, 1968; GRANIT, 1970; CURTIS et al., 1971).

Whatever the mechanism, the existence of the phenomenon must be taken into account when examining the central actions of afferent impulses from any source, including muscle. Inhibitory action of presynaptic type can be exerted on a given set of primary afferent fibres by input in others, and by descending action from a variety of suprasegmental regions, including impulses in the pyramidal tract. Because of the relative simplicity of studying primary afferent fibres, and the convenience of directly recording their depolarization in dorsal root filaments and of measuring the associated excitability changes in the depolarised terminals, nearly all the evidence relates to presynaptic action on the central projections of first order afferent neurones. However, there is reason to believe that similar modulation of interneuronal axon terminals may also occur within the central nervous system.

Most primary afferent fibres appear to be involved in these presynaptic interactions, both as donors and receivers, and afferent fibres of muscle are certainly amongst these (ECCLES et al., 1961a). It appears from a large number of studies that the central terminations of most myelinated muscle afferent fibres may be depolarised by impulses in other primary afferent fibres, of muscle, skin or joint origin, and certain patterns of such interactions have been described (ECCLES, 1961, 1962, 1963, 1964; SCHMIDT, 1971). Most of the experimental evidence relates to action on and between central terminals of the large Group I fibres, but it is generally believed that those of Groups II and III are also involved, both as donors and recipients. However, the most interesting of these actions is that exerted on Ia terminals, because of their unique monosynaptic connections with motoneurones; some modulation of the effectiveness of this direct link can thereby be brought about by impulses in other input lines. Prominent amongst the donor inputs seem to be Group I fibres of both kinds, but particularly Ib fibres from flexor muscles; Ia fibres from both flexors and extensors are subject to this control

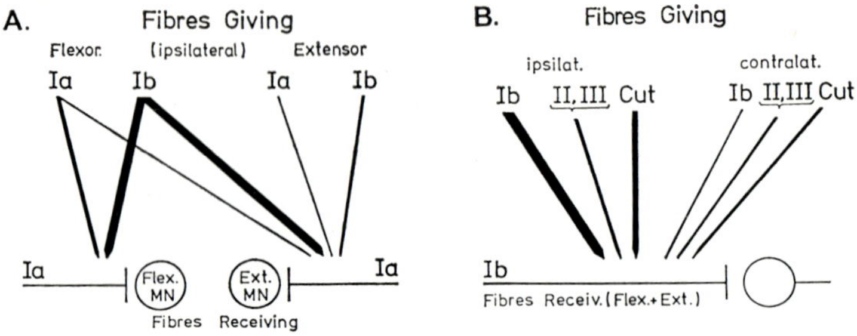

Fig. 7. Diagrammatic representation of patterns of presynaptic connections between different primary afferent fibres. A Group I presynaptic input to Ia monosynaptic collaterals supplying flexor and extensor motoneurones, all ipsilateral. B primary afferents of both ipsilateral and contralateral origin with presynaptic actions on Ib collaterals. Thickness of lines joining donor and recipient fibres indicates potency of inhibitory action. Reproduced with permission from Figs. 24 and 26 in Schmidt, R.F., Ergebn. Physiol. **63**, 20–101 (1971)

(Fig. 7A). In most of the experiments on which these conclusions are based, electrical stimulation of muscle or other nerves, usually repetitive, has been used, and the resulting input categorized as Ia or Ib on the questionable criteria of fast and slow components of the Group I volley. However, adequate stimulation of muscle stretch receptors, probably of all types, can evoke depolarization of primary afferent fibres, including those of Group Ia (Devanandan et al., 1965; Devanandan et al., 1966). Barnes and Pompeiano (1970) have also shown that selective stimulation of spindle primary endings by vibration causes depolarization of other Ia central terminals, as shown by their increased excitability, and reduction of their monosynaptic effectiveness. Despite these observations, differences of opinion persist about the extent to which such presynaptic effects operate under normal circumstances and about their precise mechanism (e.g., Kellerth, 1968; Granit, 1970; Curtis et al., 1971).

Ib central terminals of both flexor and extensor muscles are also subject to depolarization by impulses in a wide range of other primary afferents, including Ib, Groups II and III fibres of muscle and cutaneous nerves; similar fibres of contralateral origin may also be effective. The full significance of these complex interactions remains to be determined, and they will not be discussed further. However, a summary of the patterns so far described is presented in the diagrams of Fig. 7, taken from the review by Schmidt (1971).

E. Long Spinal Reflexes

Afferent impulses in muscle nerves also participate in "long spinal" reflex interactions between forelimbs and hind limbs, as first shown for the descending actions exerted on motoneurones of both hind limbs by input from the forelimb in the acute spinal cat. Volleys in forelimb muscle nerves provoke facilitation of a number of hind limb motoneurone pools, especially those of ipsilateral extensors, and when elicited in nerves supplying distal forelimb regions, also evoke the unique short latency inhibition of ipsilateral *flexor digitorum longus* and *flexor hallucis longus* (FDHL) motoneurones. However, Group I fibres are not involved, the long spinal actions appearing only when the forelimb volley engages fibres of Groups II and III (Lloyd and McIntyre, 1948). Fibres in the Group II band appear to be solely responsible for inhibiting FDHL motoneurones, and the receptive field for their action lies distally in the forelimb. It is not known whether these fibres are from spindles or other deep receptors. Volleys in distal cutaneous afferent fibres in the Group II range are much more effective in eliciting long spinal inhibition, and this can also be evoked by natural tactile stimulation of the forepaw (Aoki and McIntyre, 1972). Group III fibres are largely responsible for the more general facilitation of hind limb motoneurones which can be evoked by volleys in nerves supplying proximal or distal muscles, but again, cutaneous nerve volleys engaging the smaller Group II and Group III fibres have a more powerful action. Earlier results on ascending long spinal reflexes (McDonald and McIntyre, 1956) suggested that hind limb fibres of Groups II and III are similarly responsible for actions on forelimb motoneurones, and not those of Group I; this has recently been confirmed by Miller et al. (personal communication from S. Miller). The significance of these "hand-foot" interactions in which some afferent fibres of muscle participate seem related to the maintenance of coordination between the four quadrupedal limbs during standing, walking and the achievement of appropriate postural adjustments during operations involving a single limb, such as its withdrawal during a flexor response.

V. Supraspinal Projections

The escalation in recent years of experimental work on ascending pathways, synaptic mechanisms and actions at supraspinal levels of impulses in afferent fibres of muscle nerves is such that no more than a brief outline of their salient features as currently understood can be attempted in this context. Most of this wealth of publication concerns the identification of spinal tracts through which information travels from muscle and other receptors to the cerebellum and brainstem, and to the cerebral cortex; the properties of their synaptic interruptions; and the projection patterns and organization of the cerebellar and cerebral receiving regions. In this account, attention will be focussed on the principal projections and pathways themselves, and only brief reference can be made to the special properties and transmitting characteristics of relevant afferent synapses; nor can the now extensive information about intrinsic organization of cerebellar and cerebral cortical mechanisms be considered. However, much of this field is covered by reviews and articles such as those of Lundberg (1964); Oscarsson (1965, 1967); Eccles (1966, 1969); Eccles et al. (1967); Brooks (1969); Jansen and Walløe (1970); Jansen et al. (1966); Matthews (1972); Rymer (1972).

A. Pathways to Cerebellum

1. Direct Paths

The existence of two important direct spinocerebellar pathways for input from receptors in the hind limbs and lower part of the body has long been recognized, namely, the dorsal and ventral spinocerebellar tracts (DSCT and VSCT); corresponding paths from the forelimbs and upper trunk have been disclosed more recently. A great deal is now known about the origin, course and distribution of the fibres comprising these tracts, the kinds of receptors providing input to the tract neurones, and the properties of their synaptic linkages. Prominent amongst these receptors are those of muscle, although other superficial and deep receptors also feed information into some of the tract neurones.

a) Dorsal Spinocerebellar Tract. This pathway, long known to take origin in cells of Clarke's column, located between spinal segments T 2–3 and L 3–4, is an ipsilateral tract containing many large fibres of rapid conduction velocity travelling superficially in the dorsolateral quadrant of the spinal cord. Reaching the ipsilateral cerebellar anterior lobe and paramedian lobule via the inferior peduncle, it terminates as mossy fibres in discrete cortical areas. It has been subjected to more detailed analysis than any other spinocerebellar input-channel, and the fact that group I fibres make powerful excitatory connections with cells of Clarke's column was demonstrated in the early days of electrophysiological analysis (Grundfest and Campbell, 1942; Lloyd and McIntyre, 1950). These findings were confirmed and greatly extended by further electrophysiological studies, particularly those of Lundberg and his colleagues in Sweden, and much of this work is summarized in reviews by Lundberg (1964) and Oscarsson (1965). These and other studies (e.g., McIntyre and Mark, 1960; Eccles et al., 1961 c; Jansen and Rudjord, 1965) have shown that excitatory synaptic connections of great security are made with many tract neurones by Group Ia fibres from spindle primary endings in muscles of the hind limb. Convergence onto single Clarke's cells is mainly restricted to Ia fibres from a single muscle or synergic group, so that tract units of this type can signal to the cerebellum information about spindle responses in a restricted receptive field consisting of single muscles or small muscle groups of like function. Group II fibres of spindle secondary ending origin also make excitatory connections, probably monosynaptic, with tract cells, some of which also receive Group Ia inputs (Laporte et al., 1956; McIntyre and Mark, 1960; Eccles et al., 1961 c). Such tract units are therefore able to signal powerfully to the cerebellum the occurrence of brief, rapid length-changes in a muscle or synergic muscle group by virtue of the Ia receptors' dynamic sensitivity, together with information about maintained muscle length through the more static input from secondary endings.

Other less commonly encountered DSCT units respond to input in Ib fibres, again usually from a single muscle or muscle group, so that this component of the pathway can signal information about tension developing in different portions of the limb musculature. It was claimed some years ago that excitatory convergence of Ia and Ib fibres on the same DSCT cell is relatively frequent (Eccles et al., 1961 c; Lundberg, 1964). However, the criterion of "Ia–Ib" deflections in electrically evoked Group I volleys was the basis for this conclusion, and it is now clear from experiments with adequate stimulation that such bizarre convergence rarely, if ever, occurs (Lundberg and Winsbury, 1960 b; McIntyre and Mark, 1960). The apparent Ib input to such cells most likely consists of low threshold Group II fibres. To quote Oscarsson (1965): "It can be concluded that the proprioceptive subdivision of the DSCT contains two channels of information, one for muscle spindles and one for tendon organs, and that the spatial discrimination is good." To this may be added a further input from joint receptors (LINDSTRÖM and TAKATA, 1972).

A further, exteroceptive division of the DSCT, itself divided into sub-groups, has been described, receiving input mainly from cutaneous afferents (LUNDBERG, 1964; OSCARSSON, 1965), and will not be discussed here.

In addition to the synaptic linkages with primary afferent neurones, excitatory or inhibitory input from other sources also impinges on DSCT units. Tonic background discharge is usual, partly sustained by interneuronal bombardment since firing continues after interruption of afferent input by dorsal root section (HOLMQVIST et al., 1956; KOSTYUK, 1969). Inhibitory influences also impinge on DSCT neurones, for example, as a result of stretching muscles other than those supplying excitatory stretch receptor input to the cell (especially antagonists), or by electrical stimulation of their nerves. Both pre- and postsynaptic inhibitory mechanisms appear to be involved (ECCLES et al., 1961c; OSCARSSON, 1965).

Potency of the excitatory linkages made by Group I collaterals with DSCT cells is attested not only by the minimal need for summation and near-absence of subliminal fringe at these synapses, but also by their ability to transmit faithfully remarkably high frequencies of presynaptic repetitive firing, in some units at rates exceeding 500/sec (McINTYRE and MARK, 1960; LUNDBERG, 1964; OSCARSSON, 1965). Such high transmissive efficacy is common at synaptic interruptions of major afferent pathways, and contrasts strikingly with the much lower potency of collaterals of the same Group Ia fibres on motoneurones. The giant presynaptic boutons on Clarke's cells first demonstrated by SZENTÁGOTHAI and ALBERT (1955) are doubtless the morphological entities providing this strong synaptic drive; their ultrastructure, together with other details of the complex synaptic input to Clarke's cells, is indicated in Fig. 8, modified from RÉTHELYI (1970). That individual presynaptic Ia impulses can exert unusually large excitatory action on Clarke's cells has been strikingly demonstrated by EIDE et al. (1969). Many unitary excitatory synaptic potentials (EPSP's) in these cells during weak muscle stretch greatly exceed in amplitude those of the corresponding Ia unitary EPSP's of motoneurones, ranging from about one to several mV in the former as compared with 0.02–0.6 (average 0.1 mV) for the latter (MENDELL and HENNEMAN, 1971).

b) Ventral Spinocerebellar Tract. Although the ventral spinocerebellar tract has also been subjected to much experimentation, the picture of its organization is less clear than that of the DSCT. There is even some uncertainty about the location of its cells of origin, although it is agreed that they extend more caudally than those of Clarke's column. From electrophysiological experiments, HUBBARD and OSCARSSON (1962) described the location of these cells as being in the lateral portions of the base and neck of the dorsal horn and intermediate zone, but recently the spinal "border cells" of the ventral horn have been identified as giving origin to a large proportion of VSCT axons, in agreement with the histological study of COOPER and SHERRINGTON (1940); the majority of these border cells, located by microelectrode, were invaded antidromically from cerebellar cortical stimulation (BURKE et al., 1971).

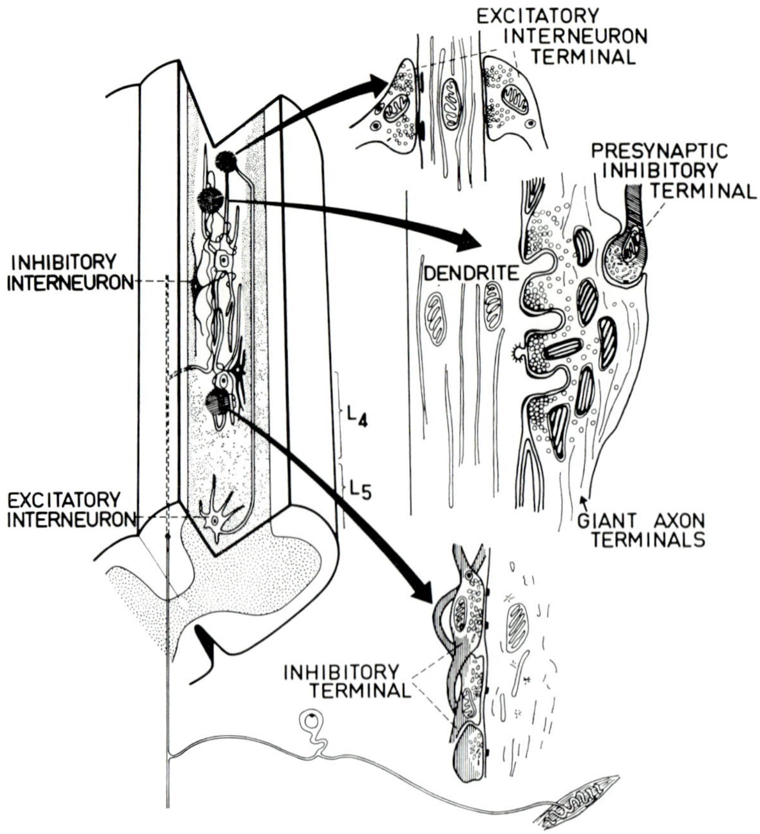

Fig. 8. Diagram showing on the left, the general arrangement of Clarke's nucleus with Ia ascending collaterals forming giant presynaptic terminals on dendrites of Clarke's cells, and the possible arrangement of inhibitory and excitatory interneurones in the nucleus; on the right, some ultrastructural details of the synaptic articulations. Redrawn from Fig. 20 in Réthelyi, M., Exp. Brain Res. **11**, 159–174 (1970), and reproduced with permission

The VSCT axons ascend in the contralateral lateral and ventral columns, enter the cerebellum through the superior peduncle and by a second decussation many terminate in the anterior lobe on the same side as the cells of origin; some, however, end on the opposite side (Oscarsson, 1965; Lundberg, 1964). The projection fields of single VSCT units on the cerebellar cortex are larger than those of DSCT fibres, as judged by the areas from which antidromic firing can be evoked. Many VSCT axons are of rapid conduction velocity, at least as fast as those of the DSCT. Until recently, the major primary afferent input of muscle origin to VSCT units has been considered to be from tendon organs, with widespread convergence from a number of muscles, either belonging to a synergic muscle group, or extending over several myotatic units, in contrast to the more discrete arrangement for DSCT units (Oscarsson, 1957, 1965). Inhibitory as well as excitatory actions from low threshold muscle receptors have also been observed, but not in any simple reciprocal pattern. Many VSCT cells, including those receiving converging excitatory Ib input, are also affected by impulses in high threshold afferents (Groups II and III) of muscle, joint and cutaneous origin (OSCARSSON, 1957, 1965; ECCLES et al., 1961b). The action of this high threshold input is predominantly inhibitory on some VSCT units, and excitatory in others; irregular background firing is seen only in those with predominantly excitatory input. The excitatory linkage of Group I fibres with many VSCT cells contrasts with the DSCT synapses in that their individual actions seem to be relatively weak and brief, so that considerable convergent input is necessary for impulse-generation. A further feature is that temporal facilitation of postsynaptic excitation occurs during repetitive presynaptic stimulation, as for the corticospinal input to primate motoneurones; thus the VSCT synapses can transmit rapid repetitive input, despite the relative ineffectiveness of single volleys (OSCARSSON, 1957; ECCLES et al., 1961b).

Until recently, Group Ia fibres have been considered to play little or no part in the input to VSCT neurones, so that the tract's proprioceptive role seemed largely restricted to the signalling of tension changes in different hind limb muscle groups involved in particular postures or movements. However, recent work provides evidence for extensive participation of spindle primary impulses, at least in the input to the spinal border cell component (LUNDBERG and WEIGHT, 1971). Both excitatory and inhibitory actions are described. Although the Ia excitatory actions are relatively weak, it is puzzling that earlier work on tract discharge failed to find even subliminal Ia excitation. It is possible that once again difficulty has been caused by continued reliance on electrical threshold differences to distinguish Ia and Ib actions. Be that as it may, it now seems necessary to include a significant contribution from muscle spindle afferents to the VSCT input, so that the earlier view of this tract's proprioceptive role being mainly the signalling of muscle tension is no longer adequate. Its functional complexity is compounded by the existence of extensive descending supraspinal control mechanisms (OSCARSSON, 1965; LUNDBERG and WEIGHT, 1971).

The heterogeneous nature of the converging input of VSCT cells makes their responses to natural stimulation of muscle more complex than those of DSCT units. Strong stretch or active contraction of muscles has predominantly inhibitory action on many VSCT neurones, despite their ready excitation by electrically evoked impulses in presumed Ib fibres of a number of muscle nerves. Such inhibition can be evoked by natural stimulation of many muscles in the limb. However, some VSCT units with predominantly excitatory input can be excited by natural stimulation of the kind which elicits tendon organ discharge. Observations such as these make it difficult to discern the significance in the intact body of VSCT-mediated information. Because of the existence of strong excitatory links between some descending tract systems (e.g., the vestibulo-spinal) and VSCT neurones (LUNDBERG and WEIGHT, 1971), it may turn out that one mode of VSCT signalling is "negative", in the sense that a particular input pattern might reduce or modulate background firing of VSCT neurones maintained by descending drive. However, further progress in understanding the normal role of pathways such as the VSCT must await the refinement of methods for recording unit activity in the intact, unanaesthetised animal.

The forelimb equivalents of the DSCT and VSCT are, respectively, the cuneocerebellar and rostrocerebellar tracts; both project ipsilaterally. Cell bodies of the cuneocerebellar tract fibres are in the external cuneate nucleus, and reach the cerebellum through the inferior peduncle; location of the rostrocerebellar tract cells is not precisely known, but their axons enter the cerebellum through both the inferior and superior peduncles. In general, the properties of these pathways resemble those of their hind limb equivalents (Oscarsson, 1965, 1967).

2. Indirect Paths

In recent years, the number of reports describing additional indirect spino-cerebellar routes has escalated sharply, but so far it does not seem possible to assign specific functions to all these, especially as little is yet known of the receptors which may transmit information through them. The use of Group III strength stimulation of peripheral nerves in many of these studies adds to the difficulties of interpretation, and so far the contribution of identified muscle receptors is unclear. However, brief mention must be made of spino-olivary pathways, because the unique cerebellar climbing fibre input is of olivary origin, and there is evidence that signals from muscle stretch receptors may contribute to this (Eccles, 1966). The ventral spino-olivary tract lies in the contralateral ventral funiculus, and corresponds to the classical spino-olivary path (Brodal et al., 1950). By way of this pathway, responses in the dorsal accessory olive to brief repetitive stimulation of cutaneous, joint and muscle nerves of the hind limb have been observed, and in some instances, near-threshold Group I volleys in muscle nerves have been effective; in others, inclusion of group II impulses in the muscle afferent volley was necessary to yield olivary responses (Armstrong et al., 1967; Oscarsson, 1968). However, the nature of the receptors served by the lower threshold afferent fibres is not established, and they could even be extra-muscular in location. The cerebellar responses are ipsilateral to the stimulus, because of re-crossing of the olivocerebellar fibres. Impulses in the smaller afferent fibres, elicited by stimulating muscle or other nerves at Group III strength, are most effective for yielding climbing fibre responses through this pathway, and appear to be solely responsible for similar responses through various other spino-olivo-cerebellar routes and the spino-reticulocerebellar path, and are regarded as flexion reflex afferents (Oscarsson, 1967; Miller and Oscarsson, 1970; Grant et al., 1966). The functional significance of these ascending systems may relate to the supraspinal signalling of the "state of play" in segmental interneuronally mediated reflex patterns, as postulated for the "bilateral ventral flexor reflex tract" (Lundberg and Oscarsson, 1962; Lundberg, 1964); indeed, identity of the latter tract with the spinal portion of the ventral spino-olivocerebellar pathway has been proposed (Grant and Oscarsson, 1965).

B. Thalamo-Cortical Projection

As mentioned in the introduction, attention has swung back in recent years to the question of input from muscle receptors to the cerebral cortex, a matter of obvious importance to the problem of kinaesthesis and the cortical control of movement. Earlier experiments in barbiturate-anaesthetised preparations had revealed cortical evoked potentials only if the stimulus to a muscle nerve engaged

Group III or Group II afferent fibres; Group I volleys seemed to be quite ineffective (MOUNTCASTLE et al., 1952; MCINTYRE, 1953, 1962). Because afferent volleys in skin and joint nerves readily elicited cortical potentials in the same preparations, it was concluded that there was no comparable simple route for impulses in fibres of spindle primary endings and tendon organs to reach the cerebral cortex. Most of these experiments were concerned with muscle afferents of the hind limb, but AMASSIAN and BERLIN (1958) observed cortical responses to volleys in low threshold muscle afferent fibres of the cat's forelimb. Using input-output technique, OSCARSSON and ROSÉN (1963) confirmed this finding, and further demonstrated that spindle (Group Ia) fibres are involved, though participation of tendon organs could not be excluded. The pathway for these responses was shown to be in the dorsal column, cuneate nucleus, medial lemniscus, thalamus and a restricted area of SI cortex between the cruciate sulcus and post-cruciate dimple (ROSÉN, 1970). This region is believed to be the feline counterpart of Area 3a in primates, to which signals from Group I muscle afferents have been shown to project by a short latency path (PHILLIPS et al., 1971). A number of authors have also described in the primate projections of somatic afferent impulses to the pre-Rolandic cortex through unknown paths not including the post-central receiving areas, and these may include signals from muscle stretch-receptors (e.g., ALBE-FESSARD et al., 1964, 1965; ALBE-FESSARD and LIEBESKIND, 1966).

In the cat, cortical responses to Group I volleys from muscles of the hind limb were not seen by a number of investigators (MOUNTCASTLE et al., 1952; MCINTYRE 1953, 1962; OSCARSSON and ROSÉN, 1963). However, using chloralose anaesthesia and computer averaging, LANDGREN and SILFVENIUS (1969) were able to demonstrate responses to hind limb Group I volleys in two cortical areas, one medial and rostral to the post-cruciate dimple, the other on the medial aspect of the post-sigmoid gyrus, regions close to but not identical with the SI projection loci for cutaneous volleys. Responses were found upon stimulation of various extensor and flexor muscle groups of the contralateral hind limb, appearing with stimuli just above threshold for the nerves; this is in marked contrast with the results of earlier work using barbiturate anaesthesia, and it seems likely that the synaptic interruptions of the hind limb Group I pathway are more susceptible to blockade by anaesthesia than are those for cutaneous hind limb volleys or Group I input from the forelimb. Landgren and Silfvenius furthermore concluded from input-output and other experiments that tendon organs as well as spindle primary impulses from the hind limb project to the cortex, and by largely independent paths. However, they relied on presumed threshold differences between Ia and Ib fibres, and on the debatable criterion of fast and slow deflections in recorded Group I volleys, so that cortical responses attributed by them to recruitment of Ib input with stimuli exceeding 1.5 times threshold could have been evoked by low threshold Group II impulses. Additional response was evoked by further increasing stimulus strength to between 3 and 10 times threshold, which would recruit most of the higher threshold Group II fibres, and still stronger stimuli engaging fibres of Group III elicited further and still more delayed components in the cortical response. The responses were not reduced by transection of the dorsal columns, in agreement with other studies (GARDNER and NOER, 1952; MCINTYRE, 1953, 1962).

Thus the pathway for impulses in Group I and smaller muscle afferent fibres of the hind limb can not lie in the dorsal column, the tract regarded so long by clinicians as being the major route for transmission of signals necessary for conscious awareness of position and movement. The pathway instead appears to lie in the dorsal part of the lateral funiculus on the side of stimulation, that is, in approximately the same region through which run fibres of the dorsal spino-cerebellar tract; however, the cortical path can not include the cerebellum, as the evoked potentials still appear after cerebellectomy (Landgren and Silfvenius, 1969). The same authors have also provided convincing evidence that the spinal fibres of the cortical projection path for hind limb Group I input end synaptically in a brainstem nucleus just rostral to the *nucleus gracilis*, the *nucleus Z* of Brodal and Pompeiano (1957), and that from this, fibres ascend to the thalamus and contralateral cerebral cortex (Landgren and Silfvenius, 1971; Silfvenius, 1972).

Thus, despite earlier negative results, it is now clear that information originating in low threshold muscle stretch receptors of both forelimbs and hind limbs can reach the cerebral cortex by fairly direct routes in both cat and monkey, in addition to signals transmitted in fibres of high threshold and possible nociceptive function. The question immediately arises of the significance to normal function of the cortical input from stretch-receptors. Are these an important component of the input underlying kinaesthesis, that is, the conscious awareness of position and movement? Earlier negative conclusions based on the apparent absence of Group I cortical projection carried the implication that any input eliciting cortical potentials must be expected to evoke some sensation, whatever its receptors of origin. The minimum input (a single afferent impulse) may elicit a cortical response (McIntyre et al., 1967), but it is not known whether this is appreciated consciously; Libet et al. (1964) have shown that duration as well as magnitude of a disturbance in the human cortex must exceed a certain minimum in order to evoke a sensation. Cortical projection of impulses from muscle stretch-receptors thus need not imply their effectiveness in eliciting sensation. Indeed, there is a good deal of evidence to the contrary (Provins, 1958; Merton, 1964; Gelfan and Carter, 1967). Such cortical projection could fulfil an unconscious role in providing feedback from receptors of muscle to the cortical and subcortical systems responsible for programming voluntary movements. The failure of repetitive Group I volleys to produce EEG arousal in unanaesthetised cats (Giaquinto et al., 1963) might be regarded as further evidence suggesting that muscle stretch-receptor input alone is without effect on the cerebral processes underlying conscious awareness. In these experiments, increasing the stimulus to include fibres of Group II could lead to arousal, but the impulses responsible may not have been of spindle origin.

However, acceptance of the notion that impulses from muscle stretch-receptors, especially those in Group I fibres, have in isolation little or no impact on consciousness need not deny them a role in kinaesthesis when combined with other input such as that from joints. Under most physiological conditions, signals from muscle stretch-receptors comprise only part of the total afferent input from the region, which together provides information about meaningful events such as movements. The recent experiments in human subjects by Goodwin et al.

(1972), in which vibratory stimulation of biceps or triceps tendons led to errors in sensing the degree of elbow-joint angulation, suggest that spindle Ia input normally participates in the making of such judgements. The recent paper by GRANIT (1972) gives a thoughtful and pertinent discussion of these matters, and the extent to which use of the old term "muscular sense" may still be appropriate.

VI. Muscle Afferent Central Actions in Sub-Mammalian Vertebrates

Remarkably little is known about central actions of muscle afferent impulses in sub-mammalian vertebrates, despite the wealth of structural and functional studies on the receptors themselves, especially in certain reptiles and anurans. According to Barker (Chapter 1, p. 96), complex receptors such as muscle spindles are not found in aquatic vertebrates, the anuran amphibia being the most primitive forms to be so equipped. Skeletal muscle in birds and at least some marsupials (e.g., the Australian brush-tailed possum *Trichosurus vulpecula*) is well endowed with spindles and tendon organs, and their principal functional characteristics have been examined (DORWARD, 1970; RYMER, 1968); they show general similarity to those of mammals. In *Trichosurus*, monosynaptic firing of motoneurones can be elicited by volleys in low threshold muscle afferent fibres, presumably from muscle spindles, but the linkage appears to be less powerful than in cat or primate (HORE and PORTER, 1971). Ascending tract systems in this marsupial are similar to those of the cat (MAGNI and OSCARSSON, 1962).

In the duck, nothing is known about segmental reflex actions of impulses in muscle afferent fibres. Their central processes have a very limited course in the dorsal column, and presumably terminate on neurones in and adjacent to the segment of entry (DORWARD, 1966). Low threshold afferent input from muscle nerves of the hind limb elicits monosynaptic discharge of a crossed ascending tract, presumably the equivalent of the VSCT, but there appears to be no ipsilateral route corresponding to the DSCT. However, ascending pathways from the forelimb resemble those in mammals (OSCARSSON et al., 1963). Perhaps this reflects a need for greater precision in the supraspinal control of flight muscles.

In frogs and toads, direct connections with motoneurones are made by collaterals of some dorsal root fibres, but their excitatory effectiveness is much weaker than in the mammal (FADIGA and BROOKHART, 1960; HOLEMANS et al., 1966). Overt stretch reflexes are not readily evoked and are commonly elicitable only in the presence of background facilitation (SASSA, 1921; MARX, 1950; CHAMBERS and SIMCOCK, 1960). However, FUKAMI (1961) recorded monosynaptic EPSP's from toad spinal motoneurones in response to stimulation of muscle nerves, demonstrating that at least some motoneurones receive excitatory terminals of fibres from muscle receptors, in all probability spindles. In most instances, excitation was subthreshold for cell-firing. Hyperpolarizing synaptic potentials of longer latency were also seen in some motoneurones.

As described by Hunt (Chapter 2, p. 222), reptilian muscle spindles have been extensively studied. In lizards of the genus *Tiliqua*, the presence of a relatively powerful stretch reflex with phasic and tonic components has been demonstrated in muscles of the hind limb (Kenins et al., 1970). The reflex includes a monosynaptic component, as shown by latency measurements of the initial discharge and of the intracellularly recorded synaptic potentials elicited by muscle stretch (Kenins, 1972). The receptors responsible for this reflex can only be those of muscle spindles, of which there are two distinct types in reptilian muscle (Proske, 1969; Fukami, 1970); it is not known whether both or only one of these participate in the stretch reflex. The central action of spindle afferent discharge in anurans and reptiles is of special interest because the motor innervation of intrafusal fibres is by collaterals from the extrafusal motor supply. This arrangement could provide an economic way of maintaining spindle firing during muscle shortening, and through a stretch reflex, automatically increase motor output when the contracting muscle encounters resistance. As in some instances of "alpha gamma" linkage in mammals, this could be a significant factor in ensuring the satisfactory performance of a centrally programmed motor act, despite the occurrence of obstacles during its execution. The positive feedback element in such a system could be considerable, depending upon the potency and extent of spindle afferent autogenetic excitatory links, and mechanisms must exist which restrict such increased output to the immediate task of overcoming the extra load.

VII. Concluding Comments

The task of presenting a comprehensive account of the impact and significance of muscle afferent input at all levels of the central nervous system is indeed formidable, and there are obvious omissions in the account given in the preceding pages. For example, the afferent equipment of muscles supplied by cranial nerves has received scarcely a mention, although these muscles and their neural circuits present a number of unusual features. Nor has the important question been discussed of muscle afferent participation in cardio-respiratory reflexes, a matter of obvious concern to the complex adjustments in muscular exercise, though understanding of its nature is as yet fragmentary and unclear. These concluding remarks survey, in greatly over-simplified and partly speculative outline, the range of central roles played by muscle afferent input as presently known.

Input signals providing information about length, tension and changes therein of individual muscles and muscle groups impinge at all times at various central nervous levels. In conjunction with input from joints and other receptors, they can be regarded as providing a dynamic display of the moment-to-moment state of the whole bodily skeleto-muscular equipment which determines posture and movement, a display which helps to mould appropriately the precise pattern of motor outflow commands required for maintenance of stance or achievement of motor goals. In addition, signals from unknown receptors in active muscles participate in the general automatic cardio-respiratory adjustments which ensure adequacy of their own blood perfusion. Should excessive stimulation or ischaemia threaten muscles or their attachments, or actual injury occur, signals from nociceptors provoke powerful protective withdrawal responses through spinal mechanisms; also, through supraspinally transmitted action, conscious and emotionally potent awareness of the threat compels behavioural adjustments appropriate for recovery or repair.

Signals from spindle and tendon-organ length and tension monitors at the spinal segmental level have relatively restricted, short latency reflex actions chiefly affecting individual muscles or muscle groups of like function and contributing to their harmonious interplay, such as the reciprocal inhibition of antagonists. At the hind-brain level, the same primary signals also exert actions by way of rapidly conducting ascending paths, some units of which transmit information about events in discrete components of the musculature, others provide a broader picture of the significant events in a whole limb or region. At cerebellar level, such information from the whole body musculature, as well as from other receptors including those of joints, skin, eye and ear, together with those from the labyrinth, can be brought together for computation and comparison with command patterns from the cerebral cortex arriving through the cortico-pontine system. The details of how these operations are carried out and the correct adjustments made cannot yet be discerned, nor can the full significance be appreciated of the forebrain projection of signals from muscle stretch receptors now known to take place, although some involvement in kinaesthesis seems likely. Finally it should be emphasized that to be meaningful most aspects of muscle receptor function need to be considered in a context of motor activity, a point underlined by the existence of two independent fusimotor systems controlling the spindle receptors themselves.

Acknowledgements

Some of the work described was supported by grants from the Medical Research Council of New Zealand and the National Health and Medical Research Council of Australia. Technical and secretarial assistance from Diana Harrison, Pat McHattie and Betty Wilson is also gratefully acknowledged.

References

ALBE-FESSARD, D., LIEBESKIND, J.: Origine des messages somato-sensitifs activant les cellules du cortex moteur chez le singe. Exp. Brain Res. **1**, 127–146 (1966).

ALBE-FESSARD, D., LIEBESKIND, J., LAMARRE, Y.: Projection au niveau du cortex somatomoteur du singe d'afférences provenant des récepteurs musculaires. C.R. Acad. Sci. (Paris) **261**, 3891–3894 (1965).

ALBE-FESSARD, D., LIEBESKIND, J., MALLART, A.: Stimulation naturelles provoquant l'activation des cellules du cortex moteur chez le singe. J. Physiol. (Paris) **56**, 525–526 (1964).

AMASSIAN, V.E., BERLIN, L.: Early cortical projections of Group I afferents in the forelimb muscle nerves of cat. J. Physiol. (Lond.) **143**, 61P (1958).

AOKI, M., MCINTYRE, A.K.: Long spinal and pyramidal actions on lumbosacral motoneurones in cats under chloralose anaesthesia. Proc. Aust. Physiol. Pharmacol. Soc. **3**, 21–22 (1972).

ARAKI, T., ECCLES, J.C., ITO, M.: Correlation of the inhibitory post-synaptic potential of motoneurones with the latency and time course of inhibition of monosynaptic reflexes. J. Physiol. (Lond.) **154**, 354–377 (1960).

ARMSTRONG, D.M., ECCLES, J.C., HARVEY, R.J., MATTHEWS, P.B.C.: Responses in the dorsal accessory olive of the cat to stimulation of hind limb afferents. J. Physiol. (Lond.) **194**, 125–145 (1968).

BARKER, D.: The structure and distribution of muscle receptors. In: Muscle Receptors, pp. 227–240. Hong Kong: Hong Kong University Press 1962.

BARKER, D.: The innervation of mammalian skeletal muscle. In: Ciba Foundation Symposium on Myotatic, Kinesthetic and Vestibular Mechanisms, pp. 3–15. London: Churchill 1967.

Barker, D., Ip, M.C., Adal, M.N.: A correlation between the receptor population of the cat's soleus muscle and the afferent fibre-diameter spectrum of the nerve supplying it. In: Muscle Receptors, pp. 257–261. Hong Kong: Hong Kong University Press 1962.

Barnes, C.D., Pompeiano, O.: Presynaptic and postsynaptic effects in the monosynaptic reflex pathway to extensor motoneurons following vibration of synergic muscles. Arch. ital. Biol. **108**, 259–294 (1970).

Bessou, P., Laporte, Y.: Activation des fibres afférentes amyéliniques d'origine musculaire. C.R. Soc. Biol. (Paris) **152**, 1587–1590 (1958).

Bessou, P., Laporte, Y.: Activation des fibres afférentes myelinisées de petit calibre d'origine musculaire (fibres du Group III). C. R. Soc. Biol. (Paris) **154**, 1093–1096 (1960).

Bessou, P., Laporte, Y.: Some observations on receptors of the soleus muscle innervated by Group III afferent fibres. J. Physiol. (Lond.) **155**, 19P (1961).

Boyd, I.A., Davey, M.R.: Composition of peripheral nerves. Edinburgh-London: Livingstone 1968.

Bradley, K., Eccles, J.C.: Analysis of the fast afferent impulses from thigh muscles. J. Physiol. (Lond.) **122**, 462–473 (1953).

Brock, L.G., Coombs, J.S., Eccles, J.C.: The recording of potentials from motoneurones with an intracellular electrode. J. Physiol. (Lond.) **117**, 431–460 (1952).

Brock, L.G., Eccles, J.C., Rall, W.: Experimental investigations on the afferent fibres in muscle nerves. Proc. roy. Soc. B **138**, 453–475 (1951).

Brodal, A., Pompeiano, O.: The origin of ascending fibres of the medial longitudinal fasciculus from the vestibular nuclei. An experimental study in the cat. Acta morph. neerl.-scand. **1**, 306–328 (1957).

Brodal, A., Walberg, F., Blackstad, T.: Termination of spinal afferents to inferior olive in cat. J. Neurophysiol. **13**, 431–454 (1950).

Brooks, V.B.: Information processing in the motorsensory cortex. In: Information Processing in the Nervous System, Leibovic, K.N., Ed., pp. 231–243. Berlin-Heidelberg-New York: Springer 1969.

Brown, M.C., Engberg, I., Matthews, P.B.C.: The relative sensitivity to vibration of muscle receptors of the cat. J. Physiol. (Lond.) **192**, 773–800 (1967).

Brown, M.C., Lawrence, D.G., Matthews, P.B.C.: Reflex inhibition by Ia afferent input of spontaneously discharging motoneurones in the decerebrate cat. J. Physiol. (Lond.) **198**, 5–7P (1968).

Burke, R., Lundberg, A., Weight, F.: Spinal border cell origin of the ventral spinocerebellar tract. Exp. Brain Res. **12**, 283–294 (1971).

Cajal, S., Ramón y: Histologie du système nerveux de l'homme et des vertébrés, Vol. 1. Paris Maloine 1909. Reprinted in Madrid: Instituto Ramón y Cajal 1952.

Cangiano, A., Lutzemberger, L.: The action of selectively activated group II muscle afferent fibers on extensor motoneurons. Brain Res. **41**, 475–478 (1972).

Chambers, R.M., Simcock, J.P.: Postural reflexes in forelimb of toad *Bufo marinus*. J. Neurophysiol. **23**, 54–61 (1960).

Conradi, S.: On motoneuron synaptology in adult cats. Acta physiol. scand. **78**, Suppl. 332, 85–105 (1970).

Coombs, J.S., Curtis, D.R., Landgren, S.: Spinal cord potentials generated by impulses in muscle and cutaneous afferent fibres. J. Neurophysiol. **19**, 452–467 (1956).

Cooper, S., Sherrington, C.S.: Gower's tract and spinal border cells. Brain **63**, 123–134 (1940).

Coppin, C.M.L., Jack, J.J.B., McIntyre, A.K.: Properties of group I afferent fibres from semitendinosus muscle in the cat. J. Physiol. (Lond.) **203**, 45–46P (1969).

Coppin, C.M.L., Jack, J.J.B., MacLennan, Catherine R.: A method for the selective electrical activation of tendon organ afferent fibres from the cat soleus muscle. J. Physiol. (Lond.) **210**, 18–20 (1970).

Creed, R.S., Denny-Brown, D., Eccles, J.C., Liddell, E.G.T., Sherrington, C.S.: Reflex activity of the spinal cord. Oxford: Clarendon Press 1932.

Critchlow, V., von Euler, C.: Intercostal muscle spindle activity and its γ motor control. J. Physiol. (Lond.) **168**, 820–847 (1963).

Curtis, D.R., Duggan, A.W., Felix, D., Johnston, G.A.R.: Bicuculline, an antagonist of GABA and synaptic inhibition in the spinal cord of the cat. Brain Res. **32**, 69–96 (1971).

Curtis, D.R., Krnjević, K., Miledi, R.: Crossed inhibition of sacral motoneurones. J. Neurophysiol. **21**, 319–326 (1958).

Denny-Brown, D.: On inhibition as a reflex accompaniment of the tendon jerk and of other forms of active muscular response. Proc. roy. Soc. B **103**, 321–336 (1928).

Devanandan, M.S., Eccles, R.M., Stenhouse, D.: Presynaptic inhibition evoked by muscle contraction. J. Physiol. (Lond.) **185**, 471–485 (1966).

Devavandan, M.S., Eccles, R.M., Yokota, T.: Depolarization of afferent terminals evoked by muscle stretch. J. Physiol. (Lond.) **179**, 417–429 (1965).

Dorward, P.K.: Responses of mechanoreceptors and intraspinal projection of their afferent fibres in the domestic duck. Ph.D. Thesis, Monash University, Australia 1966.

Dorward, P.K.: Response characteristics of muscle afferents in the domestic duck. J. Physiol. (Lond.) **211**, 1–17 (1970).

Eccles, J.C.: The mechanism of synaptic transmission. Ergebn. Physiol. **51**, 299–430 (1961).

Eccles, J.C.: Central connexions of muscle afferents fibres. In: Muscle Receptors, pp. 81–101. Hong Kong: Hong Kong University Press 1962.

Eccles, J.C.: Postsynaptic and presynaptic inhibitory actions in the spinal cord. In: Progress in Brain Research, Vol. 1.: Brain Mechanisms, pp. 1–18 Amsterdam: Elsevier 1963.

Eccles, J.C.: The Physiology of Synapses. Berlin-Göttingen-Heidelberg-New York: Springer 1964.

Eccles, J.C.: Functional organization of the cerebellum in relation to its role in motor control. In: Muscular Afferents and Motor Control, pp. 19–36. Stockholm: Almqvist and Wiksell; New-York-London-Sydney: John Wiley 1966.

Eccles, J.C.: The dynamic loop hypothesis of movement control. In: Information Processing in the Nervous System, Leibovic, K.N., Ed., pp. 245–269. Berlin-Heidelberg-New York: Springer 1969.

Eccles, J.C., Eccles, R.M., Lundberg, A.: Synaptic actions on motoneurones in relation to the two components of the group I muscle afferent volley. J. Physiol. (Lond.) **136**, 527–546 (1957a).

Eccles, J.C., Eccles, R.M., Lundberg, A.: The convergence of monosynaptic excitatory afferents on to many different species of alpha motoneurones. J. Physiol. (Lond.) **137**, 22–50 (1957b).

Eccles, J.C., Eccles, R.M., Lundberg, A.: Synaptic actions on motoneurones caused by impulses in Golgi tendon organ afferents. J. Physiol. (Lond.) **138**, 227–252 (1957c).

Eccles, J.C., Eccles, R.M., Lundberg, A.: Types of neurone in and around the intermediate nucleus of the lumbosacral cord. J. Physiol. (Lond.) **154**, 89–114 (1960).

Eccles, J.C., Eccles, R.M., Magni, F.: Central inhibitory action attributable to presynaptic depolarization produced by muscle afferent volleys. J. Physiol. (Lond.) **159**, 147–166 (1961a).

Eccles, J.C., Fatt, P., Landgren, S.: Central pathway for direct inhibitory action of impulses in largest afferent nerve fibres to muscle. J. Neurophysiol. **19**, 75–98 (1956).

Eccles, J.C., Fatt, P., Landgren, S., Winsbury, G.J.: Spinal cord potentials generated by volley in the large muscle afferents. J. Physiol. (Lond.) **125**, 590–606 (1954).

Eccles, J.C., Hubbard, J.I., Oscarsson, O.: Intracellular recording from cells of the ventral spinocerebellar tract. J. Physiol. (Lond.) **158**, 486–516 (1961b).

Eccles, J.C., Ito, M., Szentágothai, J.: The Cerebellum as a Neuronal Machine. Berlin-Heidelberg-New York: Springer 1967.

Eccles, J.C., Oscarsson, O., Willis, W.D.: Synaptic action of Group I and II afferent fibres of muscle on the cells of the dorsal spinocerebellar tract. J. Physiol. (Lond.) **158**, 517–543 (1961c).

Eccles, J.C., Pritchard, J.J.: The action potential of motoneurones. J. Physiol. (Lond.) **89**, 43–45P (1937).

Eccles, J.C., Sherrington, C.S.: Numbers and contraction-values of individual motor-units examined in some muscles of the limb. Proc. roy. Soc. B **106**, 326–357 (1930).

Eccles, R.M., Lundberg, A.: Integrative pattern of Ia synaptic actions on motoneurones of hip and knee muscles. J. Physiol. (Lond.) **144**, 271–298 (1958a).

Eccles, R.M., Lundberg, A.: The synaptic linkage of 'direct' inhibition. Acta physiol. scand. **43**, 204–215 (1958b).

Eccles, R.M., Lundberg, A.: Synaptic actions in motoneurones by afferents which may evoke the flexion reflex. Arch. ital. Biol. **97**, 199–221 (1959a).

Eccles, R.M., Lundberg, A.: Supraspinal control of interneurones mediating spinal reflexes. J. Physiol. (Lond.) **147**, 565–584 (1959b).

Eide, E., Fedina, L., Jansen, J., Lundberg, A., Vyklický, L.: Unitary components in the activation of Clarke's column neurones. Acta physiol. scand. **77**, 145–158 (1969).

Erlanger, J., Gasser, H.S.: Electrical Signs of Nervous Activity. Philadelphia: University of Pensylvania Press 1937.

Fadiga, E., Brookhart, J.M.: Monosynaptic activation of different portions of the motoneuron membrane. Amer. J. Physiol. **198**, 693–703 (1960).

Frank, K., Sprague, J.M.: Direct contralateral inhibition in the lower sacral spinal cord. Exp. Neurol. **1**, 28–43 (1959).

Fukami, Y.: Postsynaptic potentials in toad's spinal motoneurons due to muscle afferent volleys. Jap. J. Physiol. **11**, 596–604 (1961).

Fukami, Y.: Tonic and phasic muscle spindles in snake. J. Neurophysiol. **33**, 28–35 (1970).

Gardner, E., Noer, R.: Projection of afferent fibers from muscles and joints to the cerebral cortex of the cat. Amer. J. Physiol. **168**, 437–441 (1952).

Gasser, H.S., Grundfest, H.: Axon diameters in relation to the spike dimensions and the conduction velocity in mammalian A fibers. Amer. J. Physiol. **127**, 393–414 (1939).

Gelfan, S., Carter, S.: Muscle sense in man. Exp. Neurol. **18**, 469–473 (1967).

Giaquinto, S., Pompeiano, O., Swett, J.E.: EEG and behavioral effects of fore- and hind-limb muscular afferent volleys in unrestrained cats. Arch. ital. Biol. **101**, 133–148 (1963).

Glees, P., Soler, J.: Fibre content of the posterior column and synaptic connections of nucleus gracilis. Z. Zellforsch. **36**, 381–400 (1951).

Goodwin, G.M., McCloskey, D I., Matthews, P.B.C.: Proprioceptive illusions induced by muscle vibration: contribution by muscle spindles to perception? Science **175**, 1382–1384 (1972).

Granit, R.: Reflex self-regulation of muscle contraction and autogenetic inhibition. J. Neurophysiol. **13**, 351–372 (1950).

Granit, R.: Reflexes to stretch and contribution of antagonists around ankle joint. J. Neurophysiol. **15**, 269–279 (1952).

Granit, R.: The Basis of Motor Control. London-New York: Academic Press 1970.

Granit, R.: Constant errors in the execution and appreciation of movement. Brain **95**, 649–660 (1972).

Grant, G., Oscarsson, O.: Mass discharges evoked in the olivocerebellar tract on stimulation of muscle and skin nerves. Exp. Brain Res. **1**, 329–337 (1966).

Grant, G., Oscarsson, O., Rosén, I.: Functional organization of the spinoreticulocerebellar path with identification of its spinal component. Exp. Brain Res. **1**, 306–319 (1966).

Grillner S.: Supraspinal and segmental control of static and dynamic a-motoneurones in the cat. Acta physiol. scand. **76**, Suppl. 327 (1969).

Grillner, S.: Is the tonic stretch reflex dependent upon Group II excitation? Acta physiol. scand. **78**, 431–432 (1970).

Grundfest, H., Campbell, B.: Origin, conduction and termination of impulses in dorsal spino-cerebellar tracts of cats. J. Neurophysiol. **5**, 275–294 (1942).

Hagbarth, K.E., Wohlfart, G.: The number of muscle spindles in certain muscles in cat in relation to the composition of the muscle nerves. Acta anat. (Basel) **15**, 85–104 (1952).

Hoffmann, P.: Untersuchungen über die Eigenreflexe (Sehnenreflexe) menschlicher Muskeln. Berlin: Springer 1922.

Holemans, K.C., Meij, H.S., Meyer, B.J.: The existence of a monosynaptic reflex arc in the spinal cord of the frog. Exp. Neurol. **14**, 175–186 (1966).

Holmqvist, B.: Crossed spinal reflex actions evoked by volleys in somatic afferents. Acta physiol. scand. **52**, Suppl. 181 (1961).

Holmqvist, B., Lundberg, A.: Differential supraspinal control of synaptic actions evoked by volleys in the flexion reflex afferents in alpha motoneurones. Acta physiol. scand. **54**, Suppl. 186 (1961).

Holmqvist, B., Lundberg, A., Oscarsson, O.: Functional organization of the dorsal spino-cerebellar tract. V. Further experiments on convergence of excitatory and inhibitory actions. Acta physiol. scand. **38**, 76–90 (1956).

HONGO, T., JANKOWSKA, E., LUNDBERG, A.: The rubrospinal tract. II. Facilitation of interneuronal transmission in reflex paths to motoneurones. Exp. Brain Res. **7**, 365–391 (1969).

HONGO, T., JANKOWSKA, E., LUNDBERG, A.: The rubrospinal tract. IV. Effects on interneurones. Exp. Brain Res. **15**, 54–78 (1972).

HORE, J., PORTER, R.: Pyramidal and extrapyramidal influences on some hindlimb motoneuron populations of the arboreal brush-tailed possum *Trichosurus vulpecula*. J. Neurophysiol. **35**, 112–121 (1972).

HOUK, J., HENNEMAN, E.: Responses of Golgi tendon organs to active contractions of the soleus muscle of the cat. J. Neurophysiol. **30**, 466–481 (1967).

HUBBARD, J.I., OSCARSSON, O.: Localization of the cell bodies of the ventral spino-cerebellar tract in lumbar segments of the cat. J. comp. Neurol. **118**, 199–204 (1962).

HULTBORN, H.: Convergence on interneurones in the reciprocal Ia inhibitory pathway to motoneurones. Acta physiol. scand. **85**, Suppl. 375 (1972).

HULTBORN, H., JANKOWSKA, E., LINDSTRÖM, S.: Recurrent inhibition from motor axon collaterals of transmission in the Ia inhibitory pathway to motoneurones. J. Physiol. (Lond.) **215**, 591–612 (1971).

HUNT, C.C.: The reflex activity of mammalian small-nerve fibres. J. Physiol. (Lond.) **115**, 456–469 (1951).

HUNT, C.C.: The effect of stretch receptors from muscle on the discharge of motoneurones. J. Physiol. (Lond.) **117**, 359–379 (1952).

HUNT, C.C.: Diameter and function of afferent fibres from muscle. Abstr. XIX int. Congr. physiol. Sci., pp. 485–486. Montreal 1953.

HUNT, C.C.: Relation of function to diameter in afferent fibers of muscle nerves. J. gen. Physiol. **38**, 117–131 (1954).

HUNT, C.C.: Monosynaptic reflex response of spinal motoneurons to graded afferent stimulation. J. gen. Physiol. **38**, 813–852 (1955).

HUNT, C.C.: On the nature of vibration receptors in the hind limb of the cat. J. Physiol. (Lond.) **155**, 175–186 (1961).

HUNT, C.C., KUFFLER, S.W.: Stretch receptor discharges during muscle contraction. J. Physiol. (Lond.) **113**, 298–315 (1951).

HUNT, C.C., MCINTYRE, A.K.: Characteristics of responses from receptors from the flexor longus digitorum muscle and the adjoining interosseous region of the cat. J. Physiol. (Lond.) **153**, 74–87 (1960a).

HUNT, C.C., MCINTYRE, A.K.: An analysis of fibre diameter and receptor characteristics of myelinated cutaneous afferent fibres in cat. J. Physiol. (Lond.) **153**, 99–112 (1960b).

HUNT, C.C., PAINTAL, A.S.: Spinal reflex regulation of fusimotor neurones. J. Physiol. (Lond.) **143**, 195–212 (1958).

HUNT, C.C., PERL, E.R.: Spinal reflex mechanisms concerned with skeletal muscle. Physiol. Rev. **40**, 538–579 (1960).

HURSH, J.B.: Conduction velocity and diameter of nerve fibers. Amer. J. Physiol. **127**, 131–139 (1939).

IGGO, A.: Non-myelinated afferent fibres from mammalian skeletal muscle. J. Physiol. (Lond.) **155**, 52–53P (1961).

ILLIS, L.: The relative densities of monosynaptic pathways to the cell bodies and dendrites of the cat ventral horn. J. neurol. Sci. **4**, 259–270 (1967).

JACK, J.J.B.: Intensity of direct inhibitory action as a function of volley size. Proc. Univ. Otago med. Sch. **35**, 33–35 (1957).

JACK, J.J.B., MCINTYRE, A.K.: Unpublished observations, 1957.

JACK, J.J.B., MAC LENNAN, CATHERINE R.: The lack of an electrical threshold discrimination between group Ia and group Ib fibres in the nerve to the cat peroneus longus muscle. J. Physiol. (Lond.) **212**, 35–36P (1971).

JACK, J.J.B., MILLER, S., PORTER, R.: The different time courses of minimal EPSPs in spinal motoneurones. J. Physiol. (Lond.) **191**, 112–113P (1967).

JACK, J.J.B., MILLER, S., PORTER, R., REDMAN, S.: The distribution of group Ia synapses on lumbosacral motoneurones in the cat. In: Excitatory Synaptic Mechanisms, pp. 199–205. Oslo: Universitetsforlaget 1970.

Jack, J.J.B., Miller, S., Porter, R., Redman, S.: The time course of minimal excitatory post-synaptic potentials evoked in spinal motoneurones by group Ia afferent fibres. J. Physiol. (Lond.) **215**, 353–380 (1971).

Jankowska, E., Lindström, S.: Morphology of interneurones mediating Ia reciprocal inhibition of motoneurones in the spinal cord of the cat. J. Physiol. (Lond.) **226**, 805–823 (1972).

Jankowska, E., Roberts, W.J.: Synaptic actions of single interneurones mediating reciprocal Ia inhibition of motoneurones. J. Physiol. (Lond.) **222**, 623–642 (1972).

Jansen, J.K.S., Nicolaysen, K., Rudjord, T.: Discharge pattern of neurons of the dorsal spinocerebellar tract activated by static extension of primary endings of muscle spindles. J. Neurophysiol. **29**, 1061–1086 (1966).

Jansen, J.K.S., Rudjord, T.: Dorsal spinocerebellar tract: response pattern of nerve fibers to muscle stretch. Science **149**, 1109–1111 (1965).

Jansen, J.K.S., Walløe, L.: Signal transmission between successive neurons in the dorsal spinocerebellar pathway. In: The Neurosciences — Second Study Program, pp. 617–629. New York: Rockefeller University Press 1970.

Job, C.: Über autogene Inhibition und Reflexumkehr bei spinalisierten und decerebrierten Katzen. Pflügers Arch. ges. Physiol. **256**, 406–418 (1953).

Jolly, W.A.: On the time relations of the knee-jerk and simple reflexes. Quart. J. exp. Physiol. **4**, 67–87 (1911).

Kellerth, J.O.: Aspects on the relative significance of pre- and postsynaptic inhibition in the spinal cord. In: Structure and Function of Inhibitory Neuronal Mechanisms, pp. 197–212. Oxford: Pergamon Press 1968.

Kenins, P.: Central pathway of a reptilian stretch reflex. Proc. Aust. Physiol. Pharmacol. Soc. **3**, 134 (1972).

Kenins, P., McIntyre, A.K., Proske, U.: Stretch-evoked reflex response in the lizard *Tiliqua nigrolutea*. Proc. Aust. Physiol. Pharmacol. Soc. **1**, 61–62 (1970).

Kostyuk, P.G.: On the functions of dorsal spino-cerebellar tract in cat. In: Neurobiology of Cerebellar Evolution and Development, pp. 539–548. Chicago: American Medical Association 1969.

Kuffler, S.W., Hunt, C.C.: The mammalian small nerve fibres: a system for efferent nervous regulation of muscle spindle discharge. Res. Publ. Ass. nerv. ment. Dis. **30**, 24–47 (1952).

Kuffler, S.W., Vaughan Williams, E.M.: Small-nerve junctional potentials. The distribution of small motor nerves to frog skeletal muscle, and the membrane characteristics of the fibres they innervate. J. Physiol. (Lond.) **121**, 289–317 (1953).

Kuno, M.: Quantal components of excitatory synaptic potentials in spinal motoneurones. J. Physiol. (Lond.) **175**, 81–99 (1964).

Kuno, M.: Quantum aspects of central and ganglionic synaptic transmission in vertebrates. Physiol. Rev. **51**, 647–678 (1971).

Kuno, M., Perl, E.R.: Alteration of spinal reflexes by interaction with suprasegmental and dorsal root activity. J. Physiol. (Lond.) **151**, 103–122 (1960).

Landgren, S., Silfvenius, H.: Projection to cerebral cortex of Group I muscle afferents from the cat's hind limb. J. Physiol. (Lond.) **200**, 353–372 (1969).

Landgren, S., Silfvenius, H.: Nucleus Z, the medullary relay in the projection path to the cerebral cortex of Group I muscle afferents from the cat's hind limb. J. Physiol. (Lond.) **218**, 551–571 (1971).

Laporte, Y., Bessou, P.: Étude des sous-groupes lent et rapide du groupe I (fibres afférentes d'origine musculaire de grand diamètre) chez le chat. J. Physiol. (Paris) **49**, 1025–1037 (1957).

Laporte, Y., Bessou, P.: Réflexes ipsi-latéraux d'origine exclusivement amyélinique chez le Chat. C.R. Soc. Biol. (Paris) **152**, 161–164 (1958).

Laporte, Y., Bessou, P.: Modifications d'excitabilité de motoneurones homonymes provoquées par l'activation physiologique de fibres afférentes d'origine musculaire du groupe II. J. Physiol. (Paris) **51**, 897–908 (1959).

Laporte, Y., Boëunder, A.: Réflexe de flexion ipsilatéral provoqué par les fibres amyéliniques chez le chat spinal. C.R. Soc. Biol. (Paris) **149**, 808–810 (1955).

Laporte, Y., Lloyd, D.P.C.: Nature and significance of the reflex connections established by large afferent fibers of muscular origin. Amer. J. Physiol. **169**, 609–621 (1952).

LAPORTE, Y., LUNDBERG, A., OSCARSSON, O.: Functional organization of the dorsal spinocerebellar tract in the cat. II. Single fibre recording in Flechsig's fasciculus on electrical stimulation of various peripheral nerves. Acta physiol. scand. **36**, 188–203 (1956).

LEWIS, M. McD., PORTER, R.: Lack of involvement of fusimotor activation in movements of the foot produced by electrical stimulation of monkey cerebral cortex. J. Physiol. (Lond.) **212**, 707–717 (1971).

LIBET, B., ALBERTS, W.W., WRIGHT, E.W., JR., DELATTRE, L.D., LEVIN, G., FEINSTEIN, B.: Production of threshold levels of conscious sensation by electrical stimulation of human somatosensory cortex. J. Neurophysiol. **27**, 546–578 (1964).

LIDDELL, E.G.T.: The Discovery of Reflexes. Oxford: Clarendon Press 1960.

LIDDELL, E.G.T., SHERRINGTON, C.S.: Reflexes in response to stretch (myotatic reflexes). Proc. roy. Soc. B **96**, 212–242 (1924).

LIDDELL, E.G.T., SHERRINGTON, C.S.: Further observations on myotatic reflexes. Proc. roy. Soc. B **97**, 267–283 (1925).

LINDSTRÖM, S., TAKATA, M.: Monosynaptic excitation of dorsal spinocerebellar tract neurones from low threshold joint afferents. Acta physiol. scand. **84**, 430–432 (1972).

LLOYD, D.P.C.: Reflex action in relation to pattern and peripheral source of afferent stimulation. J. Neurophysiol. **6**, 111–120 (1943a).

LLOYD, D.P.C.: Neuron patterns controlling transmission of ipsilateral hind limb reflexes in cat. J. Neurophysiol. **6**, 293–315 (1943b).

LLOYD, D.P.C.: Conduction and synaptic transmission of the reflex response to stretch in spinal cats. J. Neurophysiol. **6**, 317–326 (1943c).

LLOYD, D.P.C.: Facilitation and inhibition of spinal motoneurons. J. Neurophysiol. **9**, 421–438 (1946a).

LLOYD, D.P.C.: Integrative pattern of excitation and inhibition in two-neuron reflex arcs. J. Neurophysiol. **9**, 439–444 (1946b).

LLOYD, D.P.C.: On reflex actions of muscular origin. Res. Publ. Ass. nerv. ment. Dis. **30**, 48–67 (1952).

LLOYD, D.P.C., CHANG, H.-T.: Afferent fibers in muscle nerves. J. Neurophysiol. **11**, 199-208 (1948).

LLOYD, D.P.C., McINTYRE, A.K.: Analysis of forelimb-hindlimb reflex activity in acutely decapitate cats. J. Neurophysiol. **11**, 455–470 (1948).

LLOYD, D.P.C., McINTYRE, A.K.: Dorsal column conduction of Group I muscle afferent impulses and their relay through Clarke's column. J. Neurophysiol. **13**, 39–54 (1950).

LLOYD, D.P.C., WILSON, V.J.: Functional organization in the terminal segments of the spinal cord with a consideration of central excitatory and inhibitory latencies in monosynaptic reflex systems. J. gen. Physiol. **42**, 1219–1231 (1959).

LUNDBERG, A.: Ascending spinal hindlimb pathways in the cat. In: Progress in Brain Research, Vol. 12: Physiology of Spinal Neurons, pp. 135–163. Amsterdam: Elsevier 1964.

LUNDBERG, A.: Integration in the reflex pathway. In: Muscular Afferents and Motor Control, pp. 275–305. Stockholm: Almqvist and Wiksell; New York-London-Sydney: Wiley 1966.

LUNDBERG, A.: The supraspinal control of transmission in spinal reflex pathways. In: Recent Advances in Clinical Neurophysiology. Electroenceph. clin. Neurophysiol. Suppl. 25, pp. 35–46. Amsterdam: Elsevier 1967.

LUNDBERG, A.: Convergence of excitatory and inhibitory action on interneurones in the spinal cord. In: The Interneurone, pp. 231–265. Los Angeles: University of California Press 1969a.

LUNDBERG, A.: Reflex control of stepping. The Norwegian Academy of Science and letters, Nansen Memorial Lecture. Oslo: Universitetsforlaget 1969b.

LUNDBERG, A.: The excitatory control of the Ia inhibitory pathway. In: Excitatory Synaptic Mechanisms, pp. 333–340. Oslo: Universitetsforlaget 1970.

LUNDBERG, A., OSCARSSON, O.: Two ascending spinal pathways in the ventral part of the cord. Acta physiol. scand. **54**, 270–286 (1962).

LUNDBERG, A., VOORHOEVE, P.: Effects from the pyramidal tract on spinal reflex arcs. Acta physiol. scand. **56**, 201–219 (1962).

LUNDBERG, A., WEIGHT, F.: Functional organization of connexions to the ventral spinocerebellar tract. Exp. Brain Res. **12**, 295–316 (1971).

LUNDBERG, A., WINSBURY, G.: Selective adequate activation of large afferents from muscle spindles and Golgi tendon organs. Acta physiol. scand. **49**, 155–164 (1960a).

Lundberg, A., Winsbury, G.: Functional organization of the dorsal spino-cerebellar tract in the cat. VI. Further experiments on excitation from tendon organ and muscle spindle afferents. Acta physiol. scand. **49**, 165–170 (1960b).

McDonald, W.I., McIntyre, A.K.: Observations on ascending long spinal reflexes. Proc. Univ. Otago med. Sch. **34**, 5–6 (1956).

McGrath, G.J., Matthews, P.B.C.: Support for an autogenetic excitatory reflex action of the spindle secondaries from the effect of gamma blockade by procaine. J. Physiol. (Lond.) **210**, 176–177 P (1970).

McIntyre, A.K.: Afferent limb of the myotatic reflex arc. Nature (Lond.) **168**, 168–170 (1951).

McIntyre, A.K.: Cortical projection of afferent impulses in muscle nerves. Proc. Univ. Otago med. Sch. **31**, 5–6 (1953).

McIntyre, A.K.: Unpublished observations 1961.

McIntyre, A.K.: Central projection of impulses from receptors activated by muscle stretch. In: Muscle Receptors, pp. 19–29. Hong Kong: Hong Kong University Press 1962.

McIntyre, A.K.: Deep sensibility. Proc. Aust. Assoc. Neurol. **1**, 37–40 (1963).

McIntyre, A.K.: Some applications of input-output technique. In: Studies in Physiology, Curtis, D.R., McIntyre, A.K., Eds., pp. 199–206. Berlin-Heidelberg-New York: Springer 1965.

McIntyre, A.K., Holman, M.E., Veale, J.L.: Cortical responses to impulses from single Pacinian corpuscles in the cat's hind limb. Exp. Brain Res. **4**, 243–255 (1967).

McIntyre, A.K., Mark, R.F.: Synaptic linkage between afferent fibres of the cat's hind limb and ascending fibres in the dorsolateral funiculus. J. Physiol. (Lond.) **153**, 306–330 (1960).

McIntyre, A.K., Robinson, R.G.: Pathway for the jaw jerk in man. Brain **82**, 468–474 (1959).

MacLennan, Catherine R.: Studies on the selective activation of muscle receptor afferents. D. Phil. Thesis: Oxford University 1971.

MacLennan, Catherine R.: The behaviour of receptors of extramuscular and muscular origin with afferent fibres contributing to the group I and the group II of the cat tibialis anterior muscle nerve. J. Physiol. (Lond.) **222**, 90–91P (1972).

Magni, F., Oscarsson, O.: Comparison of ascending spinal tracts activated by group I muscle afferents in the phalanger, rabbit and cat. Acta physiol. scand. **54**, 37–52 (1962).

Marx, C.: Mise en évidence de l'action centrale des proprioccepteurs musculaires de la grenouille. Arch. int. Physiol. **57**, 447–451 (1950).

Matthews, B.H.C.: Nerve endings in mammalian muscle. J. Physiol. (Lond.) **78**, 1–53 (1933).

Matthews, P.B.C.: Vibration and the stretch reflex. In: Myotatic, Kinesthetic and Vestibular Mechanisms, pp. 40–50. London: Churchill 1967.

Matthews, P.B.C.: Evidence that the secondary as well as the primary endings of the muscle spindles may be responsible for the tonic stretch reflex of the decerebrate cat. J. Physiol. (Lond.) **204**, 365–393 (1969).

Matthews, P.B.C.: The origin and functional significance of the stretch reflex. In: Excitatory Synaptic Mechanisms, pp. 301–315. Oslo: Universitetsforlaget 1970.

Matthews, P.B.C.: Mammalian Muscle Receptors and their Central Actions. London: Edward Arnold 1972.

Matthews, P.B.C., Rushworth, G.: The selective effect of procaine on the stretch reflex and tendon jerk of soleus muscle when applied to its nerve. J. Physiol. (Lond.) **135**, 245–262 (1957).

Mendell, L.M., Henneman, E.: Terminals of Ia fibers: distribution within a pool of 300 homonymous motor neurons. Science **160**, 96–98 (1968).

Mendell, L.M., Henneman, E.: Terminals of single Ia fibers: location, density, and distribution within a pool of 300 homonymous motoneurons. J. Neurophysiol. **34**, 171–187 (1971).

Merton, P.A.: Speculations on the servo-control of movement. In: The Spinal Cord, pp. 247–255. London: Churchill 1953.

Merton, P.A.: Human position sense and sense of effort. Symp. Soc. exp. Biol. **18**, 387–400 (1964).

Miller, S., Oscarsson, O.: Termination and functional organization of spino-olivocerebellar paths. In: The Cerebellum in Health and Disease, pp. 172–200. St. Louis: Warren H. Green 1970.

MOUNTCASTLE, V.B., COVIAN, M.R., HARRISON, C.R.: The central representation of some forms of deep sensibility. Res. Publ. Ass. nerv. ment. Dis. **30**, 339–370 (1952).

O'LEARY, J., HEINBECKER, P., BISHOP, G.H.: Analysis of function of a nerve to muscle. Amer. J. Physiol. **110**, 636–658 (1935).

OSCARSSON, O.: Functional organization of the ventral spino-cerebellar tract in the cat. Acta physiol. scand. **42**, Suppl. 146 (1957).

OSCARSSON, O.: Functional organization of the spino- and cuneocerebellar tracts. Physiol. Rev. **45**, 495–522 (1965).

OSCARSSON, O.: Functional significance of information channels from the spinal cord to the cerebellum. In: Neurophysiological Basis of Normal and Abnormal Motor Activities, pp. 93–117. New York: Raven Press 1967.

OSCARSSON, O.: Termination and functional organization of the ventral spino-olivocerebellar path. J. Physiol. (Lond.) **196**, 453–478 (1968).

OSCARSSON, O., ROSÉN, I.: Projection to cerebral cortex of large muscle-spindle afferents in forelimb nerves of the cat. J. Physiol. (Lond.) **169**, 924–945 (1963).

OSCARSSON, O., ROSÉN, I., UDDENBERG, U.: Organization of ascending tracts in the spinal cord of the duck. Acta physiol. scand. **59**, 143–153 (1963).

PAINTAL, A.S.: Functional analysis of Group III afferent fibres of mammalian muscles. J. Physiol. (Lond.) **152**, 250–270 (1960).

PAINTAL, A.S.: Participation by pressure-pain receptors of mammalian muscles in the flexion reflex. J. Physiol. (Lond.) **156**, 498–514 (1961).

PAINTAL, A.S.: Responses and reflex effects of pressure-pain receptors of mammalian muscles. In: Muscle Receptors, pp. 133–142. Hong Kong: Hong Kong University Press 1962.

PERL, E.R.: Crossed reflexes of cutaneous origin. Amer. J. Physiol. **188**, 609–615 (1957).

PERL, E.R.: Crossed reflex effects evoked by activity in myelinated afferent fibers of muscle. J. Neurophysiol. **21**, 101–122 (1958).

PERL, E.R.: Effects of muscle stretch on excitability of contralateral motoneurones. J. Physiol. (Lond.) **145**, 193–203 (1959).

PHILLIPS, C.G., POWELL, T.P.S., WIESENDANGER, M.: Projection from low-threshold muscle afferents of hand and forearm to area 3a of baboon's cortex. J. Physiol. (Lond.) **217**, 419–446 (1971).

PROSKE, U.: An electrophysiological analysis of responses from lizard muscle spindles. J. Physiol. (Lond.) **205**, 289–304 (1969).

PROSKE, U., LEWIS, D.M.: The effects of muscle stretch and vibration on fusimotor activity in the lightly anaesthetised cat. Brain Res. **46**, 55–69 (1972).

PROVINS, K.A.: The effect of peripheral nerve block on the appreciation and execution of finger movements. J. Physiol. (Lond.) **143**, 55–67 (1958).

RALL, W.: Experimental monosynaptic input-output relations in the mammalian spinal cord. J. cell. comp. Physiol. **46**, 413–437 (1955).

RENSHAW, B.: Activity in the simplest spinal reflex pathways. J. Neurophysiol. **3**, 373–387 (1940).

RENSHAW, B.: Reflex discharges in branches of the crural nerve. J. Neurophysiol. **5**, 487–498 (1942).

RÉTHELYI, M.: Ultrastructural synaptology of Clarke's column. Exp. Brain Res. **11**, 159–174 (1970).

REXED, B., THERMAN, P.O.: Calibre spectra of motor and sensory nerve fibres to flexor and extensor muscle. J. Neurophysiol. **11**, 133–139 (1948).

ROSÉN, I.: Projection of forelimb group I muscle afferents to the cat cerebral cortex. Dissertation, University of Lund: Lund 1970.

RYMER, W.Z.: Characteristics of muscle receptors in the phalanger (*Trichosurus vulpecula*). Aust. J. exp. Biol. med. Sci. **46**, 4 (1968).

RYMER, W.Z.: A comparison between muscle spindle receptors and dorsal spinocerebellar tract cells. Ph. D. Thesis, Monash University, Australia 1972.

SASSA, K.: A proprioceptive reflex and clonus as described in the spinal frog. Brain **44**, 308–311 (1921).

SCHMIDT, R.F.: Presynaptic inhibition in the vertebrate central nervous system. Ergebn. Physiol. **63**, 20–101 (1971).

Sears, T.A.: Efferent discharge in alpha and fusimotor fibres of intercostal nerves of the cat. J. Physiol. (Lond.) **174**, 295–315 (1964).

Sherrington, C.S.: On the anatomical constitution of nerves of skeletal muscles; with remarks on recurrent fibres in the ventral spinal nerve root. J. Physiol. (Lond.) **17**, 211–258 (1894).

Sherrington, C.S.: The muscular sense. In: Schäfer's Textbook of Physiology, Vol. 2, pp. 1002–1025. Edinburgh-London: Pentland 1900.

Sherrington, C.S.: On plastic tonus and proprioceptive reflexes. Quart. J. exp. Physiol. **2**, 109–156 (1909).

Sherrington, C.S.: Flexion-reflex of the limb, crossed extension-reflex and reflex stepping and standing. J. Physiol. (Lond.) **40**, 28–121 (1910).

Silfvenius, H.: Projections to the cat cerebral cortex from fore- and hind limb group I muscle afferents. Umeå University Medical Dissertations, No. 4, 1972.

Skoglund, S.: Anatomical and physiological studies of knee joint innervation in cat. Acta physiol. scand. **36**, Suppl. 124 (1956)

Sprague, J.M.: The distribution of dorsal root fibres on motor cells in the lumbosacral spinal cord of the cat, and the site of excitatory and inhibitory terminals in monosynaptic pathways. Proc. roy. Soc. B **140**, 534–556 (1958).

Stacey, M.J.: Free nerve endings in skeletal muscle of the cat. J. Anat. (Lond.) **105**, 231–254 (1969).

Sterling, P., Kuypers, H.G.J.M.: Anatomical organization of the brachial spinal cord of the cat. I. The distribution of dorsal root fibers. Brain Res. **4**, 1–15 (1967).

Stuart, D.G., Mosher, C.G., Gerlach, R.L.: Properties and central connections of Golgi tendon organs with special reference to locomotion. In: Research in Muscle Development and the Muscle Spindle, pp. 437–462. Amsterdam: Excerpta Medica 1972.

Stuart, D.G., Mosher, C.G., Gerlach, R.L., Reinking, R.M.: Selective activation of Ia afferents by transient muscle stretch. Exp. Brain Res. **10**, 477–487 (1970).

Stuart, D.G., Willis, W.D., jr., Reinking, R.M.: Stretch-evoked excitatory postsynaptic potentials in motoneurons. Brain Res. **33**, 115–125 (1971).

Sumner, A.J.: Properties of IA and IB afferent fibres serving stretch receptors of the cat's medial gastrocnemius muscle. Proc. Univ. Otago med. Sch. **39**, 3–5 (1961).

Szentágothai, J.: Anatomical considerations of monosynaptic reflex arcs. J. Neurophysiol. **11**, 445–454 (1948).

Szentágothai, J.: The anatomical basis of synaptic transmission of excitation and inhibition in motor neurons. Acta morph. Acad. Sci. hung. **8**, 287–309 (1958).

Szentágothai, J.: Synaptic architecture of the spinal motoneuron pool. In: Recent Advances in Clinical Neurophysiology. Electroenceph. clin. Neurophysiol. Suppl. 25, pp. 4–19. Amsterdam: Elsevier 1967.

Szentágothai, J., Albert, A.: The synaptology of Clarke's column. Acta morph. Acad. Sci. hung. **5**, 43–51 (1955).

Takeuchi, A., Takeuchi, N.: Electrical changes in pre- and postsynaptic axons of the giant synapse of *Loligo*. J. gen. Physiol. **45**, 1181–1193 (1962).

Terzuolo, C.A., Llinás, R.: Distribution of synaptic inputs in the spinal motoneurone and its functional significance. In: Muscular Afferents and Motor Control, pp. 373–384. Stockholm: Almqvist and Wiksell; New York-London-Sydney: Wiley 1966.

Vallbo, Å.B.: Muscle spindle response at the onset of isometric voluntary contractions in man. Time difference between fusimotor and skelemotor effects. J. Physiol. (Lond.) **218**, 405–431 (1971).

Wall, P.D.: Two transmission systems for skin sensations. In: Sensory Communication, pp. 475–496. New York-London: M.I.T. Press and Wiley 1961.

Westbury, D.R.: A study of stretch and vibration reflexes of the cat by intracellular recording from motoneurones. J. Physiol. (Lond.) **226**, 37–56 (1972).

Wilson, V.J., Kato, M.: Excitation of extensor motoneurons by group II afferent fibers in ipsilateral muscle nerves. J. Neurophysiol. **28**, 545–554 (1965).

Wilson, V.J., Lloyd, D.P.C.: Bilateral spinal excitatory and inhibitory actions. Amer. J. Physiol. **187**, 641 (1956).

Author Index

Page numbers in *italics* refer to the bibliography

Subject Index